TOWARD A SUSTAINABLE WHALING REGIME

TOWARD A
SUSTAINABLE
WHALING
REGIME

EDITED BY

Robert L. Friedheim

UNIVERSITY OF WASHINGTON PRESS

Seattle and London

CANADIAN CIRCUMPOLAR INSTITUTE (CCI) PRESS

Edmonton

Library of Congress Cataloging-in-Publication Data
Toward a sustainable whaling regime / edited by Robert L. Friedheim.
p. cm.
Includes bibliographical references.
ISBN 0-295-98088-5 (alk. paper)
1. Whaling—Management—International cooperation.
2. Whaling—Management—Political aspects.
3. Sustainable fisheries—International cooperation.
4. International Whaling Commission.
I. Friedheim, Robert L.
SH381.T682001   333.95′9517—DC21   00-050315

Canadian Cataloguing in Publication Data
Main entry under title:
Toward a sustainable whaling regime
(Studies in whaling ; no. 5)
(Circumpolar research series, ISSN 0838-133X ; no. 8)
ISBN 1-896445-18-7
1. Whaling—Management—International cooperation.
2. Whaling—Management—Political aspects.
3. International Whaling Commission.
I. Friedheim, Robert L.
II. Series: Studies in whaling ; no. 5.
III. Series: Circumpolar research series ; no. 8.
SH381.T68   2001   333.95′9517   C00-911364-9

CONTENTS

Foreword by John A. Knauss   *vii*

Acknowledgments   *ix*

Introduction: The IWC as a Contested Regime   *3*
ROBERT L. FRIEDHEIM

PART I / CRITIQUING THE PERFORMANCE
OF THE WHALING REGIME

1 / A New Whaling Agreement and International Law   *51*
WILLIAM T. BURKE

2 / Whales, the IWC, and the Rule of Law   *80*
JON L. JACOBSON

3 / Science and the IWC   *105*
WILLIAM ARON

4 / Is Money the Root of the Problem?
Cultural Conflict in the IWC   *123*
MILTON M. R. FREEMAN

Contents

5 / Food Security, Food Hegemony,
and Charismatic Animals   *147*
RUSSEL LAWRENCE BARSH

PART II / EXPLAINING THE POLITICS
OF THE REGIME

6 / Distorting Global Governance:
Membership, Voting, and the IWC   *183*
ELIZABETH DeSOMBRE

7 / Negotiating in the IWC Environment   *200*
ROBERT L. FRIEDHEIM

8 / The Whaling Regime: "Good" Institutions but "Bad" Politics?   *235*
STEINAR ANDRESEN

PART III / TESTING OUR ARGUMENTS
AND FINDING A SOLUTION

9 / Summing Up: Whaling and Its Critics   *269*
CHRISTOPHER D. STONE

10 / Whale Sausage: Why the Whaling Regime
Does Not Need to Be Fixed   *292*
DAVID G. VICTOR

11 / Fixing the Whaling Regime: A Proposal   *311*
ROBERT L. FRIEDHEIM

Bibliography   *337*
Contributors   *367*
Index   *371*

FOREWORD

During the peak whaling years, from 1930 to 1965, an average of thirty thousand whales a year were killed, mostly from around Antarctica. Today, the yearly take is less than a thousand under a moratorium put into force by the International Whaling Commission (IWC) in 1986. Limited whaling continues by indigenous populations, for research, and by exception as allowed under the rules of International Convention for the Regulation of Whaling (ICRW). Prior to the moratorium a number of species had been placed in a "protection" category and were no longer allowed to be taken, but there continued to be deep concern that IWC regulations were inadequate, and it would be only a matter of time before most, if not all, species of whales would be driven to near extinction.

It is now widely believed that the so-called Revised Management Procedure, accepted by the IWC in 1995, is adequate to the task of both rebuilding the stocks of endangered species and maintaining healthy stocks of those species that would be allowed to be taken under these new regulations. It seems unlikely, however, that one can expect the moratorium on commercial whaling to be lifted anytime soon.

The authors of this book address these issues from a wide range of perspectives. Why did the IWC not impose adequate limits on whaling when the stocks were still relatively plentiful? How has the economics of whaling changed now that the principal use of whales is for food and not oil? What are the problems and implications of attempting to maintain an indefinite moratorium on the killing of whales under a treaty whose stated purpose is to "make possible the orderly development of the whaling industry"? What are the political implications of operating within an organization whose only criterion for membership is paying dues, thus making it possible for those with a stake in the outcome to sponsor like-minded states for membership?

My introduction to these issues came with my appointment as the United States Commissioner to the IWC, a position I held during most of the Bush administration of 1989–93. I was commissioner, not because of any expertise in these matters, but because I was the administrator of the National Oceanic and Atmospheric Administration (NOAA), and I was told that serving as whaling commissioner was part of the job. The learning curve was steep, but I had an excellent staff, and it did not take long to perceive the ambiguities, the inconsistencies, and the sometimes Alice-in-Wonderland nature of the IWC and what passed for negotiations.

Most of the authors of this book detail the problems with the IWC and the ICRW, and some suggest solutions. One, David Victor, admits to all of the problems but concludes that attempting to implement any of the proposed cures may only make the situation worse. I confess to some sympathy for that position.

During my time as commissioner, and since, the IWC has managed to find ways to relieve major pressure points. The moratorium is being preserved, but some whales are being taken. The management regime is a significant improvement over what the IWC has had previously and is presently being tested by Norway, who is taking whales near its coast under an exception built into the ICRW. Our knowledge of whale ecology is steadily improving. Major political and philosophical differences continue to impede significant efforts to design a compromise, but all groups understand one another, and a precarious *status quo* has been maintained for some years.

However, I am concerned for the future. As an example of good faith international negotiations, the IWC is mostly a disaster. The specter of an alternative to the IWC, more sympathetic to the needs and desires of whaling nations, continues to hover. As whale stocks continue to rebound, as the world's population continues to grow, the pressure to harvest whales may also grow. For the present the IWC continues to hobble along. Whether it can continue to do so in the future, for all of the reasons so elegantly stated in this book, is less clear. For those whose interests range from the well-being of whales to the future of international civility, I believe there is good reason to be concerned.

JOHN A. KNAUSS

ACKNOWLEDGMENTS

The project behind this volume was initiated by a small group that formed in the mid-1990s to examine what, if anything, should be done to fix the whaling regime. Among its original members were Milton Freeman, Bill Aron, Bill Burke, and myself. The group was expanded incrementally as we invited people with something to say to work with us. Some of us met in Seattle, Washington, in July 1996. We also gave the first versions of several of the papers contained herein at the International Studies Association Western Regional Meeting in Eugene, Oregon, in October 1996. It is accurate to call our work a labor of love because none of us was compensated for our participation. We had no grant or subsidy, and as a result, we kept in touch by e-mail, fax, and occasional visits to each other's base of operation.

It was my task to pull it all together with few resources. It worked, but not without accumulating many debts. I have benefited from discussions with Shigeko Misaki, Tsuneo Akaha (Monterey Institute of International Studies), Gudrun Petursdottir (University of Iceland), Ted L. McDorman (University of Victoria), Ward Edwards (University of Southern California), and many others. Marshall W. Murphree, chair of the Sustainable Use Special Group (World Conservation Union), made a valiant effort to locate an expert on the elephant regime—the other "charismatic megafauna"—who might work with us. Unfortunately, we could not locate a specialist who could meet our schedule, and our volume is the poorer for not having a comparative perspective.

As we produced chapters, we needed help from others to give us perspective. Ray Gambell, secretary to the IWC, read the entire manuscript and saved us from numerous errors. He also opined that he was glad he was "not a party to any national position." All chapters also were read and commented on extensively by Jay Hastings, Esq. (Japan Fisheries Association). Jay would have joined us in the writing but for a potential conflict of interest. Nevertheless we benefited from his considerable experience with the IWC. Another faithful reader and critic was Dan Goodman, retired Canadian official experienced with the whaling regime and now working with the Institute of Cetacean Research, Tokyo. Arild Underdahl (University of Oslo), Oren Young, and Gail Osherenko (Dartmouth College)—all world-class scholars—provided the kinds of insights that one can get only from people at the pinnacle of their professions. This is also true of my colleague John Odell (University of Southern California) who helped with bargaining analysis. We also received excellent comments from Ron Mitchell (University of Oregon), a young scholar knowledgeable about ocean regimes and whaling who almost had time to join the team. Unfortunately he had other commitments. This was also the case with Harry Scheiber (Boalt Hall, University of California). We wanted Harry to join us, perhaps to turn our discourse into a dialogue, since we knew that Harry disagreed with some aspects of our "case." Unfortunately, "deaning" and scholarship are a difficult mix. Finally, we truly regret that Ed Miles (University of Washington) could not join us because of circumstances beyond his control. Ed always has something insightful to say.

No book can be successful unless the details are right. I appreciate the efforts of Warren E. Chung in preparing the bibliography, and Janice Hanks, librarian, Von Kleinschmid Library (University of Southern California), in seeing that it meets a high standard. Melissa Welebir, attorney, budding international relations expert, and editor, Kathy Mathes, and Mauricio Sanabria helped make the volume a coherent whole. Silja Omarsdottir helped prepare the index. My wife, an editor of many years' experience (some of it on the University of Washington Press staff), made sure the text flowed. Finally, all of my colleagues join me in thanking Mary Ray Worley for her outstanding copyediting; and Michael Duckworth, acquisitions editor, Marilyn Trueblood, editor, and Naomi Pascal, executive editor, all with the University of Washington Press, who helped bring to fruition both of my books for the University of Washington Press, thirty-six years apart!

ROBERT L. FRIEDHEIM

TOWARD A
SUSTAINABLE
WHALING
REGIME

Introduction

The IWC as a Contested Regime

ROBERT L. FRIEDHEIM

THE PROBLEM

Although the International Whaling Commission (IWC) has survived for more than fifty years, its present is highly contested and its future may be in doubt. It is so contested that accusations of murder (and, sometimes, blood-colored liquid) are flung at delegates from member states who wish to resume whaling, and accusations of racism are flung at representatives of those states who wish to consolidate the present preservation regime. It has become so ugly that there seems virtually no possibility of cooperation between the contending parties, no chance of reaching a consensus that would allow the commission to go forward and improve the fate of the great whales and possibly also allow for improved management of smaller cetaceans. As John Knauss, retired administrator of the National Oceanic and Atmospheric Administration and U.S. commissioner to the IWC, says in the foreword to this volume, "As an example of good faith international negotiations, the IWC is mostly a disaster."

When both sides demonize each other, it is very difficult to get a dialogue going.[1] The problem is a very difficult one to "fix." Yet try we must, however difficult the task, and even though the prospects are dim for the immediate present we must prepare for the future when circumstances may change. In the meantime, opinions are sharply divided on whether there is anything to reform, and especially whether any reform can be found that would be acceptable to all major stakeholders.

The contest is about *which general approach* the International Whaling Commission should take to manage the great baleen and toothed whales of the world's oceans—sustainable use or preservation, and *how* that approach should be implemented. It is a problem of both substance and process.

Through much of the twentieth century, from the first modern whaling treaty of 1931 on through the establishment of the International Whaling Commission in 1946 to the present, whales and whaling have been on the international agenda. After the age of oil-fired lamps and hoopskirts was over and the demand for whales dropped after the discovery of oil in Pennsylvania and the development of kerosene, and later, electric light, the take of whales from the world's oceans (now almost exclusively for food) was still too high and could not be sustained. Major states came together to sign treaties to regulate whaling, mostly on the basis of biological criteria. The idea of sustainability was built into the treaties, but achievement of that goal eluded them—sometimes because key states refused to cooperate (Germany and Japan in the earlier treaties); sometimes because participants refused to accept individual national quotas and each competed for a larger share of an overall quota; sometimes because they used the wrong criteria for establishing overall quotas (such as blue whale units, to be explained later); sometimes because the quotas were simply set so high, even though participants may have known that they could not be sustained; and sometimes because of cheating, that is, the number of whales caught was not accurately reported. But the goal—sustainability—remained constant and was recognized in the 1946 International Convention for the Regulation of Whaling (ICRW), the founding document of the IWC. The two linked purposes of the IWC are (1) conservation of the stocks, and (2) orderly development of the whaling industry.[2]

Some observers, including three contributors to this volume (William Aron, William Burke, and Milton Freeman), contend that, despite many failures in the past when the short-run interest of the whaling industries of major states dominated IWC decision making,[3] the commission was on the verge of success in developing a regime based on sustainable principles through a New Management Procedure (NMP).[4] But the NMP, endorsed by the commission in 1974, may have been developed too late. In 1972 the United Nations Conference on the Environment held in Stockholm demonstrated that the world community had been careless in its use of "natural capital" and had fired the enthusiasm of a rapidly enlarging world environmental community to do better. Where better to do better than with whales and whaling?

Television programs and movies of the 1960s through the 1990s featured cetaceans—Flipper the dolphin and Shamu and Willy the whales—as special creatures. They were almost human in their responses. They were distinctive in their size, supposedly slow in reproduction, and their very existence, it was widely proclaimed, was severely threatened. Not only did they have to be saved from extinction, but because they were "special," they had to be saved from human exploitation. They, along with elephants, became the "poster boys" of a number of major environmental organizations. A scientific wag characterized them as "charismatic megafauna."

Because of the IWC's spotty record of dealing on time to prevent significant reductions in the stocks of some major whale species,[5] and because major developed states no longer had much of an economic interest in sustaining a whaling industry, and because after 1972 a number of important environmental organizations (most obviously Greenpeace) that operated beyond national borders now had the domestic political influence and emotional backing of mass members in important developed states and could successfully sway governments, a drastic change in approach was made in the IWC. In 1982 a moratorium (to be enforced in the 1985–86 season) was imposed on all commercial whaling. Since, in theory, it was imposed to provide a pause until a new plan was developed to put whaling on a more scientifically acceptable basis, it was supposed to have been temporary and was due to be reviewed in 1990. But it has become de facto a permanent ban on commercial whaling, and the plan needed to lift it—the "Revised Management Scheme" (the management measures needed to supplement a Revised Management Procedure [RMP], which dealt with the scientific issues)—has thus far not been approved.

If the moratorium is sustained de facto or de jure, it must be considered a watershed event, a rare example of a fundamental shift in nature and purpose of an international resource management regime. The moratorium was supplemented by a Southern Ocean sanctuary in 1994, which was obviously directed at one member state who wished to resume commercial harvests in the Southern Ocean—Japan. If the moratorium and Southern Ocean sanctuary are maintained, it appears likely that another watershed has been reached.

While maintaining the moratorium, a number of major member states recognized that it was no longer possible to defend a complete whaling ban on the grounds that the conservation rules imposed under the earlier NMP (and now updated in the RMP) were not working or capable of working well, and they shifted their rationale to moral and ethical grounds inside as well as outside the IWC. In addition, the commission closely scrutinizes the whaling quotas of indigenous people (allowed in the basic IWC action document called the Schedule) to a point that some of them complain of harassment, and the commission has refused to grant a quota to artisanal whalers who are technically not indigenous under the IWC definition but who share many similar characteristics. Finally, some of the core dominant states in the commission are trying to push it into exercising control over small cetaceans, which are arguably outside its jurisdiction.[6]

Those members who want whaling to be restored protested that there was no evidence that a moratorium was needed to assess the state of the stocks that might be subject to exploitation and that the existing management scheme was adequate to control the limited whaling being requested. But states proposing the resumption of limited whaling opined that, with the imposition of the whaling moratorium, the majority should act on what was learned—that the stocks already placed in the highest IWC category as Protection Stocks (which include right, gray, humpback, blue, fin, sei, and sperm whales) are mostly recovering and that the one whale stock slated for taking—the minke whale—is quite abundant. For example, there are 760,000 minke whales in the Southern Ocean, and the Scientific Committee indicated that a take of two thousand whales would not endanger the survival, much less the stability, of the stock. Moreover, no one is proposing, or is likely in the foreseeable future to propose, that whales already protected before 1982 be placed in the next lowest category (Initial Management Stocks) or much less in the lowest category of IWC management (Sustained Management Stocks) and subject to hunting.

The states proposing the restoration of whaling were completely rejected in all of their requests—no lifting the moratorium, no Southern Ocean minke quota, no quotas for small nearshore whaling efforts off the coast of Japan, no whaling allowed in the North Atlantic or Arctic by Norway or Iceland, vigorous debate and close scrutiny of indigenous quotas, debates and scrutiny of "humane killing," even though the obvious objective of opponents of whaling was no killing at all. In addition, those in the minority were told that no quotas could be awarded until a new Revised Management Scheme (the rules for observing, supervising, and controlling the hunt) was put in place to flesh out the catch algorithm developed in the New Management Procedure. But when, for example, Norway submitted a draft supervision plan, opponents condemned it as inadequate. However, anti-whaling forces never submitted recommendations for fixing the holes in the plan or a substitute plan of their own, and, assuming that no improvements would lead to approval, pro-whaling forces did not press on with their own amendments. By delay, proponents of a permanent preservation regime hope to turn a de facto arrangement into a de jure international legal edifice.

In short, states wishing to resume whaling under sustainable conditions were expected to surrender to a new morality as expressed by a majority of developed states that, except for the United States, which had to protect its indigenous whalers, had no direct economic or cultural interest in continuing the taking of whales. Indeed, those who wished to consume whales as food were told that whales had more economic value as objects of tourist awe—the objects of whale watchers. Whalers were told that what they wanted was immoral, and they would have to be shamed into not asking for the right to kill special creatures. The implication was that not only were present whalers—in the phrase used in current American movies—"natural born killers," but so were their ancestors.

The whaling moratorium raises the most profound of human rights issues—the right to maintain a set of cultural practices as long as those practices do not lead to the overharvesting of the creatures they depend upon. Indigenous and artisanal whalers are told that "We cannot restore every traditional culture."[7] But that begs the question of whether the peoples and governments of major developed states have the right to destroy a people's culture. Shouldn't the great and powerful be a bit more careful of imposing what they claim are universal rules on others, especially those who do not share their values? All individuals and governments have the right to *attempt to persuade* others to change their ways. But do they, or should they, have the right to *coerce* them?

The quarrels are just as vehement on the procedural side. Did the new majority of states committed to preservation move the commission in such a direction by *means* that were legally permissible and political just? The issues here are quite serious. After all, the IWC is the International Whal*ing* [my emphasis] Commission, not the International Whale Commission. Two contributors to this book, Jon Jacobson and William Burke, deny that the decisions reached meet important tests of international legality. As with many legal issues, there are lawyers on the other side who will defend the legality of the actions, claiming that whatever the "hard law" requirements they are superceded by new "soft law" obligations. Often these "soft law" obligations are manifested, the majority claims, in a stream of recommendations that cajole, warn, and advise the states that whale or wish to revive whaling to change their ways. The minority point out that IWC recommendations are not enforceable in international law.

Clearly the stated regime goal of fostering the "orderly development of the whaling industry" has been dropped from the considerations of the commission, and the majority states refuse to put the legal issues to any third-party scrutiny or intervention.[8] Whether the new majority, in taking actions, met a test of having their measures "based on scientific findings,"[9] as required by the founding convention, is also questionable.

The new majority achieved its victories by means that raise interesting questions for participating states as well as students of international relations. For some time, a number of parties in the IWC have used tactics such as threatening reductions in trade quotas, or boycotts against states who vote against them in the IWC, but lately the anti-whaling forces—both governments (the new majority of "like-minded states") and nongovernmental organizations (NGOs)—have been more successful. Elizabeth DeSombre characterizes this process as "bullying and bribing." While it may get those who use such tactics what they want in the short run, it leaves a sour taste in the mouth of those on the receiving end as well as outside observers, raises fundamental issues of procedural justice, and may in the long run discourage the formation of other international environmental and resource regimes.

Other procedural issues also raise important questions that may come back to haunt the international community. One of them is the manipulation of membership. The only requirement for membership in the IWC is status as a state under international law, making it easy for states and NGOs with financial resources to get the votes needed to claim a majority position. It has long been known that some of the major contestants—NGOs as well as states—have paid membership dues for states that promise to vote with them ("bribing" in DeSombre's view).[10] However, "stuffing" the commission with "friends" may discourage states that disagree with a manipulated majority from joining or remaining in organizations open to all. Why should a state allow a majority of states within an organization, including many that have no special interest in the issues highly meaningful to that state, to make decisions that are binding upon that state? This could be viewed as a questionable precedent, as Christopher Stone, a contributor to this volume, has noted elsewhere.[11]

Moreover, the new majority in the IWC seems to act as if majority rule creates obligations for a minority that override their sovereign rights. The

ICRW is unusual among environmental agreements in that it allows a three-fourths majority to make decisions on important issues. But the ICRW's concession to sovereignty is to allow those states that disagree to lodge an "objection,"[12] and if they do so, they are not bound. This allowed the founders to give the organization some forward momentum without challenging the sovereign rights of specific members. Since the right of objection was used with some frequency by Brazil, Iceland, Japan, Norway, Peru, the USSR (and its successor, the Russian Federation) on key management measures, the new majority felt frustrated.[13] Sovereignty is supposed to be respected, and it is in the formal sense by the commission. Its Schedule (which records objections as well as moratoriums, etc.) makes clear who is and who is not bound. The official acts of the commission clearly demonstrate that the IWC is not an international environmental legislature.

But that limitation has not stopped the new majority from trying to force or shame members of the minority to drop their objections. The United States, for one, seemingly claims that the dissenting states are bound anyway, because in exercising their objection they are "reducing the effectiveness of an international environmental agreement," and therefore subjecting themselves to punishment under U.S. domestic law.[14]

For supporters of a complete ban on the taking of whales, the means are legitimate and merely one aspect of world democracy in action. From their perspective it was about time that an international organization responded to the people of the world who represented a new morality. Since they represented a majority,[15] the minority was bound to acquiesce to the will of the majority; sovereignty must give way to a new order, especially since sovereignty has become less potent in a world dependent on a globalized economy. What they are trying to do is establish a new universal norm, one that "qualif[ies] state sovereignty" on world environmental issues.[16] To be truly universal, the rules should apply to all people. In short, sovereignty is going anyway, and what is happening in the IWC is a mere indicator of the direction the world community is taking. The preservation norm has gone through a "tipping point" and has "cascaded" throughout the world community, and all that needs to be done is to have it "internalized" by the peoples and governments of the world.[17] That is being accomplished by a stout defense of the new preservation norm in the IWC.

Perhaps one reason for the rigidity of some of the preservation forces is not only that they consider the decision just, but that it is a great victory for those who wish to promote international environmental democracy and reduce the ability of states to dissent. After all, at a meeting at the Hague in 1989 they had been defeated in creating an international environmental legislative body that would have had "power to impose new environmental regulations and binding legal sanctions on any country that failed to carry them out."[18]

A further indication of the effect democratic practices have increasingly had on world environmental affairs is how NGOs have operated in the IWC since the Stockholm conference. In 1978, six years after Stockholm, adoption of an American-sponsored resolution allowed NGOs much greater direct participation in IWC proceedings. NGOs have taken vigorous advantage of their new rights, demonstrating, arguing, preparing position papers, presenting their consultants' research results, advising delegations, participating in some delegations, and even heading the delegations of some small states. They forced a new openness on the IWC. It is inconceivable that a backroom deal could escape their knowledge, and they may have sufficient clout to prevent one from being adopted by the member states.

One could perhaps see this aspect of the vigorous participation of NGOs as a good thing. The proceedings of the IWC are more open. But they are also more chaotic and more conflictual.[19] At last count some ninety NGOs participate in the annual two-week meeting of the IWC and its committees and working groups. The meeting has taken on a circus atmosphere, sometimes making it difficult to accomplish the work of the commission. The original burst of new NGO participants after 1978 came mostly from anti-whaling environmental groups. More recently groups representing whalers and indigenous groups have more vigorously represented their positions. Democracy, as Elizabeth DeSombre, Steinar Andresen, Robert Friedheim, and others among this volume's contributors will show, has its price.

The existence of a seemingly powerful majority and a weaker outvoted minority might imply that the majority has won all of its objectives, the minority lost all of its rights. That is not quite true. The majority has won the big votes, created a moratorium they refuse to lift, and augmented it with a Southern Ocean sanctuary, but they have not been able to force or convince the minority to give up all of their rights to whale. In short, preservation as a new norm has not been internalized by key states to whom it is intended to apply. The current situation is that of stalemate with neither side willing to defer to the demands of the other side.

This stalemate has come about because there are "loopholes" (as seen by the anti-whaling forces; sovereign rights as seen by pro-whaling states) in the

I C R W that allow the minority to persevere. Norway registered an objection to the moratorium and has resumed whaling for minke whales, using the algorithm developed for the Revised Management Procedure as its management guide, and it is now taking about five to six hundred whales a year. Japan too entered an objection to the moratorium but was convinced by the United States to lift it for a quid pro quo of being allocated a fishing quota in the U.S. two-hundred-mile Exclusive Economic Zone only to find that the United States eliminated all foreign fishing in its Exclusive Economic Zone. Japan felt betrayed; consequently, for this and other reasons, they exercise their right to conduct scientific whaling in the Southern Ocean and North Pacific and take some four hundred whales a year. This infuriates the antiwhaling forces, which claim that "scientific whaling" is just a dodge to get around the moratorium and sanctuary.[20] But Japan has a legal right to take whales for study, and indeed an obligation under the I C R W to process and use them.[21]

Other "whaling states"—Canada and Iceland—responded in a different way: they dropped out. Canada, because her constitution contains a provision guaranteeing her native peoples a right to take wildlife, allows limited hunting of whales under her own jurisdiction. Iceland, which threatens to resume commercial whaling but (as of this writing) has not, believes that if it does resume the hunt, it could do so with legal impunity under the jurisdiction of the North Atlantic Marine Mammal Commission (N A M M C O), created as a potential rival (or, perhaps, complement) to the I W C. Adherence to N A M M C O standards would fulfill a state's obligation under Articles 65 and 120 of the UN Law of the Sea Convention to "work through the appropriate international organizations" (note the plural).[22]

The I W C actually controls very little of the whaling that is conducted around the world today. In addition to the hunts conducted under the auspices of the Japanese and Norwegian governments (about a thousand minke whales between them), Canada's native take, the Faeroe Islanders' hunts for pilot whales, and Greenland natives' hunts for the narwhal and beluga (both Greenland and the Faeroe Islands are under Danish sovereignty but have home rule) are also beyond I W C control. It is known that there are cetaceans in the waters off the coasts of about a hundred states, but only about thirty-five to forty states (some landlocked) are members of the commission. How many allow their fishermen to take whales is unknown, but states such as the Philippines (a former member of the I W C) and Indonesia,[23] where it is known that whaling occurs, are careful to remain outside the jurisdiction of the commission. A stalemate indeed! The like-minded states control the IWC, but they do not control whaling.

THE CONTESTANTS

It is convenient to slip into language that groups disparate individuals, organizations, and governments into categories (e.g., pro- and anti-whaling) because they share one or more (albeit important) attributes. The shared attributes allow them to form a coalition, but the strength of that coalition depends more on other attributes they do, or do not, share.[24] To understand the whaling contretemps, we must briefly look at the major components of the main contending groups.

The more successful coalition calls itself (on the government side) the "like-minded states." Its principal members are the United States, the United Kingdom, France, Germany, Australia, and New Zealand, with significant support from the Netherlands and Italy. The coalition can usually command a core vote of about twenty to twenty-three states from a total membership that has varied from thirty-five to thirty-nine in recent years. It usually can pick up enough swing votes to allow it to control IWC decisions, even three-quarter majority decisions.[25] The preservation position these states espouse is supported by Greenpeace, World Wildlife Fund, International Fund for Animal Welfare, Center for Marine Conservation, Environmental Investigation Agency, International Wildlife Coalition, and a host of smaller, single-issue organizations such as Whale and Dolphin Conservation Society, International Dolphin Watch, and others. The principal like-minded states meet between and in preparation for IWC meetings and plan their common strategy.

Since the preservation forces can no longer defend a moratorium or even the Southern Ocean sanctuary on scientific grounds, they have turned to exclusively ethical arguments. The group's ideological glue is provided by deep ecologists such as Arne Naess, Bill Devall, and George Sessions, who wish to "cultivate a deep consciousness and awareness of organic unity, of the wholistic nature of the ecological webs in which every individual is enmeshed"[26] and animal rights advocates who consider it an ethical imperative to preserve all animals. They claim their arguments are biocentric (meaning that they affirm the value in every living thing and assert that people have an obligation to take that value into consideration in all of their actions).[27] They are joined by other animal rights advocates and groups who proclaim whales "special creatures." Finally, there is strong support for a preservation regime from some managers in government departments responsible for ocean, environment, and (more recently) conservation, as well as students of the history of ocean resource management. They fear that past failures in resource management augur poorly for our ability to manage "sustainably" in the future and that the best means of preventing hunting from getting out of control is to allow no hunting.[28]

For strict animal rights advocates, the present regime is merely a first step on the path leading to Eden, an evolutionary move toward an inviolable right to life for all creatures.[29] For them, animals have rights equal to, and perhaps superior to, human beings. Indeed, for some animal rights advocates, notions such as human rights are morally and legally wrong because they are "species chauvinistic."[30]

Other defenders of animal rights claim that some creatures are more privileged than the rest, especially whales, other cetaceans, and elephants. These animals have larger brains and, it is claimed, create poetry, communicate in complex ways, feel pain, and enjoy family life. Moreover, they have long gestation periods and reproduce slowly—too slowly to remain sustainable if hunting is allowed.[31]

Defenders of the whale's right to life cling fiercely to the present regime. Believing they are riding the forward wave of a consensus on a new universal norm, they work vociferously toward strengthening the present regime and applying it to cetaceans not already under I W C management, even in waters where the UN Law of the Sea Convention grants coastal states "sovereign rights for the purpose of exploring and exploiting, conserving and managing the natural resources."[32]

Animal rights activists' claim to a new morality is a powerful call to action that resonates in many urban developed societies. It appeals to substantial numbers of white, middle- to upper-middle-class, well-educated individuals who increasingly are pursuing postmaterialist goals.[33] The cry to end whaling engages their sense of guilt for bad behavior of long-gone compatriots. This guilt about the past is the flip side of the current Brundtland report–based concern with "inter-generational responsibility," whereby those in the present are obligated to show concern for what they leave their children and grandchildren.[34] Many are citizen activists who participate actively in direct-action movements[35] and nongovernmental organizations, some of which have played important roles in international environmental politics[36] and in the proceedings of the I W C.

For the organizations to which many of these committed individuals belong, the emphasis they have placed on the "sacredness" of their charismatic megafauna has created a major dilemma. Even if, as a result of new knowledge or better understanding, they might consider it sensible to shift their positions, they often feel constrained because of a fear of losing their funding and their membership base. When an individual or organization sincerely believes it is right, and the facts show weakness in their case, it is very difficult to admit that a position should be changed.

But there are pragmatic as well as idealistic reasons to try to end whaling. Some devotees of animal rights who also value utility[37] claim that they could be persuaded to sanction the killing of some whales if the proponents of whaling could show that the usefulness of killing whales, in terms of benefit to humanity, outweighs the ethical costs. For them, the defenders of whaling have not made the case. There are adequate substitutes for whales, and therefore ethics count more than utility. After all, the governments of Japan, Norway, Canada, and other developed states whose citizens still kill whales easily can provide substitutes—beef, for example. Whales are fungible, and if taking them is prohibited, potentially those who lose that right can be compensated with other benefits.[38]

For other individuals, especially those in the resource management, environment, or conservation departments of major governments, NGOs, or academies with knowledge or experience in ocean resource management, the major attraction of a preservation regime is that it has been relatively easy to enforce. It is either/or—either whale products are off the world market, a fact easily discerned, or whale products are on the world market. If whale products were to remain on the world market, managers would require much more sophisticated methods of tracking to account for the species taken, the location of catch, and the amount of product (meat, blubber, bones). Moreover, there would be many more opportunities to "cheat" and overharvest. This variety of pragmatist shares a deep suspicion that no compliance rules will be foolproof and that the Revised Management Scheme might prove as full of holes as past management schemes. After all, "overfishing is notoriously resistant to traditional resource management approaches."[39]

Deep down this quarrel that pits state against state, and at IWC meetings individual against individual, is based on the perceptions of individual human beings, with all of their biases. Moreover, some individuals with either real-world ocean management experience or historical memory look with doubt at states that propose the restoration of whaling and claim these states do not have "clean hands" and are therefore untrustworthy.[40] This accusation is especially directed at Japan, with whom U.S. fisheries managers and fishermen have been locked in conflict since the U.S. occupation forces at the end of the Second World War encouraged Japan to resume high seas fishing to feed its people. Ever since, U.S. stakeholders have been trying to push Japanese fishermen back to waters closer to their own shores. This is an unstated subtext to the noise generated at IWC meetings.

The coalition supporting the restoration of whaling under controlled conditions is much smaller, and its members are, on the whole, politically weaker and more dispersed than their opponents. They often have difficulty coordinating their strategies in the IWC.

Only two state members of the pro-whaling group are of major importance in general world politics—Japan, for its economic might (the world's second largest economy), and Norway for its wealth per capita and its environmental record. As mentioned, Canada and Iceland have dropped out of the IWC, but (along with Namibia and Zimbabwe) they are observers at the IWC meetings. Smaller states that often vote with Japan and Norway are more interested in maintaining rights of indigenous whalers or creating rights for artisanal claimants. They include most of the Caribbean state members of the IWC (St. Vincent and the Grenadines, St. Lucia, St. Kitts and Nevis, Grenada, Antigua and Barbuda, and Dominica) and sometimes China, Korea, the Russian Federation, Denmark (on behalf of Greenland), and the Solomon Islands. Occasionally, depending on the issue, the pro-whaling group can rally some of the swing voters. Some members of this group meet as a caucus between and in preparation for IWC meetings, but they do not have a successful record of developing a united coalition strategy. Norway, for example, is criticized by some of the others for going its own way and taking care of its own interests without regard to the needs of other coalition members.

The pro-whaling coalition, as noted, has been supported by increasing numbers of NGOs in recent years. Among them are the Japan Whaling Association, Japan Small-Type Whaling Association, World Council of Whalers, High North Alliance, International Wildlife Management Consortium, Inuit Circumpolar Conference, the European Bureau for Conservation and Development, and the Norwegian Whaler's Union.

The position of the pro-whalers, as might be expected, is based on a combination of defense of principles and defense of self-interest, which are inextricably intertwined. These get mixed together in the espousal of sustainability or sustainable use as a standard for human conduct, the cultural rights of peoples to establish their own food standards and defend their right of access to food, the defense of states' legal rights under international law, and a resistance by the smaller and (for the most part) weaker to the coercion of the great and powerful.

The responses of communities and states to pro-whalers' demands for animal rights range from incredulity that the major states would let their positions be captured by fanatics who espouse an abstruse idea that "the life of a sentient being takes precedence over the economic or cultural interest of human beings"[41] to outrage that they are being made to pay for the sins of others. After all, the idea of animal rights comes from peoples whose ancestors were principally responsible for devastating whale stocks when it was in their interest to do so. While Japan and Norway were among the major whaling states, they by no means bear the full responsibility for the decline of those stocks of whales that were hunted to near extinction, and as they see it, their present proposals or actions concerning the resumption of whaling present no dangers to whale survival. Others were never part of the pelagic hunt but took whales from waters near their coasts. Some were by IWC rules "indigenous," others artisanal, and the distinction between them is quite artificial. But all feel under attack.

Peoples in isolated coastal communities depend upon whales not only as an economic necessity, but also because the whale hunt gives symbolic meaning to their lives. They do not want to give up distinctive features of their cultures in order to be consistent with the postmodernist values of metropolitan societies. They do not want to be told what is or is not acceptable food. Given their situation, they do not believe that one size should fit all; there ought to be space for human variety as long as what they do does no harm to the ecosystem.[42]

Whalers see themselves as being demonized and under extreme attack by those who are interested in tearing down their societies. More than once Japanese leaders and media couched the quarrel as a struggle between the world's "meat eaters," especially Anglo-Saxons, and "fish eaters," meaning (since they recognize that the whale is not a fish) that they fear that their entire ocean-based food system and taste preference is under attack. This accounts for a significant part of their stubbornness in defending whaling despite the fact that, in self-interested economic terms, a restored controlled and limited hunt would not even represent a recognizable blip in their gross national product or any other measure of economic well-being. Even in the case of Norway, while there is a measure of economic self-interest in restoring whaling, there are also value considerations. While it is in the political and economic interest of any Norwegian government to keep the relatively small number of people of Norway's north happy, the income generated by whaling has only regional impact, and Norway probably could afford to forgo it. But Norwegians generally see themselves as living in a frontier society that exists under harsh physical conditions and that depends upon the extraction of raw materials. Their self-image is at stake. This along with an exemplary (except for whaling) environmental record!

Symbols are potent. They become "easy objects upon which to displace private emotions, especially strong anxieties and hopes,"[43] indeed strong anxieties about the past and hopes for the future that have led to an intolerant and intemperate present.

WHY THE SUBJECT OF THIS VOLUME IS IMPORTANT

The whaling issue may be largely "symbolic politics," but successful resolution is important not only for whales and whalers but for the international community. After all, the IWC's official acts have been treated as authoritative. Despite concerns expressed by anti-whaling states and NGOs that whalers are "cheating" on the new preservation regime, no member state, either while remaining a member or after dropping out, has simply ignored the formal IWC rules made by legitimate decision-making processes. No state has simply gone off and said, in effect, "to hell with the IWC!" and resumed whaling on its own authority in ways that would violate IWC rules. That is a commission achievement that should not be lost. The IWC should be salvaged, but important issues must be tackled.

Every discussion and formal proceeding about whales and whaling is suffused with, and often dominated by, a morality claim for the position espoused. However, we will try to concentrate on "ideas" that can be subjected to verification and logic. None of the authors claims to be a professional philosopher, so we do not tackle the perfect abstract notion of justice, or even one applicable to this situation. Nevertheless, this issue has been prone to strong claims of moral correctness, to securing justice for people and animals, to seeing that equitable behavior is adhered to in both process and outcome. The issue attests to the importance of core beliefs in resolving this type of conflict. Both sides claim moral correctness. It is difficult to apply rational assessment procedures on *either* claim—one claiming "to hold a more emotional, spiritual attitude toward the whale," the other claiming to support traditional beliefs often central to the defender's core sense of being.[44] Both are "spiritual" and therefore mysterious, "half shades and indeterminate outlines."[45]

The authors cannot—and probably should not even try to—convince people to give up core beliefs. Our dilemma is how to get people to recommend a rationally defensible outcome without asking stakeholders to give up their core values. We can only note that something similar was accomplished previously on an even larger problem.

In the sixteenth century, Europe was engulfed in religious wars. People believed that, since the right to rule came directly from God to the king, one could be ruled only by a coreligionist. The sovereign *secular* state was invented in part to demonstrate that all people within a territory could be treated equally as citizens, effectively separating church from state.[46] Loyalty to the state could be owed regardless of religion. In short, to achieve civil peace it was necessary to put aside some core beliefs. Can the stakeholders involved in the whaling issue put aside their particularistic notions and rally around an idea such as the survivability of whales in a world now more obviously interconnected?

One of the tragedies of the way this issue has been framed is that it prevents the contestants from seeing that they do share certain basic premises and they might be able to use that insight for finding a consensus-based solution. They might not have to put aside everything. There is an "overlapping rather than a strict consensus"[47] on the fact that whales should never be exploited to the point of seriously diminishing the stocks, much less driving them to extinction. No one proposes that whales should be treated as common pool resources. Perhaps these shared premises can be built upon.

JIHAD VERSUS McWORLD, OR CENTRALIZATION VERSUS DECENTRALIZATION

Vast changes in the world economy and political system in the last part of the twentieth century have lead to a situation in which connectedness, and perhaps integration, and, certainly, significant degrees of cultural homogenization have occurred in many regions. The Cold War is over, capitalism and free trade are triumphant, and as many observers have noted, borders of nation-states are no longer sacrosanct. The world is now safe for what

Benjamin R. Barber calls "McWorld," a social system in which the cultural standards of the West (and particularly the United States, whose main exports many believe are "sex, drugs, and rock and roll") dominate, no national government can truly control activities within its own territory, and multinational corporations do as they please.[48] But the victorious forces of change have engendered their own opposition—"Jihad"—which, Barber notes, may have begun as "a simple search for local identity, some set of common personal attributes to hold out against the numbing and neutering uniformities of industrial modernization and the colonizing culture of McWorld," and has sometimes become a full-scale revolt against the standards imposed, sometimes implicitly, sometimes deliberately, by the world's great postindustrial political powers. McWorld versus Jihad provides a contextual nexus for the whale quarrel.

Both major forces involved in the whaling fight, attempting to confront the unalloyed consumerism of McWorld, invoke aspects of Jihad—one relies upon a mystical notion of a higher responsibility of human beings toward animals, the other upon the right of small, remote, or culturally different communities to live by standards they design and not by standards imposed upon them by metropolitan powers. But the forces of the animal rights jihadists have been able to take advantage of McWorld's loosening of national borders and more centralized decision making on issues of worldwide importance. They have done this through control of multilateral organizations. It may be paradoxical that Jihad can exist inside McWorld, but Barber demonstrates this is not unusual.

One of the key features of the whaling quarrel is a strong underlying pro-whaling jihadist perception that through control of the central decision-making mechanisms of the IWC (backed by U.S. enforcement), major states are trying to impose their will—and their values—upon smaller or culturally different states. A strong sense that centralized decision making threatens the liberties of smaller or different states runs through the issue. It raises important questions in the minds of some who, despite the fact that the world is more homogenized and decisions on some important issues more centralized, still recognize that the world system is still largely anarchic and that it may not be in their interest to cooperate in future multilateral efforts.[49] As important a test as whaling is of the viability of multilaterally established rules, what is happening within the IWC is only one example of a more general revolt against dominant powers imposing their will on others.[50] In a June 1997 meeting of the Convention on International Trade in

Endangered Species (CITES) in Harare, Zimbabwe, three African states that have managed their elephant herds but were banned from selling ivory from culled animals to help pay for their efforts argued and successfully lobbied to overturn the rigid ban on international ivory sales. Perhaps the partial accommodation of the needs of these states is a sign that those in control of international resource management issues have become more sophisticated in their approach and can take care of local needs and sensibilities, and this is a sign that a serious threat to global environmental governance can be avoided. One hopes so, but the jury is still out.

CLARIFYING THE CRITERION OF "SUSTAINABILITY" AS THE BASIS OF HUMAN USE OF NATURAL RESOURCES

Some observers of international environmental affairs fear that abandoning a sustainability standard in the IWC means repudiating a concept vital to international environmental management. Unless human beings forswear extracting economic value from nature, they need a principle to guide them on how far their exploitation may go. The only candidate is sustainability, as the Brundtland Commission Report recognized, but it is not an easy concept to implement. The Brundtland Commission, established by the United Nations General Assembly in 1983 to propose long-term environmental strategies for achieving sustainable development, posited that "sustainable development is development that meets the needs of the present without compromising the ability of future generations to meet their needs."[51] It is a useful general guide (to make qualitative improvements of some dimension but not to allow increases in extraction of natural capital)—but the specifics of how far exploitation may continue requires much more precise definition to fit the different types of exploitation. Unfortunately, several years ago, Baranzini and Pillet noted at least twenty different interpretations of the meaning of *sustainability*—and this in the economics literature alone![52]

A working definition of sustainability for whales and whaling must be tailor-made, specific to the problem of maintaining healthy stocks of wild animals while exploiting them. As noted, the International Convention for the Regulation of Whaling (1946) is one of the oldest international environmental instruments. It was remarkably prescient and "Brundtland"-like in its attempt to define its goal: "that the whale stocks are susceptible of natural increases if whaling is properly regulated, and that increases in the size of whale stocks will permit increases in the number of whales which may be captured without endangering those natural resources."[53] But the IWC went further in its Schedule by developing categories for establishing permissible and impermissible activities. It defined Sustained Management Stocks, Initial Management Stocks, and Protection Stocks in terms of percentages above and below a maximum sustainable yield (MSY).[54] How these were defined—especially since MSY has been a controversial concept in the biological and resource economic communities for some time—may be faulted, and their implementation can be heavily faulted, but the approach of establishing trip wires for permitted levels of exploitation is fundamental to all wildlife management including the risk-averse Revised Management Procedure developed by the IWC Scientific Committee. It is an approach being used in the recent Straddling Stocks Agreement, in which scientists are asked to develop "reference points" to help make the "precautionary principle" (i.e., reducing the risk of extensive environmental damage in advance of scientific knowledge if doing so is cost-effective) of practical use.[55] It is a concept we cannot abandon without finding a better replacement.

THE ROLE OF SCIENCE IN INTERNATIONAL ENVIRONMENTAL MANAGEMENT

The whaling case is also an important test of the use of science in international environmental discourse. Interpretation of the facts in numerous environmental situations that might have lead to serious quarrels have been reduced to manageable issues because the parties have agreed to be guided by the findings and advice of scientists knowledgeable about the problem. In some cases, in which there has been consensus concerning cause of and solution to a problem, the findings and recommendations of a community of scientists (characterized as an "epistemic community") have forced political decision makers to make decisions consistent with the scientists' views.[56] Scientists have been treated as authority figures.

Even if one recognizes that science is not divorced from politics and culture,[57] and that all recommendations from involved scientists have moral and political baggage, it is difficult to find a substitute that would allow all sides to defer to "expertise" and avoid making a decision based exclusively on moral claims. The views expressed above are "modernist," I admit, and not wholly "objective." But one of the purposes of science is to reduce mystery. Science may not inevitably lead to social progress (a la Vannevar Bush), but science must still play an important role in choice.[58] Those who wish to go in another direction have the burden of demonstrating that a better basis of dialogue exists.

How science has been used in the IWC—and not just recently—is a dismal story.[59] Lately, there has been a near consensus in the community of whale scientists concerning the facts and proper scientific interpretation of the case, but with silent dissents by a limited number of well-known scientists. These scientists did not speak up in the Scientific Committee and therefore did not block formal consensus, but their doubts were cited by anti-whaling political forces as the reason for being skeptical about the worth of scientific recommendations, and they were used to undermine the consequences of scientific consensus.[60] The majority of states on the commission ignored the advice of the Scientific Committee, praised its report on the RMP, and initially refused to accept it. This rejection led to the resignation of the Scientific Committee's chair, Philip Hammond.[61] In a later session, the commission finally did adopt the RMP, but it still refuses to implement it.

More "horror" stories about the role of scientists in the politics of the IWC could be told—but to little purpose. The problem is to find an acceptable future role for science in the management of cetaceans.[62] There are wider implications. Some "activists" pick and choose "facts" about whaling that they prefer and ignore others, so they should not complain when their politicians and general publics also pick and choose what they prefer to believe about the nature of a potential problem, for example, global climate change. Consider the difficulty President Clinton encountered in trying to convince the American public to take seriously and be willing to sacrifice to avoid the worst consequences of a human-induced change in global climate. He tried to make the case with three Nobel laureates and four other scientists at his side.[63]

Although in December 1997 the United States helped negotiate and signed the Kyoto Protocol to the Framework Convention on Climate Change, its implementation is strewn with roadblocks, especially from those who either honestly or dishonestly exploit the (narrowing) dissensus within the scientific community concerning climate change.[64] If scientists' luster as authority figures is dimmed today, how helpful will their potentially nonauthoritative voices be in the future?

THE RIGHTS OF SMALLHOLDERS AND INDIGENOUS PEOPLE

Although we can treat the problems being encountered by smallholders and indigenous people in the IWC as part of the larger problem of McWorld versus Jihad, they deserve separate treatment because of the importance of the issue: as we enter the twenty-first century, can major developed states with sense and sensitivity address the problems of peoples, especially those living in remote regions, who wish to maintain their cultures and also enjoy the benefits of the modern material world? Often, as well as their cultures, their physical existence is dependent upon their ability to live off wildlife. While we in developed states pay lip service to the idea of cultural diversity, we act by a set of our own culturally determined rules to impose conditions on peoples in areas where, to misquote Ira Gershwin, "the living ain't easy." We expect them to respond as we would. Why shouldn't they accept beef from their government in return for not whaling or sealing? I heard a Canadian Inuit elder say, "I eat the beef, but when I go outside I am cold." Is he cold? Or is it his perception of being cold that is important? Or doesn't it matter what the explanation is if he is being coerced into eating beef?

Indigenous people have special standing in the ICRW. But the way the IWC attempts to manage indigenous hunts is insulting and perpetuates serious problems. The ICRW borrowed from a 1931 treaty restrictions that do not permit indigenous people "to deliver the products of their whaling to any third person."[65] That definition of total noncommercial involvement is characteristic of an era in which indigenous people were—at best—treated with romantic fervor as "noble savages." But a total absence of exchange was never a characteristic of most peoples, even those in isolated areas.[66] It cuts them off from the material benefits of interacting with the rest of the world. This is not to say that aboriginal hunts should not be regulated. Indeed, the Alaskan Inuit wish to use the latest in navigation equipment and harpoons to reduce risk[67] and increase the probability of success of the hunt, so their hunting capabilities are greater than their ancestors' were and, if abused, could do harm. We must strike a reasonable balance between the rights of peoples with special needs and the principles of sustainability.

Smallholders or artisanal whalers who are not considered indigenous in IWC discussions (even if their ancestors have lived in their current location a very long time) fare even worse under the ICRW; they have no special rights. This group includes peoples of northern Norway, Iceland, and Japan. Although the pilot whalers of the Faeroe Islands and Japan are not subject to the ICRW, they are under similar pressures. For them as well as for indigenous people, we must raise the question of whether the activities of the IWC are reducing human cultural diversity at the same time the world is concerned with the reduction in biological diversity.

FOOD SECURITY

There are approximately 6 billion people living today. If the rate of increase slows down as various countries develop (but citizens' expectations of material rewards increase), in the mid-twenty-first century we can expect approximately 7.8 billion people to be in need of food, shelter, and recreation.[68] If fertility remains high, we will have 12.5 billion people. As the number of people and their needs grow, so does our worry over whether there is enough space and resources to go around—even if the major states are willing to treat the needs of third- and fourth-world people equitably, which they show no signs of doing. There is a fear that the planet's future is one of increasing scarcity and therefore increasing conflict over access to or control over resources, particularly in the poorer regions of the world.[69]

Food security is a matter of vital importance and has a much greater role in the thinking of leaders of countries on the margin than of countries of abundance with an expectation of continuing plenty.[70] Ronald Inglehart, after a thorough survey of the development of attitudes relating to political, economic, and social change in forty-three countries, concludes that "The root *cause* of the Postmodern value shift has been the gradual withering away of the value systems that emerged under conditions of scarcity, and the spread of security values among a growing segment of the publics of these societies."[71] Since most people in developed states share a "subjective" sense of well-being, it is hardly surprising that their leaders have no "feel" for the feelings of those who have recently faced hunger—or will face hunger in the future.[72]

Could whale meat be a major contributor in any effort to stave off hunger in a future world that pessimists such as Lester Brown are convinced will not have sufficient resources to properly feed all the peoples of the world?[73] Not likely, even if uncontrolled hunting were to be restored. But it might be able to help, especially in particular vulnerable communities, if we do not box humankind in a corner and forbid taking restored whale stocks on a sustainable basis. And is the real unspoken "food security" issue the one raised by Christopher Stone, that in the twenty-first century whales may be seen as serious competitors for—predators upon—dwindling stocks of ocean fish?

"CHARISMATIC MEGAFAUNA"

A distinctive feature of the problem of managing whales—and elephants— is that many people in developed states think of them as "special creatures" because of their size or purported intelligence and therefore assert that they require a special regime not necessarily congruent with regimes for other forms of wildlife.[74] Often, humanlike attributes such as the ability to "talk" and "sing," compose poetry, and have complex thoughts, the quality of being majestic or—if they are porpoises or orcas—of being "cute," are assigned to animals that are now being characterized as "charismatic mega-fauna." But should they be exempt from the normal rules being developed under the notion of sustainable use? Should local peoples in remote or less-developed regions that have interacted with these creatures for ages be forced to change their relationship with the megafauna of their regions because of rules made for them by certain segments of the international community?

This issue is important because, whether just or not, a predominant de-cision-making role will be played by major developed states in developing and enforcing rules related to the management of the international envi-ronment. How their leaders think will be influenced by what their publics think. As they become more postmodern, less concerned about scarcity, more influenced by television, we will be living in a world even more dom-inated by the symbols television conveys. Sometimes leaders of developed states, in private, will admit that what they are doing in relation to charis-matic megafauna makes little sense. But while in office, they fear that buck-ing the messages of *Free Willy* and *Flipper* will cost them votes. A very pow-erful political constraint has been created.

More general is the question of where the thinking of leaders and publics of developed states concerning "nature" will lead us. As it is, we often think of "nature" as "naive reality." But nature is a human construct, and a trou-bling one. As William Cronon put it, "recent scholarship has clearly demon-strated that the natural world is more dynamic, far more changeable, and far more entangled with human history than popular beliefs about the 'balance of nature' have typically acknowledged." Nature is not so natural as it seems. We cannot describe nature without our values and assumptions influencing our understanding.[75] We need to manage something that is always changing, and exempting some of the creatures of the system from the rules, that is, fixing forever the way we manage them, does not seem to be a sensible way to proceed in dealing with a dynamic reality.[76]

A LAW-BASED APPROACH AND GLOBAL GOVERNANCE

The whaling case is an important test of the usefulness of a formal rule or law-based approach to international resource management, or more generally global governance. When it could be assumed that the international system was an "anarchic" one—that is, it lacked a central authority—then consent was the fundamental requirement for binding sovereign states to a joint solution to a shared problem. Consent meant that states were expected to enforce their commitments within their territories, and largely they did. Many successful treaties were mostly "self-enforcing." Formally, the structure of the system is still primarily anarchic, but, as we have seen, the ability of nation-states to control their territories has been slipping as a result of the economic, political, and social consequences of globalization. Consent—or withholding of consent—may not mean what it used to mean. Has world "integration" gone so far that states that have withheld their consent are bound anyway? If it has proved impossible to gain formal consent to a change in international resource or wildlife law but rules of "soft law" have been developed, are dissenting states bound?[77]

Trying to coerce a state to enforce rules to which that state has not consented troubles some international lawyers, our colleagues among them. It calls into question notions of contract as a basis of human behavior. International treaties are akin to contracts; states are expected to follow the rule "*pacta sunt servanda*," meaning that states must abide by their contracts or written agreements with other states. So that events are not frozen in time, international law also has recognized a notion of "progressive development," but how far can progressive development go to change and modernize international rules if those being ruled do not consent to the change in the rules? This is a fundamental dilemma to those who believe in a law-based approach to managing change in a rapidly changing world.

The problem of consent has even wider implications. Gaining consent has always been difficult. It is difficult one-on-one—that is, when two states face each other across a negotiating table. It is even more difficult when three or more states must agree before action is taken. A multilateral environment that is consent-based puts considerable leverage in the hands of the intransigent.[78] A multilateral environment in which a majority rules is much easier to bring to decision, but if it chooses to come to decision and enforce its mandate on dissenters, it violates our usual notion of sovereignty. Since many of the problems relating to the environment are inherently transboundary and regional (and sometimes global) in their impacts, the consent of many is needed to effectively deal with a large class of problems. It is little wonder that we hope for a situation in which international "legislation" is possible. But the consequences of altering the formal rules governing international process to force compliance by the intransigent are serious for the notion of international governance. If, despite the weakening of sovereignty, states can successfully resist the demands of a majority, what benefit has majority rule created, at least in terms of solving the problem? It can only exacerbate conflict to create "empty law."[79] Multilateral decision making is further complicated by the seemingly tighter linkage today between key domestic considerations, and often key domestic decisions, and international actions. Whaling may also be an example of "intermestic" policy in which bargaining at the international level may be mere posturing to satisfy one or more domestic constituencies.[80] Therefore, it is possible that those concerned with what has been occurring in the IWC are spinning their wheels, when the real action is elsewhere and the real decisions will be made by decision makers having little directly to do with the IWC.

THE IWC AND OUR FUTURE ABILITY TO RELY UPON INTERNATIONAL REGIMES

If lawyers are concerned with the rule of law, political scientists worry about a parallel problem—the viability of the underlying structures that allow for cooperation in the international system—international regimes. Typically, international regimes (defined as social institutions made up of agreed-upon principles, norms, rules, and decision-making procedures that govern the action of specific actors in specific issue areas[81]) attempt to solve coordination problems.[82] In a world that still is anarchic, with independent decision making still possible by so-called sovereign states, we cannot solve problems that transcend national borders without cooperation through coordination. While we rely heavily upon such institutions, they are fragile, since it is possible for members to defect or cheat. In fact, as the world becomes more globalized, we rely ever more heavily upon international regimes. The future of the IWC is an important test case of the world's ability to develop rules and rights based on shared norms in an issue area. A failure in the area of whales and whaling might have a ripple effect. What is at stake is not only whether the affected states will remain within the IWC but whether the IWC can demonstrate an ability to establish rules that will be obeyed, whether it can develop effective compliance mechanisms, and whether the necessary entrepreneurial leadership will emerge to get the IWC out of its present state of stasis.[83]

More generally, the whaling case provides a good test of theories of regime design, regime establishment, and implementation.

WHAT WE ARE TRYING TO DO

Although the purpose of this volume is to bring light to the subject of the management of whales and whaling, we still will emit some heat because the authors favor one case over the other. This book is an extended discourse on the problems of a preservation regime being fostered by the International Whaling Commission, and it considers whether anything ought to be done about those problems, and what might be done. While a dialogue would be preferred, there seems as yet little basis for one. Therefore we have set out to show our concerns with the current path down which the IWC seems determined to go. Since we present arguments on behalf of a sustainable-use approach to the problem of whales and whaling, we hope that those not acquainted with this side of the argument, or those opposed, will view our arguments with an open mind. We will try to reciprocate.

We will proceed first by presenting a set of essays on the legal, people, and scientific problems associated with a preservation regime. In short, the concerns here are mostly substantive. This will be followed by essays on the problems of process—how the regime was put in place and how it is being sustained, in short, the politics. We finish by tackling two challenges: First, are our arguments as good as we think they are? Second, what should be done about the IWC—if anything? The first of these concluding essays forces us to examine whether our analyses are as balanced and honest as we can make them; the second and third tackle the question of where the world community should go from here.

We may fail to impose "neatness" on all these diverse and complex issues. We are international lawyers, ocean scientists, anthropologists, and political scientists and therefore bring different professional perspectives to the analysis, but our interests and expertise overlap. William Burke and Jon Jacobson, given their long experience with fishery management, address the management system for whales as well as the formal legal requirements. Russel Barsh comments on food security, smallholder rights, *and* international treaties. Milton Freeman knows not only the cultures of peoples of whaling communities but also the culture of metropolitan societies. He has also observed the IWC in action many times and can speak of the politics with authority. William Aron, scientist and former U.S. commissioner to the IWC, has as much to say about commission procedure as whale science. Steinar Andresen, another close observer of the IWC, comments not only about the decision-making process within the IWC but also about the effectiveness of various regime arrangements. Elizabeth DeSombre, David Victor, and Robert Friedheim bring knowledge of other regimes to the whaling imbroglio, and Christopher Stone offers a helpful overview of all the arguments presented in *Toward a Sustainable Whaling Regime.*

All authors of this volume chose to participate because they think that there are problems with the present regime. In this sense, we are united, but no ideological or other filter was applied for participation. While we agree that there are problems, we do not necessarily agree as to the severity of the problems, and we disagree on what, if anything, should be done. In one of the concluding chapters, David Victor proposes that nothing be done because change that improves the outcome for one side could be achieved only at the expense of the other side. Therefore, we should leave well enough alone. In the final chapter, I propose a course of action not because I disagree with Victor's insight, but because it is evident that the parties will not leave the status quo alone and therefore we must think about how to bring about change; let us hope that it will be change through consensus.

Appendix. Whales and Whaling—A Short History

Some brave readers, if they have followed us to this point, might now be scratching their heads about the sequence of events alluded to and the concepts invoked. To help make the rest of this volume more understandable we offer below a very short history of attempts at managing whaling.

EARLY ATTEMPTS TO MANAGE WHALING

The history of industrial whaling is a record of extraordinarily rapacious behavior. Little concern was shown for the survival of the largest mammals this world has known. European and American whalers from the sixteenth to the twentieth centuries hunted one species after another to the point of extinction, usually switching to smaller species when the populations of larger animals were so decimated that hunting them was no longer commercially viable.[1] Whale oil was in demand, since it was the major source of lighting fuel before Colonel Edwin Drake discovered oil in Pennsylvania in 1859. Japan joined the high seas hunt after the Meiji Restoration removed the prohibition against leaving the home islands in 1868. Japanese whalers were mostly interested in hunting whales for food.[2]

Millennia before the major states of the world developed distant-water whaling fleets, peoples in many other areas of the world used whales and other cetaceans for food. Often these were communities of "aboriginal" peoples, especially the Inuit people of the far north. Even when these fisherfolk did not fit the aboriginal label, they were artisanal or small-type exploiters.[3] They usually live in remote locations and even today operate at or near subsistence and have rarely overexploited cetaceans.[4] Whaling is crucial to their livelihood, and it is integral to their culture.[5]

Whaling needed little regulation when the technical capability of the whalers was low. The hunters' desire to take every animal spotted was counterbalanced by limits in hunting equipment, navigation equipment, vessel speed, and ability to process animals into useful products. Those limitations changed with the advent of steam-powered vessels and the 1884 invention by the Norwegians of an explosive grenade harpoon fired from a cannon mounted on a fast catcher vessel.[6] It was inevitable that by the 1930s serious overharvesting would occur, so serious that the concerned states were willing to enter into an international convention to regulate whaling.

A 1931 convention was the first "modern" effort at managing whales. It was characteristic of early attempts to manage common property resources. It tried to regulate the taking of whales without seeking to solve the open entry problem. That is, there was little incentive for a number of parties to cooperate to not overexploit a resource if one or more of them could take what they pleased while the majority agreed to act with restraint. Those who would accept restrictions while others did not were "suckers." The signatories, using a "species" approach, attempted to regulate by restricting hunting to baleen whales only. It also regulated by other "biological" standards. The convention established "seasons," exempted the taking of females accompanied by calves, and established size limits. It required the collection of statistics so that regulation could be put on a more "scientific" basis. The signatories tried to use an international convention not so much to solve the problem of overharvesting but the problem of overproduction. Those who predominantly influenced attempts to regulate whaling during the 1930s were the industry managers.[7] Little or no infrastructure at the international level was created that could carry the load of managing the necessary rules.

The 1931 agreement proved to be inadequate to the task of managing whaling. It was extended by a 1937 agreement signed by nine whaling states. That agreement used management devices similar to its predecessor's. These were supplemented by a protocol signed in 1938 that banned the taking of humpback whales and created a whale sanctuary in the Pacific sector of Antarctic waters.

Although significant portions of the major whaling fleets were sunk during the Second World War, the resumption of whaling was anticipated in a 1944 agreement promoted by the Whaling Committee of the International Council for the Exploration of the Sea. An overall quota—a measure that had eluded earlier negotiators—was worked out at approximately two-thirds of the prewar catch. However, it was to be measured in a new unit, the notorious blue whale unit (B W U). Since most whalers from the Allied states were still interested in whale oil and the largest whales produced the most oil, the total catch quota was measured in "blue whale" equivalents. That is, one blue whale equaled two fin whales or two and a half humpbacks or six sei whales. The impact should have been predictable—every whaling nation rushed to take as many of the B W Us as they could.

There was soon more competition from Japan, which was interested mostly in providing scarce protein to feed its war-ravaged population. Over objections of other states, the Supreme Commander for the Allied Powers allowed Japan to resume Antarctic whaling on a "temporary" basis in 1946–47. Japan returned to Antarctic whaling in full force. By 1965 Japanese whalers were taking nearly twenty-seven thousand whales a year. Until 1963, the Japanese consumed more whale meat than any other type of meat.[8] The Soviet Union was also a major whaling state after the Second World War, largely for the same reasons as Japan—whales were high-quality protein and the cost, compared with the equivalent protein from land sources, was low. To this day we do not know how many whales Soviet whalers took before they dropped out of pelagic whaling. The numbers are likely to be very large, but during more than seventy years of the Soviet Union, the totals were deliberately underreported to the IWC.[9]

THE INTERNATIONAL CONVENTION FOR THE
REGULATION OF WHALING

The International Convention for the Regulation of Whaling was negotiated in Washington in 1946. The United States, the major physically intact developed state, was just beginning to recognize its responsibilities for postwar leadership and its obligation to assist the reconstruction of its devastated friends and former foes. While the United States was still a whaling state, the age of the Yankee whaler was over. Before too long it would be a nonwhaling state, as would many of the other former major European whaling states. Nevertheless, the ICRW was negotiated by most of its parties to protect their whaling interests. Their status changed over the years, and much of the evolution of IWC policy can be explained by the fact that many of the major players no longer had any, or very limited, commercial whaling interests to protect.

The ICRW was constructed along the lines of earlier attempts at whaling regulation. The purpose of the agreement was to "ensure proper and effective conservation and development of whale stocks" and "thus make possible the orderly development of the whaling industry."[10] To that end, the convention established the IWC. The commission was composed of one member from each of the contracting parties. Each contracting party had one vote. Decisions were to be taken by simple majority of members voting, but important substantive decisions required a positive three-fourths majority.[11] Member states could "defect." If a state notified the commission that it objected to a policy decision (technically an amendment to the Schedule), that policy decision would not be effective with respect to that government unless or until it withdrew the objection.[12] Member states also could issue permits to conduct scientific whaling and allow the whalers to "process" or use commercially the whales taken, as long as the scientific data derived was transmitted to an international data archive.[13] Compliance with the agreement was self-enforcing. Each contracting government was supposed to ensure the application of the treaty to its citizens.[14] The commission was authorized to create a secretariat (which was always kept small) and establish subordinate bodies (most importantly a Technical Committee and a Scientific Committee). The commission was authorized to perform studies and collect statistical data.[15] It was also expected to cooperate with member governments and international agencies.

The major policy tool of the IWC is found in an attached document, a Schedule through which the commission could regulate whaling by, among other measures, specifying (1) protected and unprotected species; (2) open and closed seasons; (3) open and closed waters (including sanctuaries); (4) size limits for each species; (5) time, methods, and intensity of whaling; and (6) gear restrictions.[16] These measures are to be employed "for the conservation, development and optimal utilization of whale resources" and are supposed to be based on scientific findings.[17] While the commission could ban all whaling, or whaling in a particular region, the agreement *did not* give the commission the right to restrict the number or nationality of factory ships or allocate specific quotas.[18] Its ability to limit entry was constrained. It could not allocate or determine *who* should get *what.* Like five of the eight fisheries commissions created after the Second World War, it could not divide the catch and eliminate the incentive for a whaling company (and its sponsoring state) to take as much of an overall quota as possible.[19]

During its early years, the dominant influence on the IWC was industry managers who affected the national policy of the whaling states and often participated in the IWC Technical Committee as representatives or observers.[20] The yearly catch limit (16,000 BWUs) established by the IWC, while lower than the overall yearly prewar catch (30,000 BWUs) was woefully inadequate for maintaining many species and stocks at a sustainable level. The whalers were engaging in what Ray Gambell, secretary of the IWC, called a "Whaling Olympics."[21] While the postwar limits were set after study by whale scientists, the state of the science was such that they were largely guessing at what might be a viable yield.

REFORM EFFORTS

In 1961 a new attempt was made to put whaling management on a more scientific basis. A committee of three (and later four) population dynamicists was formed to assess baleen whale stocks using more mathematically sophisticated tools than those available earlier. They recommended drastic reductions in the take. The active whaling states resisted these recommendations until 1965, and even then the whalers were given three years to adjust their catch downward to below the sustainable yield. Blue whale units were not eliminated until 1972, and shipboard observers reporting to the commission were not authorized until 1972.

The difficulties in gaining consensus among whalers, their state protectors, whale scientists, and the increasingly assertive conservationists, preservationists, and animal rights activists did not go unnoticed. In 1972 a resolution of the Stockholm Conference on the Human Environment called for a ten-year moratorium on commercial whaling, partially in response to the inability of the IWC to manage in a sustainable manner.[22] In 1974 the IWC responded with a New Management Procedure that went into effect in 1975. It purported to reorient management with a different conceptual approach. Management of whaling was to be on the basis of Maximum Sustainable Yield (MSY).[23] In theory, if the original stock size could be calculated, it should be possible to take whales when they are at 50 to 60 percent of their original abundance. This rate of predation should be sustainable over time, presuming the stocks to be exploited can be brought back to the acceptable percentage of their original numbers.

Unfortunately, the scheme proved unworkable. Data were difficult to acquire. The method was probably flawed scientifically, and it did not take the economics of the industry into account. The evident failings of the MSY scheme finally induced twenty-five of thirty-two members of the IWC at its 1982 meeting to vote for a moratorium on commercial whaling that was to take effect in the 1985–86 seasons and was to be reviewed by 1990. During that period the quota was reduced to zero[24] and the Scientific Committee was to embark upon a comprehensive assessment of whale stocks and the development of a new management procedure to replace maximum sustainable yield.

THE CURRENT SITUATION

The moratorium did not represent a mere hiatus in whaling but a significant shift in the way a new majority decided to manage whales. The new anti-whaling coalition, lead by major developed ex-whaling states, turned from a sustainability standard to a preservation standard.[25] It recruited new states to guarantee a three-fourths majority, and in 1978 it allowed more direct participation by representatives of nongovernmental organizations in many of the commission's activities. NGO numbers have now swelled to about ninety, making the environs surrounding the formal IWC meetings public, noisy, confrontational, "democratic," and definitely more chaotic.

The new majority was supposedly required to review the moratorium by 1990, but it ignored the deadline and seemingly tried to stretch out indefinitely the day it might be lifted. However, it was constrained to move ahead on the formal agenda items, particularly developing a Revised Management Procedure to be based on the best scientific evidence. This was an arduous task. After much study, five different schemes were proposed. The third on the list was finally recommended by the Scientific Committee in 1991 for implementation as a key component of the Revised Management Procedure. But it was turned back in the spring 1991 plenary session in favor of the maintenance of the commercial moratorium. As a result, the chairman, Phillip Hammond, resigned.[26]

The new majority insisted that the procedure alone was insufficient and that a Revised Management Scheme composed of detailed enforcement measures was necessary before any lifting of the moratorium could be considered. In recent sessions neither proponent nor opponent of the scheme has vigorously pushed a detailed proposal for carrying out the scheme's requirements, although Norway and Japan did submit drafts (Japan's draft submitted to the forty-ninth meeting was based on the earlier Norwegian draft).[27] But the effort to ban commercial whaling continued in the forty-sixth session—a Southern Ocean sanctuary was created. Japan demurred under its right to object to its implementation. Earlier, Iceland and Canada left the commission.

The IWC also moved vigorously to expand its agenda by moving into a variety of related questions such as humane killing[28] (although advocating "no killing" they devoted much time to this effort), and the potential effects of environmental degradation such as global climate change on whales.[29] (It is well known that nearshore pollution caused by onshore activities can create significant hazards for cetaceans; but this gets into the sticky question of whether the commission has any real powers in the two hundred Exclusive Economic Zones of coastal states where the coastal states are "sovereign for the purpose of exploring and exploiting . . . the natural resources."[30]) Some state members are also trying to move the commission into controlling small cetaceans, even though it is questionable whether the commission has any legal mandate here. They do have a mandate to allocate whale species of interest to commercial whalers for taking by indigenous people. Indigenous people claim that all the commission does is harass them.[31]

While "commercial" whaling was rigidly banned, limited whaling continues, some authorized, some not, although all types of whaling remain under vehement attack by anti-whaling forces. For example, the United States has successfully requested a quota of bowhead whales for Alaskan Inuit whaling villages and gray whales for its Makah Indians in Washington State (whose rights to whale are guaranteed under an 1854 treaty), but lately some of its allies, in the effort to end all whaling, threaten to turn on it to reduce or force an end to aboriginal whaling.[32] As might be expected, the commission refused Japan's request to establish a new category of permitted whaling—"small-type coastal whaling."[33] A more controversial form of exception is the ICRW provision allowing "scientific whaling." Iceland, Norway, and Japan have exercised their rights under that provision. All are condemned by anti-whaling foes. As Greenpeace put it, "scientific whaling is almost universally regarded as nothing more than commercial whaling under a different name."[34] While meat of the whales killed has been sold commercially to help pay expenses of the expeditions, Japan, for one, denies that scientific whaling expeditions even recover their costs. They state that high-quality investigative work has been done to support the effort to put a new management procedure on a scientific basis. Norway under its Article V right to object to a decision of the commission concerning the Schedule and exempt itself from its application, has decided to resume commercial whaling, claiming that their scientific surveys have indicated no shortage of minke whales in North Atlantic waters.[35] They set their quotas in accordance with the Revised Management Procedure.[36]

Little progress in resolving the major issues was evident in the commission's sessions until the forty-ninth annual meeting in Monaco, and even then, the effort by Michael Canny, the Irish commissioner and newly elected IWC chair, to begin a process that might break the deadlock, was greeted with suspicion by the major pro- and anti-whaling states' delegates. Canny, seemingly in response to host Prince Rainier III's warning that continued intransigence by anti-whaling states might backfire and cause the breakup of the commission,[37] proposed allowing a limited form of local whaling inside the two hundred Exclusive Economic Zones (EEZs) of coastal states, the termination of lethal scientific whaling (which cannot be done legally without formal amendment of the ICRW), and creation of a whale sanctuary in all oceans outside of EEZs. In some respects, he tried to bypass the core problem of completing the Revised Management Scheme. Only when the Revised Management Scheme was adopted could the commission consider the question of the resumption of commercial whaling. Delay was, and is, the name of the game. As Canny himself expected, although his proposal was informal and not in writing, opposition was immediate and vociferous. Soon after the session, animal rights campaigners descended upon Dublin to persuade the Irish government to drop the proposal.[38]

Canny tried hard to get I W C members to treat his proposal as a "negotiating text," not a negotiated text. That is, it was to be a beginning point for a serious negotiation whose purpose was to find a negotiated solution to the I W C conflict over whales, whaling, and the future role of the I W C. Unfortunately it included the mix of proposals mentioned above, which would have required both sides to make painful compromises on core issues. Neither camp budged at the Oman fiftieth anniversary meeting of the commission nor at the subsequent fifty-first meeting, and therefore stalemate continues.

Whether any new attempt to find a negotiated solution could succeed depends heavily on the attitude of the United States. Its initial response was not promising. It was one of the most vehement opponents of Canny's probe. It remains the "guarantor" of the commission and acts as its policeman under the Pelly amendment, claiming the United States can enact unilateral trade sanctions against states that, in its opinion, "diminish . . . the effectiveness of an international fisheries conservation program." But when the United States has used its big stick in the tuna-dolphin imbroglio, it was twice declared in violation of world trade laws.[39] On whaling, Pelly has been used only to warn, to put diplomatic pressure on, not to impose severe economic sanctions. Its effectiveness in frightening states from taking unilateral or regional action is diminished. Norway resumed commercial whaling without U.S. trade sanctions.[40] The North Atlantic Marine Mammal Commission, created as a threat to the I W C, has not yet made decisions or recommendations relating to species managed by the I W C. However, it stands ready to do so and to become an "appropriate" international organization for these species under Article 65 of the United Nations Convention on the Law of the Sea if the I W C does not act soon. In a recent meeting of the C I T E S organization, whose mandate is to control trade in endangered species, there was a strenuous effort to remove whales and elephants from a listing that declared them endangered. It succeeded in the case of elephants and almost did in the case of whales, with a majority in favor (57–51) but not the two-thirds majority necessary to pass.[41] Some observers claim that the tide is turning albeit slowly not only in C I T E S but also the I W C. They claim they can foresee the day when a ban on all but indigenous whaling will be lifted, and when indigenous whalers will be treated reasonably.

NOTES

1. Fred Charles Ikle, an important analyst of international negotiations, looked at the role of emotions in negotiation in a recent essay. He noted that there were three types: (1) emotions animated by something in the future; (2) emotions animated by something in the past; and (3) emotions animated by something in the immediate present. While all three types of emotional involvement are present in the IWC case, emotions animated by the immediate present are the most salient type. Ikle analyzed the first two and skipped the third case of high emotions caused by face-to-face interactions, doubtlessly because it is too difficult to get people to overcome the personal animus they feel for people on the other side of the table. Yet overcoming such feelings is necessary to create a prospect for a successful negotiated outcome. Fred Charles Ikle, "The Role of Emotions in International Negotiations," in *International Negotiation: Actors, Structure, Process, Values*, ed. Peter Berton, Hiroshi Kimura, and I. William Zartman (New York: St. Martin's, 1999), 335–50.

2. Preface, *International Convention for the Regulation of Whaling*, 2 December 1946, 62 Stat. 1716, 161 U.N.T.S. 74 (hereinafter ICRW).

3. M. J. Peterson. "Whalers, Cetologists, Environmentalists, and the International Management of Whaling," *International Organization* 46, no. 1 (winter 1992): 149–53.

4. William Aron, William Burke, and Milton Freeman, "Flouting the Convention," *Atlantic Monthly*, May 1999, 22.

5. Jørgen Wettestad and Steinar Andresen, "The Effectiveness of International Resource Cooperation: Some Preliminary Findings," *International Challenges* 11, no. 3 (1991): 55ff.

6. It is interesting to note that while the majority "like-minded" states have no qualms about simply voting down every request made for a restoration of whaling of whale stocks clearly under the jurisdiction of the IWC, they are forced to "address the issues in a consensual manner" on small cetaceans that are beyond the jurisdiction of the commission. "Draft Resolution on Small Cetaceans," Agenda Item 9, *Proceedings of the 46th Meeting of the IWC*, 25 May 1994 (IWC/46/54).

7. Harry N. Scheiber, "Historical Memory, Cultural Claims, and Environmental Ethics in the Jurisprudence of Whaling Regulation," *Ocean and Coastal Management* 38, no. 1 (1998): 35.

8. "Resolution on Legal Matters Related to the Adoption of the Southern Ocean Sanctuary," Agenda Item 13, *Proceedings of the 47th Meeting of the IWC*, May–June 1995, (IWC/47/45).

9. Article V(2)(c), ICRW.

10. What other promises may have been extracted are more difficult to document.

11. Christopher Stone, "Legal and Moral Issues in the Taking of Minke Whales," in *Report: International Legal Workshop, Sixth Annual Whaling Symposium,* ed. Robert L. Friedheim (Tokyo: Institute of Cetacean Research, 1996), xix.

12. Article V(3)(a)(b), ICRW.

13. Schedule, ICRW, as amended by the commission at the 51st annual meeting, 1999, and replacing that dated January 1999.

14. This is the so-called Pelly amendment to the *U.S. Magnuson Fisheries Act.* Public Law 92-219, 85 Stat. 786 (1971): 34; Public Law 96-61, 3(a), 93 Stat. 407 (1979).

15. The new "majority," working through NGOs, claim they represent the peoples of the world, but, of course, they have not been elected to office, nor are they publicly licensed. Accountability is an important issue here, as more-sensitive supporters of NGO activity recognize. For example, Mary Risely, "Environmentally Responsible Global Governance," in *New Directions in International Environmental Negotiation,* ed. Lawrence E. Susskind and William Moomaw (Cambridge, Mass.: Program on Negotiation, 1999), 8:120.

16. Paul Wapner, "Reorienting State Sovereignty: Rights and Responsibilities in the Environmental Age," in *The Greening of Sovereignty in World Politics,* ed. Karen Litfin (Cambridge: MIT Press, 1998), 287.

17. Martha Finnamore and Kathryn Sikkink, "International Norm Dynamics and Political Change," *International Organization* 52, no. 4 (autumn 1998): 888.

18. Lawrence E. Susskind, *Environmental Diplomacy: Negotiating More Effective Global Agreements* (New York: Oxford University Press, 1994), 21.

19. This is observable not only in the IWC but also in a number of war and peace types of international crises. See Bernard I. Finel and Kristin M. Lord, "The Surprising Logic of Transparency, " *International Studies Quarterly* 43, no. 2 (1999): 3, 315–39.

20. Japan also entered an objection to the Southern Ocean sanctuary, but only for minke whales; therefore it might be argued that Japan has accepted the sanctuary.

21. Article XIII(1)(2), ICRW.

22. Articles 65, 120, *United Nations Convention on the Law of the Sea* (New York: United Nations, 1983). While it might withstand legal scrutiny, it might not survive U.S. Pelly amendment retaliation and NGO fish boycotts, so Iceland remains cautious.

23. About thirty sperm whales are taken each year by Indonesian artisanal fishermen, mostly from Lamalera Island. Whether the fishermen would be classified as indigenous is an interesting but academic question. For a description of their whaling practices, see Jeffrey Gettleman, "Drawn to the Sea, to Tradition, to Danger," *Los Angeles Times,* 29 August 1999, A26.

24. The strength of a coalition will depend upon what kind of crosscutting or overlapping cleavages its members possess. Robert Axelrod, *Conflict of Interest: A*

Theory of Divergent Goals with Applications to Politics (Chicago: Markham, 1969), 158–64.

25. These swing-voting states include Sweden, South Africa, Switzerland, Ireland, Mexico, Monaco, and Oman.

26. John S. Dryzek, *The Politics of the Earth: Environmental Discourses* (New York: Oxford University Press, 1997), 156.

27. Peter S. Wenz, *Environmental Justice* (Albany: State University of New York Press, 1988), 273. For a useful review of environmental ethics, see Kara L. Lamb, "Ethical Discourse: An Exploration of Theories in Environmental Ethics," in *Handbook of Global Environmental Policy and Administration,* ed. Dennis L. Soden and Brent S. Steel (New York: Marcel Dekker, 1999), 243–59.

28. Although their remarks were directed at fisheries management, the following statement also captures the attitudes of some ocean resource management critics: "Large levels of natural variability mask the effects of overexploitation. Initial overexploitation is not detectable until it is severe and often irreversible. In such circumstances, assigning causes to past events is problematical, future events cannot be predicted, and even well-meaning attempts to exploit responsibly may lead to disastrous consequences." Donald Ludwig, Ray Hilborn, Carol Walters, "Uncertainty, Resource Exploitation, and Conservation: Lessons from History," *Science* 260 (2 April 1993): 17.

29. The animal rights literature is large. A classic statement of the animal rights case can be found in Peter Singer, *Animal Liberation: A New Ethics for Our Treatment of Animals* (New York: Avon, 1975).

30. Anthony D'Amato, "Agora: What Obligation Does Our Generation Owe to the Next? An Approach to Environmental Responsibility," *American Journal of International Law* 84, no. 1 (January 1990): 195.

31. See the references in D'Amato, "Agora," and Anthony D'Amato and Sudhir K. Chopra, "Whales: Their Emerging Right to Life," *American Journal of International Law* 85, no. 1 (January 1991): 21–62.

32. Article 56, *United Nations Convention on the Law of the Sea* (New York: United Nations, 1983).

33. As Ronald Inglehart put it: "Postmaterialists come from middle-class backgrounds, but they support change. . . . This is conducive to a decline in social class voting, as middle-class Postmaterialists move left—and working-class Materialists move to the right." Ronald Inglehart, *Modernization and Postmodernization: Cultural, Economic, and Political Change in Forty-three Societies* (Princeton, N.J.: Princeton University Press, 1997), 254. For a different view of the spread of environmental

ethics, see Willett Kempton, James S. Boster, and Jennifer A. Hartley, *Environmental Values in American Culture* (Cambridge: MIT Press, 1997).

34. Todd Sandler, "Intergenerational Public Goods," in *Global Public Goods: International Cooperation in the Twenty-first Century,* ed. Inge Kaul, Isabelle Grunberg, Marc A. Stern (New York: Oxford University Press, 1999), 20–50.

35. Bron Raymond Taylor, ed., *Ecological Resistance Movements: The Global Emergence of Radical and Popular Environmentalism* (Albany: State University of New York Press, 1995); Carolyn Merchant, ed., *Ecology* (Atlantic Highlands, N.J.: Humanities Press, 1994).

36. Thomas Princen and Matthias Finger, *Environmental NGOs in World Politics* (London: Routledge, 1994).

37. The intellectual leader of the animal rights movement, Peter Singer, is an avowed utilitarian "who judges whether acts are right or wrong by their consequences." Peter Singer, "The Singer Solution to World Poverty," *New York Times Magazine,* 5 September 1999, 61.

38. In the dispute over the resumption of whaling by the Makah tribe of the Olympic Peninsula, it was reported that a telephone magnate offered the tribe $12 million to forgo its right to whale under an 1854 treaty between the tribe and the government of the United States. Is this bribery or just compensation for a right that is fungible and therefore transferable? See *World Council of Whalers News* 5 (February 1999): 4.

39. Robert Costanza et al., "Principles for Sustainable Governance of the Oceans," *Science* 281 (10 July 1998): 198.

40. Scheiber, "Historical Memory."

41. Ali Shirvani-Mahdavi claims his position is based on Peter Singer's ideas. "Toward Legal Standing for Natural Objects: A Proposal for the Creation of the Universal Declaration of the Rights of the Environment and the Rights of Humans to a Healthy Environment," in *New Directions in International Environmental Negotiation,* ed. Lawrence E. Susskind and William Moomaw (Cambridge, Mass.: Program on Negotiation, 1999), 8:174.

42. Compare: "Indeed one of the dangers of the disappearance of isolation and the development of a single world culture is this destruction of cultural variety, as the stronger species of artifacts simply exterminate weaker ones all over the world." Kenneth Boulding, *Ecodynamics* (Beverly Hills: Sage, 1978), 81.

43. Murray Edelman, *The Symbolic Uses of Politics* (Urbana: University of Illinois Press, 1967), 164.

44. David Rothenberg, "Have a Friend for Lunch: Norwegian Radical Ecology

versus Tradition," in *Ecological Resistance Movements: The Global Emergence of Radical and Popular Environmentalism* (Albany: State University of New York Press, 1995), 212.

45. Georges Sorel, *Reflections on Violence* (New York: Collier, 1950), 144.

46. Jean Bodin, *On Sovereignty,* ed. Julian Franklin (Cambridge: Cambridge University Press, 1992); W. A. Dunning, *A History of Political Theories: From Luther to Montesquieu* (New York: Macmillan, 1905), 81–123.

47. John Rawls, *A Theory of Justice* (Cambridge: Harvard University Press, Belknap Press, 1971), 388.

48. Benjamin R. Barber, *Jihad vs. McWorld: How Globalism and Tribalism Are Reshaping the World* (New York: Ballantine, 1995), 9.

49. Peter Cowhey put it very well: "Given the role of the dominant powers within a regime for its implementation and adaptation, and given their relatively numerous options for foreign policy (relative to lesser powers), how can other countries trust the good faith of the dominant powers? This is particularly worrisome because multilateralism provides even fewer external checks on dominant powers than purely bilateral and minilateral orders do." Peter Cowhey, "Elect Locally—Order Globally: Domestic Politics and Multilateral Cooperation," in *Multilateralism Matters,* ed. John Gerard Ruggie (New York: Columbia University Press, 1993), 157. Also see Detlev F. Vagts, "Taking Treaties Less Seriously," *American Journal of International Law* 92 (1998): 458–62.

50. The clash between developed states pressuring African governments and the effect on local resource users is well documented in Nancy Peluso, "Coercing Conservation: The Politics of State Resource Control," in *The State and Social Power in Global Environmental Politics,* ed. Ronnie D. Lipschutz and Ken Conca (New York: Columbia University Press, 1993).

51. World Commission on Environment and Development, *Our Common Future* (Oxford: Oxford University Press, 1987), 43.

52. Andrea Baranzini and Gonzague Pillet, "The Physical and Biological Environment—the Sociobiology of Sustainable Development," in *Economy, Environment, and Technology,* ed. Beat Burgenmeier (Armonk, N.Y.: M. E. Sharpe, 1994), 140.

53. Preamble, ICRW.

54. Paragraph 10, Schedule, ICRW.

55. Article 6(4), Agreement for the Implementation of the Provisions of the United Nations Convention on the Law of the Sea of 10 December 1982 Relating to the Conservation and Management of Straddling Fish Stocks and Highly Migratory Fish Stocks (UN A/CONF. 164/37,), 8 September 1995.

56. Peter Haas, "Introduction: Epistemic Communities and International Policy

Coordination," *International Organization* 46, no. 1 (winter 1992): 1–36; Peter Haas, "Do Regimes Matter? Epistemic Communities and Mediterranean Pollution Control," *International Organization* 43, no. 4 (summer 1989): 377–404. For a "reflectivist" view of the epistemic community notion, see Karen T. Litfin, *Ozone Discourses: Science and Politics in Global Environmental Cooperation* (New York: Columbia University Press, 1994).

57. Sheila Jasanoff, "Science and Norms in Global Environmental Regimes," in *Earthly Goods: Environmental Change and Social Justice*, ed. Fen Osler Hampson and Judith Reppy (Ithaca: Cornell University Press, 1996), 173–97.

58. Daniel Sarewitz, "Social Change and Science Policy," *Issues in Science and Technology* 13, no. 4 (summer 1997): 31; Richard P. Feynman, "The Value of Science," in *What Do You Care What Other People Think? Further Adventures of a Curious Character* (New York: Bantam, 1989), 240–48.

59. Tore Schweder, *Intransigence, Incompetence, or Political Expediency? Dutch Scientists in the International Whaling Commission in the 1950s: Injection of Uncertainty,* SC/44/O 13 (Cambridge, England: IWC, 1992).

60. After participating in the working group providing population estimates, Justin Cooke, developer of the C procedure, did not object to the consensus-based working group report but later tried to undermine it. When asked why he did not make his views clear in the group or why he formed a "minority of one," he replied that it was a result of a "weak character." Later he was a major contributor to a report to the International Union for the Conservation of Nature that was used in the 1996 CITES meeting, which other scientists objected to as deceptive. Working Group, Scientific Committee, *Proceedings of the 48th Meeting of the IWC, 1996.* (IWC/48/4).

61. IWC, "Resignation of the Chairman of the Scientific Committee," Circular Communication to Commissioners, Contracting Governments, and Members of the Scientific Committee, 1 June 1993.

62. D. S. Butterworth, "Science and Sentimentality," *Nature* 357 (18 June 1992): 532–34.

63. "Threat of Global Warming Is for 'Real' Clinton Declares," *Los Angeles Times,* 25 July 1997, A21.

64. See *IPCC Second Assessment Synthesis of Scientific-Technical Information Relevant to Interpreting Article 2 of the UN Framework Convention on Climate Change* (Geneva: World Meteorological Organization, 1995) and *Kyoto Protocol to the United Nations Framework Convention on Climate Change* (FCCC/CP/1997/l.7).

65. Article 3(4), Convention for the Regulation of Whaling, 24 September 1931, reprinted in United States Senate, Committee on Commerce, *Treaties and Other In-*

ternational Agreements on Fisheries, Oceanographic Resources, and Wildlife to Which the United States Is Party, 93d Cong., 2d sess., 31 December 1974, 339.

66. Nicholas Peterson, "Introduction" in *Cash, Commoditisation, and Changing Foragers,* ed. Nicholas Peterson and Toshio Matsuyama, Senri Ethnological Series, no. 30 (Osaka: National Museum of Ethnology, 1991), 1–3; Milton M. R. Freeman, "The International Whaling Commission, Small-Type Whaling, and Coming to Terms with Subsistence," *Human Organization* 52, no. 3 (1993): 243–51.

67. During the spring 1997 bowhead hunt a group of 142 hunters was trapped on an ice floe that began drifting out to sea. The rescue was made easier because the Inuit whalers were using handheld global positioning systems. *Los Angeles Times,* 19 May 1997, A12.

68. Joel E. Cohen, "Population Growth and Earth's Human Carrying Capacity," *Science* 269 (21 July 1995): 341–46; Amartya Sen, "Population: Delusion and Reality," *New York Review of Books* (22 September 1994): 62ff.; John Bongaarts, "Population Policy Options in the Developing World," *Science* 263 (11 February 1994): 771–76; John Bongaarts, "Can the Growing Human Population Feed Itself?" *Scientific American,* March 1994, 36–42; Roy L. Prosterman, Tim Hanstad, and Li Ping, "Can China Feed Itself?" *Scientific American,* November 1996, 90–97.

69. Thomas Homer-Dixon and Jessica Blitt, eds., *Ecoviolence: Links among Environment, Population, and Security* (Lanham, Md.: Rowman & Littlefield, 1998; Thomas F. Homer-Dixon, Jeffrey H. Boutwell, and George W. Rathjens, "Environmental Change and Violent Conflict," *Scientific American,* February 1993, 38–47; Thomas F. Homer-Dixon, "On the Threshold: Environmental Changes as Causes of Acute Conflict," *International Security* 16, no. 2 (fall 1991), 76–116; Jessica Tuchman Matthews, "Redefining Security," *Foreign Affairs* 68, no. 2 (spring 1989): 162–77.

70. For example, Japan, an island country, has long feared dependence upon other states for control of access to its grain and protein needs. Their fears are reinforced by the insensitivity of others to their concerns. Japanese analysts point to the first of the Nixon "shocks" as an example of their vulnerability. In 1973, because of the El Niño–induced failure of the catch of Peruvian anchovita (a reduction fish ground into meal subsequently fed to American chickens), President Nixon, without consultation with Japan, banned the export of soybean meal so that the soybean meal could be fed to American chickens. Reiko Niimi, "The Problem of Food Security," in *Japan's Economic Security,* ed. Nobutoshi Akao (New York: St. Martin's, 1983), 169–96.

71. Inglehart, *Modernization,* 78.

72. See the declaration by Argentina, Australia, New Zealand, and the United States at the food security conference sponsored by the Food and Agricultural Or-

ganization of the United Nations (FAO) in Kyoto, December 1995, putting on record that the conference outcome—the Kyoto Declaration—not "affect the competency of, or change the current status in, other international organizations, including the International Whaling Commission." While it might well be that Japan and others hoped to have the outcome of the FAO conference influence events in the IWC, this is a clear example of representatives of developed states, who know no hunger and expect no hunger, failing to take the issue of food security seriously. *Interpretative Statement Made by Argentina, Australia, New Zealand, and the United States in Relation to the Kyoto Declaration and Plan of Action*, 9 December 1995.

73. Lester R. Brown, *Who Will Feed China?* (New York: Norton, 1995).

74. Margaret Klinowska, "How Brainy Are Cetaceans?" *Oceanus* 32, no. 1 (spring 1989): 19–20.

75. William Cronon, *Uncommon Ground: Rethinking the Human Place in Nature* (New York: Norton, 1995), 24. For the opposite view, that there is a boundary between nature and human beings, see Michael E. Soule and Gary Lease, *Reinventing Nature? Responses to Postmodern Deconstruction* (Washington, D.C.: Island Press, 1995).

76. Interview with Daniel Botkin in Wallace Kaufman, *No Turning Back* (New York: Basic Books, 1994), 94–96 nn. 3, 6. Also see Daniel Botkin, *Discordant Harmonies: A New Ecology for the Twenty-first Century* (New York: Oxford University Press, 1994).

77. Susskind, *Environmental Diplomacy*, 11.

78. For a discussion of consent in a multilateral context, see Robert L. Friedheim, *Negotiating the New Ocean Regime* (Columbia: University of South Carolina Press, 1993), 3–5.

79. David Victor and Julian Salt, "Keeping the Climate Treaty Relevant: An Elaboration," *International Institute for Applied Systems Analysis* (April 1995): 22–23.

80. Peter B. Evans, Harold K. Jacobson, Robert D. Putnam, *Double-Edged Diplomacy: International Bargaining and Domestic Politics* (Berkeley and Los Angeles: University of California Press, 1993).

81. Oran R. Young and Gail Osherenko, eds., *Polar Politics: Creating International Environmental Regimes* (Ithaca: Cornell University Press, 1993), 1; also see Marc A. Levy, Oran R. Young, and Michael Zurn, "The Study of International Regimes," *European Journal of International Relations* 1, no. 3, (19XX): 267–330; and the special issue of the journal *International Organization*, volume 36, 1982, especially the essays by Steven D. Krasner, "International Regimes," and Robert O. Keohane, "The Demand for International Regimes."

82. Gary D. Libecap, "The Conditions for Successful Collective Action," in *Local*

Commons and Global Interdependence, ed. Robert O. Keohane and Elinor Ostrom (Thousand Oaks, Calif.: Sage, 1995), 182–17.

83. Young and Osherenko, *Polar Politics,* 187.

APPENDIX

1. Ray Gambell, "The Management of Whales and Whaling," *Arctic* 46, no. 2 (1993): 97–107; Ray Gambell "Management of Whaling in Coastal Communities," *Whales, Seals, Fish, and Man,* ed. A. S. Blix, L. Walloe, and O. Ulltay (Elsevier, 1995), 699–708.

2. Hugh Borton, *Japan's Modern Century* (New York: Roland, 1955), 13.

3. Milton M. R. Freeman, "The International Whaling Commission, Small-Type Whaling, and Coming to Terms with Subsistence," *Human Organization* 52, no. 3 (1993): 243–51.

4. Tomoya Akimichi et al., *Small-Type Coastal Whaling in Japan: Report of an International Workshop* (Edmonton, Alberta: Boreal Institute for Northern Studies, 1988).

5. Oran R. Young et al., "Subsistence, Sustainability, and Sea Mammals: Reconstructing the International Whaling Regime," *Ocean and Coastal Management* 23 (1994): 117–27.

6. For a history of whaling, see Johan Nicolay Tønnessen and Arne Odd Johnsen, *The History of Modern Whaling* (Berkeley and Los Angeles: University of California Press, 1982).

7. M. J. Peterson, "Whalers, Cetologists, Environmentalists, and the International Management of Whaling," *International Organization* 46, no. 1 (winter 1992): 149–53.

8. Robert L. Friedheim and Tsuneo Akaha, "Antarctic Resources and International Law: Japan, the United States, and the Future of Antarctica," *Ecology Law Quarterly* 16, no. 1 (1989): 139.

9. The logs of Soviet whalers are currently being reanalyzed in order to estimate the take. The effort has been paid for by the United States. "United States Opening Statement" (IWC/46/OS USA), "Resolution on the Unreliability of Past Whaling Data" (IWC/46/60), Table 4 (IWC/46/8a), and "Intersessional Meeting of the Working Group on a Sanctuary in the Southern Ocean" 46th meeting of the IWC (IWC/46/19), 3.

10. Preamble, *International Convention for the Regulation of Whaling,* 2 December 1946, 62 Stat. 1716, 161 U.N.T.S. 74 (hereinafter ICRW).

11. Article III(2), ICRW.

12. Article V(3), I C R W.

13. Article VIII, I C R W.

14. Article IX, I C R W.

15. Article IV, I C R W.

16. Article V(1), I C R W.

17. Article V(2), I C R W.

18. Article V(2), I C R W.

19. Robert L. Friedheim, "International Organizations and the Uses of the Ocean," in *Multinational Cooperation,* ed. Robert Jordan (New York: Oxford University Press, 1972), 242–51.

20. Peterson, "Whalers," 160.

21. Gambell, "Whales and Whaling," 106.

22. For an assessment of the Stockholm conference, see Lynton Keith Caldwell, *International Environmental Policy,* 2d ed. (Durham, N.C.: Duke University Press, 1990), 21–93.

23. M S Y had been under attack by resource economists. See James Crutchfield and Giulio Pontecorvo, *The Pacific Salmon Fisheries: A Study in Irrational Conservation* (Baltimore: Johns Hopkins University Press, 1990); Francis Christy Jr. and Anthony Scott, *The Common Wealth in Ocean Fisheries* (Baltimore: Johns Hopkins University Press, 1965); Harry N. Scheiber and Chris Carr, "The Limited Entry Concept and the Pre-History of the ITQ Movement in Fisheries Management," in *Social Implications of Quota Systems Fisheries,* ed. Gisli Palsson and Gudrun Petursdottir (Copenhagen: Nordic Council of Ministers, 1997), 235–60.

24. Tables 1–3, Schedule, I C R W, January 1995.

25. For an assessment of recent I W C politics, see Robert L. Friedheim, "Moderation in the Pursuit of Justice: Explaining Japan's Failure in the International Whaling Negotiations," *Ocean Development and International Law* 27 (1996): 349–78.

26. "Resignation of the Chairman of the Scientific Committee," Circular Communication to Commissioners, Contracting Governments, and Members of the Scientific Committee, I W C, 1 June 1993.

27. Chairman's Report, *Proceedings of the 47th Meeting of the I W C,* 1995, 19–21.

28. A special workshop was held in 1997 on humane killing. Final Press Release, *Proceedings of the 48th Meeting of the I W C,* 28 June 1996.

29. A workshop was held on the subject in 1996, even though the commission noted that "the most vulnerable species to such threats might well be those reduced to levels at which the R M P, even if applied, would result in zero catches." If they knew the answers in advance, how much could a workshop contribute? Final Press Release, *Proceedings of the 48th Meeting of the I W C,* 28 June 1996.

30. Article 56(1)(a), *United Nations Convention on the Law of the Sea* (New York: United Nations, 1983).

31. "The IWC management of indigenous whaling plays an important role in this conflict. The great majority of anti-whaling member states and their affiliated NGOs have not understood, or even shown a willingness to understand, the nature of this kind of whaling. Instead the IWC has become a forum for expressing the most common misperceptions and prejudices about indigenous peoples that regrettably still prevail in many Western societies." Joint Opening Statement by the Observers for ICC and IWGIA, *Proceedings of the 46th Meeting of the IWC,* 1994 (IWC/46/OS/ICC-IWGIA).

32. The Breach Marine Protection UK together with Australian marine protection organizations threatened to sue in U.S. courts to prevent the permitted aboriginal hunt of 34 California gray whales. "U.S. Challenged on Indigenous Whaling," posted on the World Wide Web, 4 July 1997.

33. For an analysis of the Japanese "case," see Tomoya Akimichi et al., *Small-Type Coastal Whaling in Japan: Report of an International Workshop* (Edmonton, Alberta: Japan Social Science Association of Japan, Fund to Promote International Education Exchange and Boreal Institute for Northern Studies, 1988).

34. "Whales," *Greenpeace Fact Sheet* (Greenpeace International, via Greenbase, 18 September 1991).

35. Statement of the Norwegian commissioner on Draft Resolution on Northeast Atlantic Minke Whales, *Proceedings of the 48th Meeting of the IWC,* 28 June 1996, (IWC 48/41).

36. Proposal from Norway to Transfer Minke Whale *Balaenoptera acutorostrata* from Appendix I to Appendix II: Draft for Consultation with Range States, CITES, 19th Meeting, Zimbabwe, 9–20 June 1997.

37. *Los Angeles Times,* 22 October 1997, A12.

38. Amanda Brown "Campaigners to Urge Irish to Drop Whaling Proposals," via Internet.

39. Richard J. McLaughlin, "UNCLOS and the Demise of the United States' Use of Trade Sanctions to Protect Dolphins, Sea Turtles, Whales, and Other International Marine Living Resources," *Ecology Law Quarterly* 21, no. 1 (1994): 1–78.

40. Walter Gibbs, "Journal: Norwegian Whalers Say the Wind Is Turning in Their Favor," *New York Times,* 23 July 1997.

41. "Majority Vote for Downlisting Minke Whales," *HWNNews,* 20 June 1997; "CITES: Continues to Defer to IWC Decisions," *HWNNews,* 20 June 1997, posted on the World Wide Web (http://www.highnorth.no).

PART I
CRITIQUING THE PERFORMANCE
OF THE WHALING REGIME

1 / A New Whaling Agreement and International Law

WILLIAM T. BURKE

In seeking change in the regime for whales and whaling, an initial question is how that is to be done. A common view is that the International Convention for the Regulation of Whaling (ICRW) needs to be revised because it no longer reflects the shared interests of the parties. The need for a revision has long been obvious, but previous attempts have failed. This failure has led to the contortions that attempt to square the circle by application of interpretive principles that allegedly permit revision of the ICRW without its amendment[1] or by "compromise" resolutions that make no effort to disguise the attempt at treaty amendment.[2]

Although it is currently next to impossible to detect an interest shared by anti-whaling and pro-whaling states, except perhaps for continuing indigenous harvesting, the parlous state of the International Whaling Commission (IWC) suggests that it is still useful to discuss how negotiations on a new agreement might proceed in light of relevant principles of international law and agreements.[3]

At least one point is clear, if anything can be where whaling is concerned. The International Whaling Commission has no direct role to play, even if the aim of interested states is to modify the ICRW itself. Other than changing the Schedule, where the commission makes the decision, the ICRW has no provision for amendment at all, let alone one giving some role to the commission. Hence the commission itself has no authority to change the basic treaty.[4] The states parties (and any other state participating) must negotiate a new agreement.[5] Obviously the difficulties in the way of this achievement are formidable, especially in light of the great differences between those who would prohibit all whaling and those who would allow a regulated harvest of sufficiently abundant species and stocks, of which there are several.[6] While states are free to reach whatever agreement can be forged to satisfy their common interests, the general context of international law (including existing agreements) will or might affect their choices.

Some elements of this context are examined in the following discussion. Consideration is first given to the relevance for possible negotiations of current international law for whaling, including the United Nations Convention on the Law of the Sea, especially the alleged special position of the IWC under the convention and how this might affect negotiations over a new agreement. Second, the sad experience under the ICRW of parties willfully distorting or ignoring its principles and provisions is cited to suggest the utility of some obvious negotiating and drafting tactics when a new agreement is sought. Third, in light of the continuing controversy over failures to observe the ICRW, it may be helpful even now, if not urgent, to consider application of the provisions on compulsory dispute settlement in UNCLOS. A future agreement ought to take careful account of this potential and of the avoidance of problems in its adaptation to the potential disagreements. Fourth, and finally, some miscellaneous emerging or already applicable principles are briefly assessed, including the precautionary approach and its abuse by the IWC.

THE GENERAL PRINCIPLES OF INTERNATIONAL LAW

Current International Law for Whaling

Current international law regarding marine mammals (the present legal regime), including cetaceans, consists of customary principles and the provisions of relevant international agreements, including the 1982 United Nations Convention on the Law of the Sea (UNCLOS), the ICRW, and other particular agreements.[7] Under customary law, the nationals of all states are free to exploit all living marine resources, including whales, outside national jurisdiction, where flag states of these nationals are obligated to prescribe conservation measures for these operations and to cooperate for this purpose. Customary law principles govern the activities of states not party to an agreement restricting whaling activities. The vast majority of states are not party to specific restrictions on whaling accepted by an international agreement.

A very large number of states are now parties to UNCLOS. Accordingly, customary law and the conventional law represented by UNCLOS provide the regime for the vast majority of states, only a relatively few of which engage in direct harvesting of cetaceans. Many more conduct fisheries having a bycatch of marine mammals, including small cetaceans.[8]

The 1982 United Nations Convention on the Law of the Sea does not materially change customary law, which already required states to conserve high-seas resources and to cooperate with others to that end, although it does seriously constrict the area of the ocean within which the customary principle of free access applies.[9] Part V of UNCLOS establishes in Article 56 that a coastal state has sovereign rights over marine mammals within a two-hundred-mile Exclusive Economic Zone (EEZ) and in Article 65 (Marine Mammals) that it has no obligation to allow harvests of marine mammals therein irrespective of their abundance or whether there is a surplus over domestic use. On the other hand, if it chooses, the coastal state may allow harvests of whales within its EEZ, subject to the obligation (applicable to all living marine resources within national jurisdiction) that it does not allow them to be endangered by overexploitation (Article 61).

For the area beyond national jurisdiction, UNCLOS Articles 87 and 116 reiterate the long-standing principle that freedom of fishing applies on the high seas and that the nationals of all states are entitled to enjoy this freedom subject inter alia to other international agreements to which they are party, including UNCLOS.[10] The further obligations of states to take conservation measures and to cooperate with each other are provided in Articles 117–18. Article 119 sets out the conservation measures to be observed by states on the high seas and obviously envisages utilization subject to regulation.[11] However, Article 120 incorporates Article 65, which makes it clear that states and international organizations, as appropriate, are entitled to adopt stricter conservation measures than would otherwise be applicable, including prohibition of exploitation. Article 120 also provides that "States shall cooperate with a view to the conservation of marine mammals and in the case of cetaceans shall in particular work through the appropriate international organizations for their conservation, management and study."

In the context of potential negotiations for a new international agreement, the latter provision of Article 120 might be thought to have special significance. The question is whether Article 120 requires a harvesting state to participate as a member of a specific international organization established to govern whaling on a global basis. UNCLOS Articles 87 and 116

clearly allow high-seas harvesting of whales, as other marine living resources; must a state join a particular international organization with competence over whales before it may engage in whale harvesting? In carrying
out its obligation to cooperate under Article 118, is the individual whaling
state required to join a particular agency regulating whaling? Obviously, the
answers to these questions might be important for a particular state's decision to enter into negotiations, continue them, or accept an outcome.

A possible implication of the idea that by virtue of UNCLOS the IWC is
the only competent international organization for regulating whaling is that
actions to change the whaling regime must proceed by way of revising the
ICRW and that this treaty is the basis on which a future regime must rest. If
UNCLOS means that the IWC is the only competent body and most states
are bound by UNCLOS, how else can change occur?

At least two answers to this question are apparent. One is that UNCLOS
does not provide that a single agency is required nor that a state must join
a particular agency before beginning (or continuing) whaling. Another is
that even if it did, states in general are not disabled from taking action to
create a new whaling regime or regimes differing from the ICRW. The following discussion examines the assumption that only one competent
agency for whaling should be recognized. It concludes that the assumption
has no basis in international law or in UNCLOS.

Of several articles on living marine resources in general, UNCLOS contains two directly concerning whales, Articles 64 and 65 (which also apply to
the high seas under Articles 116 and 120, respectively). Neither supports the
notion that harvesting whales by the national of a state party must be pre-
ceded by joining an international regulatory agency. Article 64 on "highly
migratory species," which includes cetaceans listed in Annex I, calls for a
"coastal state and other States whose nationals fish in the region for highly
migratory species" to "cooperate *directly or through the appropriate international organization with a view to ensuring conservation and promoting the
objective of optimum utilization of such species throughout the region, both
within and beyond the exclusive economic zone*" (emphasis added).

Two major points of relevance here are that Article 64 allows for direct
interaction between states as an acceptable mode of cooperation and that it
contemplates the potential existence (and creation) of more than one international organization for marine mammals, including cetaceans. A provision allowing for conservation action through direct bilateral or multilateral action is obviously inconsistent with the proposition that states can

harvest cetaceans only if they join an international organization. Similarly the emphasis on regional organization in Article 64 is inconsistent with the idea that only a single such international organization is appropriate. Annex I to Article 64 specifically lists the families of cetaceans to which the article applies.

The provision in Article 120 that in the case of cetaceans states shall "work through appropriate international organizations for their conservation, management and study" is the principal basis for the belief that the IWC is the sole international body competent to deal with whales. The evidence for this interpretation is weak at best. The language is poorly designed if the intent was to establish an obligation to become a member of a specific international organization. There is, first, the use of the plural term "international organizations," which is difficult to square with the notion that only the IWC is meant.[12] The suggestion that the drafters had in mind that the plural form allowed for other agencies to regulate small cetaceans is convenient, but a long way from adequate explanation of a treaty that specifically lists both large and small cetaceans.[13] The terms of Article 65 would take a good deal of revision to make this distinction, and there is no record of a proposal to amend the treaty language.

Beyond this major difficulty, Article 65 was poorly designed if the intent was to establish an obligation to become a member of a specific organization. A declaration that a state shall "work through" an organization does not convey the obligation to join the organization, since "working through" can be (and is) achieved through means other than membership in the organization, including attendance at annual meetings, collaboration in working out acceptable conservation measures, timely submission of information and data, recognition and acceptance of scientific findings, voluntary observance of prescribed conservation measures, coordination with enforcement schemes, and no doubt others.

The notion of a single global body with jurisdiction over any whaling anywhere seems an inappropriate and top-heavy way to regulate activities focusing upon regional or local stocks, including cetaceans. Some states might justifiably be inclined to prefer an agency with a regional geographic orientation for this purpose. For this reason the use of the plural "international organizations" in the second sentence of Article 65 seems sensible.

Furthermore, if regard is had to the real world of international agency action in relation to interpreting UNCLOS, the appropriateness of the IWC as the single agency is highly questionable when consideration is given to its record. This point is obvious in light of the fact that under the current controlling group within the IWC, the ICRW regime has been defined, twisted may be a better word, to embody an objective not shared by all members and not even consistent with its basic charter. In addition, significant decisions have been made without regard to the procedures required by the treaty. The well-documented practice of the IWC of ignoring scientific advice, or not even requiring either advice or findings despite the mandate of the ICRW Article V, adds a great deal to doubts about this agency (as presently manipulated) as an appropriate entity for implementing an agreement that calls for the use of the best scientific evidence available in making conservation decisions. Under such circumstances, the prospect of interpreting UNCLOS to require adherence to the ICRW as a condition of entering into whaling is not only appalling but most likely self-defeating. States insisting on this view of UNCLOS cannot be taken as serious defenders of whale conservation. In its current stage of control by members who reject the basic tenet of sustainable harvesting as established by the ICRW, the IWC is not itself interested in conservation in any sense consistent with the treaty's mandate and purposes.

To underscore this point, no member state has yet provided a reasoned statement for departing from the treaty's terms. The only proffered justification is strictly political, as in the declaration of the United States that it cannot support sustainable takes of whales because there is no political support for doing so in the Congress,[14] as if this adequately explained and even justified its refusal to abide by its treaty obligations. The similarly irresponsible positions of other member states are well known.[15]

In considering how UNCLOS should be interpreted concerning an alleged membership requirement in Article 65, it is pertinent to consider how states have handled the issue in other circumstances. When it has been thought desirable to make membership in an international organization a requirement for access to a stock subject to internationally agreed-upon conservation measures, other agreements accomplish this by specific provision. For example, the 1995 Straddling Fish Stock Convention declares in Article 8(4) that "Only those States which are members of such an organization or participants in such an arrangement, or which agree to apply the conservation and management measures established by such organization or arrangement, shall have access to the fishery resources to which those measures apply."[16] Nothing of this nature or anything remotely resembling it appears in UNCLOS. Based on what UNCLOS actually provides and what it does not say, the conclusion must be that there is no requirement in it that whale harvesting necessitates membership in an existing international organization or in one to be created. That it would be highly desirable to work through such an entity is manifest, but the treaty falls short of mandating membership in such an agency. It might of course be helpful if such a regional or global agency existed.

Fortunately, for those states wishing to invoke their customary law right to harvest whales on the high seas in the North Atlantic, the creation of the North Atlantic Marine Mammal Conservation Organization (NAMMCO) provides the mechanism for cooperation in genuine conservation measures. Unless the recalcitrant states in the IWC at some stage agree to accept the goal of sustainable harvesting of whales, other regional whaling organizations devoted to this goal may be necessary, or an entirely new international organization will need to be created for this purpose by those concerned with sustainable harvesting, as opposed to its prohibition.

In this connection, the provisions of Agenda 21 are sometimes construed as if the drafters of that document were authorized to, and did, pronounce either that Article 65 mandates participation in the IWC for whale conservation or that other international organizations can proceed only under IWC guidance and overview.[17] Several objections to this position are evident. First, the United Nations Conference on Environment and Development (UNCED) participants had no standing to determine the meaning of UNCLOS or of the ICRW. Second, they had no authority to decide how individual states shall carry out their obligation to cooperate in whale conservation. Third, although UNCED participants' view about the treaty might be helpful in some contexts, assuming they actually expressed such a view, it is something of a stretch to consider the relevant passage in Agenda 21 as anointing the IWC as the single authorized agency to accomplish whale conservation. The passage is as follows: "17.61 States recognize (a) the responsibility of the International Whaling Commission for the conservation and management of whale stocks and the regulation of whaling pursuant to the 1946 International Convention for the Regulation of Whaling." No other specific marine mammal organization could fit the description in this sentence. Since only the IWC can carry out the obligations set up by the ICRW, then only the IWC *is* responsible under the ICRW. However, the ICRW binds only a relative handful of states, most of whom do not engage in whaling. Accordingly, this sentence is hardly a general statement about whaling and international organizations.

It is also relevant in considering Agenda 21 to note that participants did not accept the proposal advocated by some anti-whaling states that whales should not be subject to any exploitation at all. This choice was expressly left to individual states and to international organizations with the necessary competence and the inclination to use it. The International Whaling Commission operates on the basis of an agreement that does not support a permanent prohibition of whaling irrespective of population abundance.

In sum, whaling is permissible under general international law subject to the duty to conserve, a duty that under UNCLOS Article 64 can be discharged through direct interaction with other states or through international organizations. In choosing the latter mechanism, a state may join a particular international organization concerning whaling, including but not limited to the International Whaling Commission. A regional organization with cognizance of a particular stock of whales would also be appropriate. States might understandably not choose the IWC in light of its record (not only in recent times) for disregarding its own treaty and particularly for noticeably less-than-faithful adherence to the requirement that IWC actions are based on scientific findings.[18] For states party to the ICRW, their obligations are spelled out in that agreement.

The above considerations lead to the conclusion that the Law of the Sea treaty does not set conditions affecting negotiations for a new whaling regime. It would be a considerable stretch to establish that UNCLOS requires that changes in the current whaling regime must proceed by revising the existing multilateral agency for whaling. Member states are not required by UNCLOS to ratify the ICRW as a continuing agency and are certainly not confined to its revision in order to create an institutional regime for whaling. So far as UNCLOS is concerned, member states are free to create a new agency, revise the ICRW, or join a regional agency such as NAMMCO if it assumes authority to regulate whaling in the North Atlantic by its members.

From the standpoint of a desirable decision process, regional organizations for whaling regulation and for marine mammals in general constitute an attractive objective for several reasons. One is the utility of permitting overview of a marine mammal stock whose range is regional in scope. Overview refers to maintaining cognizance of the development of data and information and how it is employed, and of regulation and enforcement measures and processes. An entity with competence over the entire marine mammal component of a large ocean region is more likely to contribute to holistic management actions (including interactions with other entities concerned with fishery conservation) than one fragmented by distinctions irrelevant to ecosystem management. Of course, the ideal arrangement would subject all living marine resources in the region to the same institutional structure.

Regional management is also not inconsistent with creating a global entity devoted to overall cognizance and action to be sure that regional measures in toto are adequate or need to be supplemented or complemented to ensure general effectiveness and to avoid gaps in coverage and reporting. The idea of using regional panels for decision making within a larger fishery body is not new; this approach has been used in the North Atlantic Salmon Conservation Organization and in the original International Commission for Northwest Atlantic Fisheries.[19]

GENERAL PRINCIPLES OF INTERPRETATION

A significant difficulty that might confront any negotiations for a new whaling regime is to overcome the mistrust generated by the too frequent practice by IWC members of implementing the ICRW as if the terms originally adopted didn't really matter. Such recurrent behavior has been justified by pointing to the alleged "flexibility" of the convention,[20] which by some alchemy is seen by such members as New Zealand, Australia, and the United Kingdom to allow for permanent elimination of what the convention was established to accomplish, namely whaling on stocks that are regulated so as to be abundant enough to yield a safe harvest on a continuing basis. That there are such stocks is no longer subject to serious scientific question.

Under an alleged principle of flexible interpretation, the convention can be construed to authorize abolition of the main purpose for its existence while the commission devotes attention to other matters of an ancillary nature. Among the latter is the matter of seeing that killing is done in a humane fashion, despite that some members of the commission are trying their best to prohibit any commercial killing at all. If anything is to be accomplished, it will be necessary to reduce the range of "flexibility" in interpretation (that good faith use of principles of interpretation can result in flexibility is not denied) and to recognize that commitments and obligations are not so lightly dropped or ignored or expanded.

In view of past experience, a main practical problem now is how to frame an agreement that will not be so easily undermined by the machinations of members, in concert with other supportive entities, who no longer accept the purposes and methods of the new agreement and resort to questionable interpretations to escape it. If this cannot be achieved for the relatively insignificant segment of economic activity of whaling, some wonder whether vastly more important environmental problems can be dealt with through normal international decision processes. Of course, the analogy may be far-fetched since the willingness of states to abuse the treaty process in the case of whaling arises partly *because* they have no economic interest at stake and no broad constituency in favor of whaling. Other users, such as the aboriginal and small coastal whaling participants, are taken care of separately (when and if they are) and, in any event, have sometimes simply been ignored anyway.

Whether or not whaling regulation has some precedential value for other subject areas, a number of legal devices or approaches may be helpful to avoid future problems while devising a new agreement. Among these are refinements in the negotiating process, improvements in drafting, and, perhaps most important, the adoption of a dispute settlement procedure that has been lacking, or, more recently, ignored by those concerned.

Drafting and Negotiating Tactics

A first step might be to make a determined effort to avoid provisions that are vague and ambiguous, leaving excessive "wiggle room" as possible escape hatches from suddenly inconvenient commitments. Although vagueness and ambiguity are sometimes necessary and valuable negotiating tools, and minimizing the occasion for their use is essential in this context, the adoption of dispute settlement procedures may help lessen the impact of vagueness as a means of obscuring basic disagreements. Even so, should disputes later arise, precise drafting may facilitate the use of the major principle of interpretation found in the Vienna Convention on the Law of Treaties, the ordinary and natural meaning of words in their context.[21] However fragile and question-begging this principle may be as an aid to interpretation,[22] it does serve to caution parties to be wary of papering over their disagreements with highly general and excessively ambiguous terms that might later be employed to give color to unexpected and destructive differences. One might point to some provisions in UNCLOS to demonstrate how this process works.[23]

Another method for helping to avoid, or at least for alleviating, future problems in interpretation is to establish a sound legislative history of the agreement-making process. Such a record may assist in placing boundaries on future opposing and conflicting interpretations. The lack of an adequate legislative history was a serious deficiency of the law-making process employed in negotiating UNCLOS and partially accounts for the advances made in this treaty on provision for compulsory dispute settlement, a development that could contribute a great deal in the present context of achieving a new whaling agreement.

In actuality, it was not defective drafting or indeterminate terminology that led to the distortion in implementation of the ICRW—it is straightforward in expressing the purpose of conserving whales in order that species and stocks can continue to be harvested. Nothing in the ICRW is so ambiguous that it can reasonably be made to yield an interpretation precisely the reverse of the original purpose. Unfortunately, however, no drafting legerdemain can successfully prevent the regime destruction that results from the blindness of ideological conviction and moral superiority. But at least clarity and precision of expression in declaration of purpose and modalities of action might make actions contrary to or inconsistent with shared expectations more difficult to achieve.

In addition, no matter how skilled drafting may be, it is easily conceivable that differences of interpretation will arise without any origin in bad faith. The obvious step is to invoke third-party assistance in the event of dispute over divergent interpretation. There are several ways to do this but in the end there will be a need to make adequate provision for the compulsory settlement of disputes over the interpretation and application of whatever agreement can be negotiated. In this regard, the renegotiation of the ICRW or the crafting of an entirely new agreement will take place in an atmosphere far more favorable for resort to compulsory dispute settlement than any time in the past. This is due entirely to Part XV of UNCLOS, which requires the use of compulsory and binding procedures to settle disputes over the interpretation and application not only of UNCLOS itself but, in some circumstances, of other agreements dealing with subject matters "covered" by (and sometimes incorporated into) UNCLOS.[24] The possibilities of this means of coping with difficulties in implementation of a future whaling agreement may be illustrated by considering what might even now be done in the case of alleged violations of the ICRW.

This matter has not drawn much attention for a number of reasons, including a probable lack of interest by the majority of parties to the ICRW perhaps coupled with the active persuasion that their actions can easily be squared with the ICRW itself, as they interpret it, or any other relevant agreement. But the situation has recently changed with the coming into force in late 1994 of UNCLOS. The potential for review of IWC decisions may be attractive in some circumstances.

DISPUTE SETTLEMENT UNCLOS PART XV AND ANNEXES VI–VIII

States party to both the ICRW and UNCLOS, or only to the latter, differ from other nations because they have available, and are subject to, the dispute settlement provisions of UNCLOS. Future whaling agreements can anticipate the application of these parts and lessen the difficulties presented by their implementation.

UNCLOS Part XV

Part XV is most important for dispute settlement because its process is compulsory and binding when it applies, as it would for disputes over an agreement about taking whales on the high seas or about the provisions in UNCLOS. In brief, Part XV section 1 provides that the parties to the UNCLOS treaty are obligated to seek settlement of disputes over the interpretation or application of the treaty but that they are wholly free to choose the mechanism and procedures they will pursue. If the effort to settle the dispute through the chosen route is unsuccessful, any party may submit the dispute to the court or tribunal having jurisdiction under Section 2, Article 288(1), subject to Section 3, which provides limitations and exceptions to Section 2.

Under Section 3 a coastal state is not obliged to submit to dispute settlement in any dispute relating to its sovereign rights over fisheries in the exclusive economic zone or their exercise. It may, of course, agree specifically to do so in particular instances, or by itself invoking an Article 287 tribunal it might effectively waive its immunity from suit. Disputes over fisheries or whaling in the high seas area, however, are not excepted from dispute settlement and compulsory procedures are applicable.[25]

It follows from the above that states party to the ICRW that also are parties to UNCLOS are obligated to seek settlement of their disputes over the interpretation or application of Article 116 on freedom of fishing. In this instance an aggrieved whaling state can assert that it has a right to take whales on the high seas (as provided in Articles 87 and 116 of UNCLOS) and that this right can be terminated or conditioned only by obligations imposed consistent with another agreement so providing. A state party to the ICRW might claim that termination of its right under UNCLOS by imposing an obligation adopted in violation of the ICRW is also a violation of its rights under Article 116.

The fundamental point is that states have a right to harvest living resources of the high seas under customary law and under the 1982 Law of the Sea treaty. If there are to be restrictions on that right by another agreement, to which reference is made in Article 116 of the UNCLOS treaty, those restrictions can be imposed only in accordance with such agreement. Members of international organizations who impose restrictions in violation of their basic charter not only infringe that agreement but are in violation of Articles 87 and 116 of UNCLOS. A dispute between parties to UNCLOS over this interpretation of UNCLOS is subject to compulsory dispute procedures.[26]

The argument suggested here is not limited to these provisions on living resources. Disputes about the interpretation or application of other intricate provisions of the UNCLOS treaty may be complicated by treaty relationships. The UNCLOS treaty makes numerous cross-references to other agreements, subjecting parties to UNCLOS to obligations deriving from other agreements to which they are not necessarily party and, therefore, not bound by such other agreement. Disputes could arise about whether a particular state has complied with UNCLOS obligations defined under other agreements and whether noncompliance is a violation of the UNCLOS treaty. A dispute of this kind would be over the effect of the UNCLOS treaty on a party's rights or obligations thereunder, but the basic issues would or might turn on the interpretation of still another agreement. For example, the latter might provide for exceptions, exemptions, qualifications, or conditions that arguably justify or excuse noncompliance with its terms.

Article 116 of the UNCLOS treaty is a variation of this situation. It subjects a right under UNCLOS, as provided in Articles 87 and 116, to the provisions of other agreements. In the immediate instance, the freedom of fishing provided in Article 87 and the right to fish in Article 116 are subject to the rights and obligations of states party to the ICRW. Accordingly, in determining whether a right under UNCLOS is being denied or simply regulated as agreed, resort must be had to the provisions of the ICRW.

The ICRW provides for regulation of whaling for specific purposes identified in the Preamble and in accordance with requirements in Article V. The IWC is authorized to adopt regulations to regulate whaling and to include

63

them in a Schedule annexed to the basic treaty. The Schedule is subject to amendment. However, in amending the Schedule, the IWC is not free to adopt simply any restriction or condition its members may desire. The convention is careful to circumscribe this power. Schedule provisions must be such as are "necessary for the objectives and purposes" of the treaty, be based on scientific findings, and take into account the interests of consumers and the whaling industry. Regulations that do not comply with these requirements, or seek to change them, unlawfully restrict whaling. Such regulations deprive ICRW members of their right under UNCLOS to take living resources on the high seas.

The position is as follows: in accepting the UNCLOS treaty, states agree to limit their right to take whales by reference to obligations assumed under other relevant international agreements. But this is not an agreement to accept restrictions on whaling that exceed those authorized by the other international agreement. To impose restrictions by measures not consistent with the ICRW is a violation of the UNCLOS treaty itself. If other parties to the ICRW deny that a member's rights have been violated under UNCLOS, there is a dispute over the interpretation of UNCLOS within the meaning of Part XV of UNCLOS. Such a dispute is subject to compulsory and binding procedures.

Annexes VI–VIII

The specific avenues for dispute settlement arise under Part XV, Article 287, 288(2), and Annexes VI, VII and VIII.[27] Under Part XV or the annexes, the competence of a tribunal extends to disputes arising under another treaty or convention already in force concerning the subject matter "related to" or "covered by" the UNCLOS treaty. Under Annex VI, the tribunal is open not only to state parties but to other entities when all the parties to the dispute accept the jurisdiction of the tribunal. In the whaling context, one of the parties might be, for example, the International Whaling Commission in its status as the principal organ of an international organization. Whether it could participate as an entity would presumably be decided by the states parties to the ICRW, not by the commission itself since nothing in the ICRW appears to authorize it to make such a decision and its recommendation on such a matter might be without legal effect.

The significance of Annex VI may not be fully apparent. The new tribunal has the competence to entertain disputes over the interpretation or application of other treaties or conventions and the proceeding may be open not only to states parties to the UNCLOS treaty but to any parties to a dispute under the other international agreement. It is not necessary that the other agreement itself make provision for settling disputes. This can be accomplished by a supplementary agreement. Furthermore, the scope of the jurisdiction conferred upon the tribunal under Article 22 of Annex VI depends upon the agreement of the parties.[28] The parties may agree that the tribunal be limited to issuing an advisory opinion, or they may request some particular action by the tribunal that it might not otherwise be authorized to take.[29]

Under contemporary international law, therefore, if there is political willingness in the IWC to seek a third-party, objective, neutral decision maker to resolve differences or to give an advisory opinion about differences over what the IWC is authorized to do under the ICRW, there is no longer any doubt about the availability of such a tribunal.

The likelihood of this route being used in the IWC context might be remote in light of the requirement for all the parties to the dispute to agree to confer jurisdiction and to agree on its scope. The possibility of an advisory opinion may make the prospect less forbidding, but the force of even such an opinion might be discouraging to some states, whether they are pro- or anti-whaling.

Implementation of Dispute Settlement Provisions

In devising a new whaling agreement, taking account of potential problems concerning dispute settlement under the ICRW and UNCLOS may be useful. Some of the complexities may be illustrated by considering possible proceedings to resolve differences over a specific issue. There is certainly no shortage of occasions in the history of the IWC that could trigger resort to a third-party settlement procedure. Perhaps the most recent is the 1994 decision to establish a Southern Ocean sanctuary despite the absence of scientific findings and advice supporting the action. Others include the moratorium on commercial whaling itself, both in its original adoption without scientific findings and its continuation despite scientific findings that some whale stocks are sufficiently abundant to withstand a regulated harvest. Perhaps the most egregious issue arises from the clear view of some members that the original purpose of the ICRW to regulate whaling to maintain harvests is no longer relevant and binding, despite what the treaty says.[30]

The example here discussed briefly is the decision on the Southern Ocean sanctuary, which Japan claimed was adopted in violation of the requirement, inter alia, that it be supported by a scientific finding

Since there is no provision for this precise situation in either the ICRW or UNCLOS, it will need to be decided whether to attempt to make the commission itself the main party to the dispute or the commission plus members in their individual capacity or just selected members. If the IWC is to be involved, as sole party or a joint party with members, consent might require negotiation of an agreement to submit the dispute to the tribunal. The agreement would probably need to be negotiated with the states party to the ICRW, as opposed to the IWC itself, which does not appear to have authority under the ICRW to reach such an agreement nor does it appear to have control over the resources needed to implement an agreement. How many of these parties must agree is not clear; presumably a majority could agree to submit the dispute if the IWC were to be associated as a party to the dispute.

Whether or not the IWC is made a party, it might be necessary to negotiate an agreement with other states. An obvious potential difficulty is that some states may refuse to be a party, hoping to block any settlement procedure. Avoiding this potential obstacle should be a must in any new agreement.

The need in some circumstances for an agreement for parties to accept the jurisdiction of the Law of the Sea Tribunal under Annex VI introduces something of a wild card since it is impossible to predict what the agreement might include or what will be involved in its negotiation. Negotiating such an agreement might place those opposing the sanctuary decision under pressure to accept some unpalatable conditions as the price for securing use of the procedure. One condition might be to limit the competence of the tribunal to an advisory opinion, that is, one not binding on the parties (this is not necessarily unfavorable to a party challenging the sanctuary decision). Another condition might be to require the "loser," if the outcome were to be binding, to pay some or all the costs of the proceedings, which might be substantial.

It can be expected also that any agreement on jurisdiction will entail technical arrangements that might involve trade-offs. The arrangements could relate to determining who are to be parties to the procedure, what entities will have access to the tribunal (such as filing memorials or making oral arguments), whether scientific or technical experts need to be added (as provided in UNCLOS Article 289), whether to seek to have a chamber of the tribunal make the decision (as opposed to the tribunal as a whole), who should serve on the chamber or the tribunal if this question is open, the timing of the process, its location, specific details on production of documents and data, and perhaps conditions dependent upon the outcome. Some or all of these matters might appropriately be anticipated or avoided in a new agreement.

It may be desirable to consider asking for provisional measures to be taken pending the final decision. This is related to the larger issue of what remedy to seek if the final decision is favorable, a problem noted further below.

In the end, it is not possible to forecast the outcome of negotiations over submission of such an issue to the International Tribunal for the Law of the Sea since that outcome could be influenced by the attitudes and positions of the negotiating parties. For example, if this option were presented as an alternative to an action that was unpalatable, it might be met with a different response than otherwise. One such bargaining point might be reduction of discretionary financial support for the IWC, assuming such support were significant enough to make a difference to the conduct of IWC business. Another option would be some form of temporary suspension of membership in the IWC (this might be fashioned so that no disadvantage might be incurred). The most extreme alternative to dispute settlement would be withdrawal from the IWC, an action that has actually been mentioned as an alternative because of the blatant violations of the ICRW in adopting the Southern Ocean sanctuary. This would, of course, have no force unless it was perceived as a realistic possibility.

The other alternative for dispute settlement, that of invoking a compulsory and binding procedure under Part XV that requires no agreement between the parties, requires particular care in framing the issue to be resolved in a dispute settlement procedure. The goal is to be sure that the issue involves a dispute over the interpretation or application of the UNCLOS. As already suggested, this can be accomplished by arguing that the selection of the Southern Ocean sanctuary violated the ICRW and that this violation denies or will deny a member its right to take whales as provided in the UNCLOS treaty.

A possible response to the contention that the Southern Ocean sanctuary decision violates the ICRW is the assertion that the IWC acted properly under the ICRW and therefore did not violate any member's rights under

UNCLOS. The premise of this position would be that the question to be resolved is whether the IWC violated the ICRW and that this is not the proper subject matter of an action under Part XV. A secondary argument is that the IWC has already determined the issue of compliance with the ICRW and that, therefore, there can be no basis for a proceeding under Part XV.

The main counters to these contentions are that the question of conduct of parties under the ICRW must be considered by the tribunal in order to decide whether there is a violation of UNCLOS and that the tribunal is not foreclosed from that determination by the prior action of the IWC.

Another countering response is that failure to comply with the ICRW would itself be a violation of UNCLOS Article 116 and that denial of an allegation of such a violation is sufficient to establish that there is a dispute for the purpose of Part XV of UNCLOS. A tribunal would then have to consider whether actions by the IWC (or by the states composing the IWC) were consistent with the ICRW. If not, such actions would be a violation of UNCLOS.

Assuming resort to UNCLOS dispute settlement provisions is possible under the analyses presented here, it may be useful to ask what gains or losses might accrue if a tribunal upheld the position of the complaining party.

One significant gain would be to establish that a violation of the ICRW can be successfully challenged as a violation of the UNCLOS treaty. This would provide a basis for challenging other IWC actions or actions of individual state parties to the ICRW such as the statements from the United Kingdom and others that they will not support any future commercial harvests of whales, even if the best available scientific information shows that a sustainable harvest can be taken. This is tantamount to an announcement that they intend to violate the ICRW.

For purposes of the present discussion, the major potential gain from successfully challenging member actions under the ICRW is that it might provide some impetus toward parties' compliance with the provisions for the sustainable harvesting of whales. If this happened, it is not wholly inconceivable that members could agree on continued implementation of the ICRW without substantial revision.

This possibility of creating pressure toward future compliance with the ICRW is likely to be more important than any specific remedy a tribunal could offer.

If a dispute settlement tribunal did find that the IWC action violated a member's rights under the UNCLOS treaty, there is considerable question about what it could do beyond this to provide a remedy. It may be limited simply to pronouncing the violation, as opposed to entering an order that the IWC pay damages or take some other specific action. Even the International Court of Justice lacks authority to order specific performance as a remedy to support its judgments.[31] The advantage of using a tribunal pursuant to Annex VI is that its authority regarding remedies can be negotiated and thereby specified in advance. Perhaps the new whaling agreement should seek to anticipate this question of remedies when a tribunal resolves an issue. The following section considers some other elements in a new agreement.

EMERGING AND APPLICABLE PRINCIPLES

Precautionary Approach

Application of the precautionary approach is now widely considered desirable in the context of living resource management, and perhaps especially in the context of whale management, where the margin for error is narrow. Ironically, the principle is currently treated as largely irrelevant by those who both oppose commercial whaling and seem anxious to reduce if not eliminate aboriginal and small-type whaling. The moratorium on commercial whaling was adopted despite the absence of any finding by the IWC Scientific Committee that a blanket moratorium was needed. While there was uncertainty about the data on the abundance of some stocks, which justified a cautious approach, this was not uniformly true. Now, after several years of population increase and despite the development and approval in principle of a risk-averse management procedure, considered among the most rigorous developed for any marine animal, the majority of the IWC refuses to allow harvests from stocks that are judged by their own Scientific Committee to be sufficiently abundant for the purpose. So even when scientific information and findings remove serious uncertainty, the IWC majority is unwilling to act. To continue to delay resumption of whaling for particular stocks under stringent catch limitations is an abuse of the precautionary approach as well as a violation of the ICRW and UNCLOS. For the future, whether in a new organization or a revamped IWC, the Revised Management Procedure would appear to reasonably meet injunctions of caution in safeguarding whale populations.

Enforcement System

Realistic need for precaution no doubt extends to the critical requirements of an enforcement system in any new or revised regime since effective implementation is critical in the whaling context. All phases of such a system need careful attention including monitoring, surveillance, inspection, reporting, adjudication, trial, and penalty. It is common knowledge that securing the adequate performance of these operations is one of the most difficult accomplishments of international fishery management efforts.

Another downside of the current moratorium is that it formally removed the need for an enforcement system, thus there is no currently operating system upon which to build something more adequate. The previous enforcement arrangement under the IWC has been very differently assessed by observers and in any case might not provide a model in the foreseeable future when whaling operations are likely to be limited to small-type coastal whaling and aboriginal catches. For example, the kind of system needed for Japanese coastal whaling where operations are mostly within twenty-five to thirty miles of the coast might be very different from one suitable for distant water operations in the Southern Ocean or in a more remote part of the North Pacific or Indian Ocean.

One might expect these matters to be negotiable, but the present position of the parties to the ICRW will have to change before viable negotiations can take place. When the parties again become serious about a new enforcement system, recent developments will offer considerable promise. While it may not fully meet the needs of different types of whaling operations and may require modifications of various elements, the ongoing observer system in the tuna fishery in the Eastern Tropical Pacific Ocean appears to be fully effective as well as acceptable to those concerned.[32] This is a far-flung fishery, operating on the high seas outside exclusive economic zones and therefore may present some of the problems of whaling operations. Its main attribute is that it appears to have the confidence of the states concerned, including the one state, the United States, which has evidenced the most concern about the bycatch of small cetaceans in the fishery. The system has worked so well that the United States Congress voted recently to eliminate its ban on imports of tuna for those states in compliance with the International Dolphin Protection Program, which calls for an observer aboard tuna vessels to certify that no dolphins were killed or seriously injured in taking tuna.

Another factor of potential significance concerns innovations in technology for surveillance and monitoring. The fallout from the ending of the Cold War includes greater potential for civilian use of military sensing systems, both in the oceans and in space. These extend not only to vessel operations but to whales themselves. It is conceivable that these innovations may contribute to the negotiability of new agreements on enforcement by lending greater credibility and timeliness to observations and to scientific data.

The question of who will bear the cost of enforcement has apparently been a factor in recent negotiating difficulties within the IWC. Whether this is solely a product of the unwillingness of anti-whaling states to create a regime that will enable agreement to end the moratorium is unknown to this writer. Such a stance would be regrettable but simply another instance of the bankruptcy of the current situation. However this may be, in fairness costs must be shared in some proportion between those who do and do not choose to harvest whales under agreed-upon regulations. If nonwhaling states expect to participate in negotiations that fix the elements of the enforcement system, especially the observer system, then it is also equitable that they help defray part of the cost of that system. These states should not be allowed to insist on having a decisive voice in determining the ingredients of an effective system and at the same time demand that somebody else pick up the bill. Certainly it is also equitable that whaling states carry a significant part of the cost burden of an agreed-upon system. How much it is worth to each group of states to have the system can hardly be determined in advance, but the share of the costs that each bears must be sufficient that paying is not a trivial matter and can cause some pain. For the whaling states the calculation may be easier since presumably they may be able to place some value on the product, whether it is food or cultural needs. Evaluating the worth of the latter may be difficult.

On the other hand, nonwhaling states have to place a value on protection of whales as part of the marine environment, and this kind of calculation may be harder to determine. Indeed this value may be very high, in which case the cost of assuring protection should conceivably also be comparably high. Both sets of states must also take account of the shared value of maintaining diversity not only in the ecosystem usually thought of as animals and habitat, but equally of human groups and their cultural and religious needs. There is need to recognize that these considerations are involved in this issue and need to be weighed as much as other consequences.

Scientific Evidence in Management

Another principle that must be given greater weight than it has received in the current implementation of the ICRW is the need to be guided by the best scientific evidence available. This is no longer only a matter of highly desirable policy; it is also mandated by UNCLOS and should be considered to reflect general international law binding on all states. Article 119 of UNCLOS declares that in establishing conservation measures for fisheries on the high seas, states shall "take measures which are designed, on the best scientific evidence available to the States concerned." In any context other than the current situation in the IWC, where science is often irrelevant for specific decisions, it would hardly need emphasis that actions must take account of the best scientific evidence available. To allow this situation to continue will threaten not only whales but, if the credibility of the treaty implementation process is sufficiently undermined, perhaps other significant environmental problems.

NOTES

1. Patricia Birnie, "Are Twentieth-Century Marine Conservation Conventions Adaptable to Twenty-first-Century Goals and Principles?" *International Journal of Marine and Coastal Law* 12 (1997): 307, 438.

2. "The Irish commissioner [to the International Whaling Commission] introduced a proposal intended to break the deadlock between the governments opposed to a resumption of commercial whaling and those in favor. It would complete and adopt the Revised Management Scheme, designate a global sanctuary for whales, allow closely regulated and monitored coastal whaling within two-hundred-mile zones by communities with a long tradition for such activity but allow no international trade in whale products, and end scientific research catches. *1997 IWC Annual Meeting,* Final Press Release, Monte Carlo, Monaco. For the precise text, see "Discussion Document for Antigua Meeting Setting Out in Further Detail the Package of Measures for the Conservation and Management of Whales Proposed by Ireland at the IWC Meeting in Monaco in September 1997." The detailed text of this proposal differs in some respects from the above description but is the same regarding elements that are outside the scope of the ICRW, including the intrusion into international trade and the end of scientific whaling. The former issue is dealt with in the

Convention on International Trade in Endangered Species, and the latter is specifically exempt from the ICRW under Article VIII thereof. The proposal for the global sanctuary is not likely to be consistent with the Article V requirement of a scientific finding.

3. The reason for exploring some legal aspects of seeking a new whaling agreement is that other avenues for resolving differences do not seem promising. But the later portion of this paper discusses the possibilities of resolving divergent interpretations of the ICRW by resorting to dispute settlement procedures under the UN Convention on the Law of the Sea. As written in 1946, the ICRW provided for what is now labeled sustainable harvesting. As currently interpreted by some IWC members, however, the treaty supports complete prohibition of any whale mortality for commercial purposes. This interpretation might well be discredited if it were challenged in a dispute settlement proceeding.

4. In recent years, however, the commission has managed to avoid the treaty provisions by amending the Schedule in disregard of the treaty and then finding (1) that the amendment is consistent with the treaty and (2) that it has sole and exclusive authority to decide how the treaty is to be interpreted. This appears to be an indirect, and unlawful, way to amend the treaty, but this neat maneuver has not yet been challenged successfully by members who disagree with the commission action. Given the IWC majority's perceived disregard for their treaty obligations in other respects, it is difficult to expect a minority member to accept an IWC decision about its authority to amend the basic agreement.

5. Perhaps it would make no difference that representatives of states rather than commissioners as such would be involved in reaching a new whaling agreement, but it is conceivable that in some instances the views of governments may differ from those of commissioners. It may be that in some instances the issue of national whaling policy is determined routinely by the commissioner, without instructions from upper levels. High-level foreign affairs officials might have a different perspective because they have broader responsibilities. This applies particularly within a context in which a subject matter is perceived as symbolizing broader questions.

6. The IWC comprehensive assessment has apparently been completed for minke whales in the Southern Hemisphere (761,000), North Atlantic (approximately 149,000), northeast Pacific and Okhotsk Sea (25,000), fin whales in the North Atlantic (47,300), and gray whales in the eastern North Pacific (21,000) (this number was increased in 1999 to 26,635). Although these figures are the best estimates of present abundance (although the minke whale number for the Southern Hemisphere was for 1989), no catch limits have been established because the algorithm has been applied only to minke whales in the North Atlantic and Southern Hemisphere *and*

because the commission has not yet agreed to a Revised Management Scheme. This is a political decision, not a scientific one. It is widely believed that the eastern gray whale is also sufficient to support a harvest. See John A. Knauss, "The International Whaling Commission: Its Past and Possible Future," *Ocean Development and International Law* 28, no. 1 (1997): 79, 82 n. 35. See also Jon Conrad and Trond Bjørndal, "On the Resumption of Commercial Whaling: The Case of the Minke Whale in the Northeast Atlantic," *Arctic* 46, no. 2 (1993): 164; William Aron, "The Commons Revisited: Thoughts on Marine Mammal Management" *Coastal Management* 16 (1988): 99–110; Peter Weber, "Abandoned Seas: Reversing the Decline of the Oceans," *World Watch Paper* 116 (November 1993).

7. At this writing 132 states are parties to UNCLOS, thirty-four of which are also parties to the ICRW. Assuming there are 185 nation-states that are generally recognized, about one-sixth accept the ICRW and around two-thirds accept UNCLOS. Other agreements include the "Agreement on the Conservation of Cetaceans in the Black Sea, Mediterranean, and Contiguous Atlantic," *International Legal Materials* 36 (1997): 777; Agreement on the Conservation of the Small Cetaceans of the Baltic and North Seas (1992); Agreement on the Conservation and Management of Marine Mammals in the North Atlantic (1992); Convention on the Establishment of the Inter-American Tropical Tuna Commission (1949). The latter three agreements are in *The Marine Mammal Commission Compendium of Selected Treaties, International Agreements, and Other Relevant Documents on Marine Resources, Wildlife, and the Environment*, comp. Richard L. Wallace, vol. 2 (Washington, D.C.: GPO, 1995).

8. A 1994 report of the World Conservation Union suggests that at least thirty states directly harvest small cetaceans. It is difficult to estimate how many take bycatches but it is probably very large. See Randall Reeves and Stephen Leatherwood, comps., *Dolphins, Porpoises, and Whales* (Gland, Switzerland: IUCN, 1994).

9. Although not applicable to marine mammals, the 1995 Straddling Stock Agreement does change customary law affecting high seas fishing for those states party to it by establishing obligations not previously recognized. See UN, Agreement for the Implementation of the Provisions of the United Nations Convention on the Law of the Sea of 10 December 1982 Relating to the Conservation and Management of Straddling Fish Stocks and Highly Migratory Fish Stocks, 8 September 1995, UN Doc. A/CONF. 164/37.

10. That freedom of fishing applies to the take of marine mammals extends back at least to the Fur Seal Arbitration in 1890. See John B. Moore, *Digest of International Law*, vol. 1 (Washington, D.C.: GPO, 1906), 890 and following for the award. The scope of the claims based on freedom of fishing are fully developed in the sixteen-volume record of this arbitration.

11. The 27 August 1999 order by the International Tribunal for the Law of the Sea in the Southern Bluefin Tuna cases suggests (the decision is somewhat enigmatic on its precise interpretations of treaty provisions, so "suggests" is as far as one can go) that these provisions may be given significant content in particular circumstances. In this case Australia and New Zealand submitted a dispute with Japan to the tribunal requesting provisional measures (the tribunal did not have direct jurisdiction under Part XV of UNCLOS) prohibiting Japan from conducting an experimental fishing program that would have the result of increasing its total catch of southern bluefin tuna over the allocation that had been agreed upon two years before under the provisions of the Convention for the Conservation of the Southern Bluefin Tuna between these three states. Despite the fact that there was no currently agreed-upon total allowable catch for the southern bluefin tuna, the tribunal decided that Japan was forbidden to continue its program without the agreement of Australia and New Zealand unless its total catch remained within the formerly agreed-upon allocation. Whether this decision will stand is not known at this writing, since the tribunal had jurisdiction only to issue the provisional measure pending the creation and decision by an arbitral tribunal under Annex VII of UNCLOS. That decision may find that the arbitral tribunal has no jurisdiction to consider the case, which would signify, in turn, that the international tribunal had no jurisdiction to issue the provisional measures. The arbitral tribunal's decision on the merits may also differ from the international tribunal's. The writer was a consultant to counsel for Japan in this case. The tribunal's order is available at www.un.org/Depts/los/ITLOS/Order-tuna34.htm.

12. One means of avoiding this difficulty is to misread the treaty so that it refers only to "international organization." See Jose A. Yturriaga, *The International Regime of Fisheries: From UNCLOS to the Presential Sea* (The Hague: Kluwer Law International, 1997), 131, noted in William T. Burke, review of *The International Regime of Fisheries: From UNCLOS to the Presential Sea,* by Jose A. Yturriaga *American Journal of International Law* 91 (1997): 752–53. For this same misreading see also Birnie, "Twentieth-Century Marine Conservation," 314, 338. But see page 500 of that work, where the correct form is used, suggesting that the first instances are inadvertent. The most egregious distortion of Article 65 is in Peter J. Stoett, *The International Politics of Whaling* (Vancouver: University of British Columbia Press, 1997), 68, where the author simply substitutes an article of his own creation for Article 65: "Article 65 stipulates: States must abide by the regulations of the IWC, whether or not they are parties to the International Whaling Convention, except where they adopt stricter regulations for the conservation of whales." [Dan Goodman's review of the Stoett book highlights the "misrepresentation" of Article 65 as one of the "most serious and obvious errors" in the book. See Dan Goodman, review of *The International Politics*

of Whaling, by Peter J. Stoett, *Marine Policy* 21 (1997): 547.] Further search discloses that this quotation, which is not footnoted, comes from another book [Simon Lyster, *International Wildlife Law* (Cambridge: Grotius Publishing, 1985), 36] in which the author accurately quotes Article 65 but adds the sentence quoted by Stoett as "the implication" of the article. The Lyster reference is footnoted after another sentence, but the page reference is wrong. Lyster does not explain how he reaches his conclusion either.

13. It deserves to be recalled that this view is advocated by some who contend that the ICRW already provides for the regulation of small cetaceans.

14. Congressional resolutions express opposition to any resumption of commercial whaling.

15. Statements by Australia, New Zealand, and the United Kingdom that they cannot vote for a resumption of commercial whaling under any circumstances are tantamount to rejection of the basic purpose of the ICRW and the substitution of a contradictory purpose. Statements attributed to fourteen members, including those mentioned above, record their objection to any resumption of commercial whaling in the context of the Japanese request for an allocation for their small whaling villages. See *Proceedings of the 46th Meeting of the International Whaling Commission,* 1994, 18. Voting against a whale harvest solely on such a basis is a violation of the ICRW.

16. This paragraph refers to an international organization already established for the fish of a particular region. Article 8(1) provides that states shall pursue cooperation directly or through appropriate regional international organizations or arrangements. See UN, Agreement for the Implementation of the Provisions of the United Nations Convention on the Law of the Sea of 10 December 1982 relating to the Conservation and Management of Straddling Fish Stocks and Highly Migratory Fish Stocks, 8 September 1995, UN Doc. A/CONF.164/37. The convention applies to highly migratory fish stocks, not to highly migratory species in general and, therefore, is inapplicable to whaling organizations.

17. See Patricia Birnie, "UNCED and Marine Mammals," *Marine Policy* 17 (1993): 501, 514. Nothing in the ICRW or any other document, including Agenda 21, recognizes such a role for the IWC.

18. Reluctance to adhere to the ICRW might only be increased by the recent Irish proposal, which is advanced as a "compromise" between pro- and anti-whaling states. See note 2 above. This proposal would foreclose future whaling by any nation that has not done any whaling for the past five years and would, in any case, limit catching to coastal areas (not defined) under the still uncertain application of the Revised Management Procedure and under enforcement procedures that are not yet determined.

19. For discussion of the latter see William T. Burke, "Aspects of Internal Decision-Making Processes in Intergovernmental Fishery Commissions," *Washington Law Review* 43 (1967): 115, 140–42. Article 4 of the NAMMCO agreement appears to anticipate this possibility in authorizing the council to establish "appropriate management Committees and coordinate their activities." *Marine Mammal Commission Compendium* (1995): 1619. For an insightful examination of the requirements for scientific advice in an international management setting, including the value of regional approaches, see Lee Kimball, "Treaty Implementation: Scientific and Technical Advice Enters a New Stage," *Studies in Transnational Legal Policy* 28 (1996): passim.

20. Birnie is a principal advocate of this approach to interpretation. See Birnie, "Are Twentieth-Century Marine Conventions Adaptable" Birnie also cautions, however, that interpretation of the ICRW must remain consistent with its objects and purposes. This tends to lose force, however, when she finds the latter to be ambiguous and highly subjective and in need of interpretation in light of numerous events occurring subsequent to the adoption of the ICRW.

21. Article 31, Vienna Convention on the Law of Treaties: "A treaty shall be interpreted in good faith, in accordance with the ordinary meaning to be given to the terms of the treaty in their context and in light of its object and purpose." Official Records, United Nations Conference on the Law of Treaties, Documents of the Conference 293, 1969, UN Doc. A/CONF. 39/27.

22. Myres Smith McDougal, Harold D. Lasswell, and James C. Miller, *The Interpretation of Agreements and World Public Order: Principles of Content and Procedure* (New Haven, Conn.: Yale University Press, 1967), 95–97.

23. An example of how general provisions can lead to questionable interpretations is the order issued in the Southern Bluefin Tuna cases; see note 11 above. In the 1993 Southern Bluefin Tuna Convention, the three states agreed to take measures to conserve and manage the southern bluefin tuna, establishing a commission to make decisions by consensus, including the determination of the total allowable catch for each fishing season. The agreement did not provide for what was to happen if the total allowable catch was not agreed upon. Nonetheless, the parties were held to be bound by the last agreed-upon total allowable catch, after failing for two years to agree on one. This meant that Japan was bound to refrain from an experimental fishing program that yielded a catch over its allocation of the last agreed-upon total allowable catch. The tribunal's order appears to rest on the view that the cooperation required by Articles 64 and 119 of UNCLOS overrides the consensus requirement in the Southern Bluefin Tuna (SBT) Convention, and Japan is required to agree with a decision that it has the right to reject under the basic treaty. Although the tribunal

concedes that it cannot "conclusively assess the scientific evidence presented by the parties," it apparently believed the Australian/New Zealand view of this evidence and acted "to avert further deterioration of the southern bluefin tuna stock." Japanese evidence was that the stock was increasing, and its experimental fishing program was designed to lessen the uncertainty about the status of the stock.

24. The reach of the UNCLOS dispute settlement provisions is clarified by the decision of the tribunal in the SBT cases. While the basic dispute in the SBT cases arose from disputed scientific issues under the 1993 SBT Convention, the tribunal declared that the dispute involved the obligations of the parties under UNCLOS. The tribunal's order is available at www.un.org/Depts/los/ITLOS/Order-tuna34.htm.

25. UNCLOS Article 282, however, provides that if states parties that are parties to a dispute have agreed "through a general, regional or bilateral agreement or otherwise" to another binding procedure, a party may request use of that procedure and it applies in lieu of Part XV. In the SBT cases, the tribunal raised the question with the parties about their common acceptance of the compulsory jurisdiction of the International Court of Justice (ICJ). For unknown reasons, Japan expressed the view that it did not believe that the ICJ had jurisdiction but that this was a close question. Australia/New Zealand essentially passed on the question. It seems likely that the ICJ would have jurisdiction over the dispute to the exclusion of Part XV if one of the parties so requested, but none did. The tribunal's order says nothing about this aspect of its jurisdiction.

26. Parties to UNCLOS also have obligations to provide for conservation of whales and to cooperate with other states to this end. Issues arising over alleged failures to take adequate conservation measures or to cooperate with other states might also turn on disputes about the interpretation or application of UNCLOS or of another agreement. Such disputes might also precipitate invocation of UNCLOS dispute settlement provisions. Statements in the text are also applicable to such disputes.

27. Article 287 allows states to choose among four different mechanisms for dispute settlement: (1) the International Tribunal for the Law of the Sea, (2) the International Court of Justice, (2) an arbitral tribunal under Annex VII, and (4) a special arbitral tribunal under Annex VIII. Failure to designate a choice means that an arbitral tribunal under Annex VII is deemed to be accepted. Part XV determines the jurisdiction of such a tribunal.

28. "If all the parties to a treaty or convention already in force and concerning the subject-matter covered by this Convention so agree, any disputes concerning the interpretation or application of such treaty or convention may, in accordance with such agreement, be submitted to the Tribunal."

29. The text of Part XV does not make it clear that the jurisdiction of the tri-

bunals under Annexes VII and VIII is coextensive with that of the tribunal under Annex VI, although it seems likely that this is the case.

30. See note 15 above.

31. See Shabtai Rosenne, *Developments in the Law of Treaties, 1945–1986* (Cambridge: Cambridge University Press, 1989).

32. The IWC has been fully briefed on the Inter-American Tropical Tuna Commission system, but differences have still not been overcome.

2 / Whales, the IWC, and the Rule of Law

JON L. JACOBSON

Let me state right up front that I would prefer a world in which whales are not hunted and killed by humans. I would not object if the International Whaling Commission (IWC) were to be disbanded and its treaty replaced by a global agreement dedicated to the well-being of all cetaceans. For that matter, I would personally be grateful if the great whale nations of the world ocean—and by this I mean nations made up of whales, not of people—were in return to adopt a global agreement dedicated to the well-being of humankind. Some pro-whale (i.e., anti-whaling) literature, by suggesting that whales may be at least as intelligent as our own species, seems to raise the prospect that humans and cetaceans might someday meet together in a truly global conference to converse about and decide many matters of mutual interest and concern. I, for one, am not prepared to disbelieve this. Certainly by that time, if it does come to pass, the human hunting of whales will have been completely outlawed everywhere.

In the meantime, current events concerning the fortunes of whales and whalers present us with a crucial international law question of some difficulty. Summed up, that question is whether it is legally possible to use the present IWC and its convention to protect and preserve all whales from whalers without sacrificing even greater values. To me, alas, the answer is no.

In this chapter, I begin by pointing out, as others have done in more detailed analyses,[1] the apparent dissonance between, on the one hand, the whaling convention's clear original purpose of regulating whaling for the benefit of the whaling industry and its clients and, on the other, the IWC's recent actions in purporting to establish a possibly permanent moratorium on commercial whaling and an apparently permanent Southern Ocean sanctuary for whales. I then briefly examine the lawful means by which this dissonance might be eliminated, including replacement of the I W C's convention with a whale-friendly regime or, failing that, amendment or interpretation of that treaty. In view of the unlikelihood of replacement or amendment in the near future, I offer what I (and some others) consider the best case that can be made for justifying the commission's anti-whaling actions under the present I W C convention: characterizing the convention as a "constituent document," like a constitution or charter that establishes a government or decision-making body of indefinite long-term tenure, thus making it arguably eligible for an expansive evolutionary interpretation of the I W C's powers. Like other treaty "reinterpretations," however, even the expansive-interpretation case falls short, in my view, of justifying the I W C's actions under scrutiny. More important, such an overly broad reading of the convention is yet another unfortunate example, among others recently, of ignoring clear but inconvenient legal mandates and strictures. This disregard undermines the authority of all aspects of the whaling convention itself (e.g., it invites convention parties to indulge in an equally expansive interpretation of their right to engage in "scientific" or research whaling) and encourages disrespect for international conventional law in general. This outcome, I submit, is not in the interest of anyone, particularly of those, including most people concerned about the commercial killing of whales, who have been working so hard to establish and implement treaty-made law to protect the planet's environment and its biodiversity.

Perhaps it is time for those of us who are concerned about the future of whales to bite the bullet and learn how to persuade rather than manipulate. The fate of more than whales is at stake.

THE PROBLEM WITH THE IWC

The whaling treaty, formally the International Convention for the Regulation of Whaling,[2] or the I C R W, was adopted in 1946 and established the International Whaling Commission for purposes that, despite some recent rather imaginative (and perhaps sometimes disingenuous) interpretations by those admittedly seeking to end commercial whaling, are quite obvious from the language and context of the convention. The main crystal-clear objective of the treaty was, and still is, to provide for the conservation of whales for the benefit of the whaling industry and its customers. The ICRW is, indeed, one of the primary early treaty commitments to the ideal expressed today as sustainable development of the earth's natural resources. This ideal, heavily promoted by environmentalists in recent decades,[3] has probably emerged by now as a principle of customary international environmental law.

In its pursuit of the goal of sustainable development, the whaling treaty is in its essence no different from scores of international fisheries agreements of the past and present that attempt to achieve, through conservation measures, an optimal harvest of renewable marine living resources indefinitely into the future.[4] Some of these agreements have worked fairly well in pursuing this goal, and others have fallen short or, sometimes, failed miserably. Nevertheless, there has seldom been any dispute about the reality—or the worthiness—of any of these agreements' true major purpose or objective: conservation to enable sustained harvest. It would be absurd for anyone to contend that the "conservation" goal was or had become one of preservation and protection of the resources from all fishing.

Yet for the past several years, we have witnessed analysts of the ICRW apparently contend that the treaty's "conservation" objective can now be interpreted to allow the IWC (1) to prohibit all whaling indefinitely in a Southern Ocean sanctuary for the preservation and protection of the animals there, and (2) to impose a global moratorium on all commercial whaling that, while technically temporary, has certainly begun to take on an aura of permanence.[5]

Why the difference between the treatment of the ICRW and other international agreements for the conservation and management of the sea's renewable living resources? Start, of course, with the fact that, unlike fish, whales are mammals. This difference is, indeed, crucial. Mammals have much lower reproductive rates than fish, and overwhaling can therefore truly cause species extinction, while overfishing alone can probably never cause extinction. Extreme care must therefore be taken in the regulation of whaling, both to prevent extinction of whale species and to meet the objective of sustainable harvest. This difference between whales and fish has been understood for generations and is a major reason the IWC was established. Like a number of fishing agreements, however, the whaling treaty and the IWC have, over the decades, often failed badly in the pursuit of their conservation assignment and, thus, their sustainable harvest goal.[6] Yet new approaches, aided by better science and predictive technology, today show great promise for helping the I W C to realize the treaty's original purpose.[7]

But another distinction between fish and mammals apparently stands in the way of any serious attempt to implement the new plan. This distinction emphasizes the mammalian link between humans and whales and the consequent empathy many of us feel toward those creatures and other "charismatic megafauna."[8] Because of this empathy and the suspicion that whales are sentient beings who share a level of intelligence with humans,[9] an undoubtedly growing number of people (in many countries), at least in historical terms, are today opposed, sometimes adamantly, to human hunting and killing of cetaceans. At the same time, the number of people (in a few countries) who are actively engaged in whaling has probably dwindled.[10]

These relative numbers of people (and countries) have, to some extent, come to be reflected in the member states of the I C R W and the I W C. The commission is now controlled by delegations, including that of the United States, that are opposed to the practice of commercial whaling. As a result, the I W C has purported to establish a Southern Ocean sanctuary, where no whaling is allowed, in the face of persuasive scientific evidence that an adequate number of minke whales exist to allow a sustainable harvest with no danger of extinction.[11] And the commission has prolonged a moratorium, which it first imposed in 1986, on commercial whaling worldwide.[12] Representatives of some anti-whaling parties have reportedly stated that they intend to ensure that the moratorium is extended indefinitely, for the sake of the whales.[13]

Other analysts of the I C R W, notably William T. Burke, have forcefully argued that the I W C's Southern Ocean sanctuary is not authorized by the convention and is, therefore, beyond the powers of the commission to establish.[14] A permanent moratorium on commercial whaling, whether or not labeled as such, would seem to be even more clearly outside the treaty's grant of authority to the I W C. The treaty's sustainable harvest objective can, of course, never be realized if no harvest is allowed. Thus the dissonance I referred to earlier. In theory at least, there are ways to erase the dissonance. Let's look at the major options.

REPLACING THE CONVENTION

For those of us who are concerned about the commercial harvest of whales and would like to see the hunting lawfully ended, the obvious preferred legal approach would be to replace the ICRW with a new regime that would outlaw whaling altogether. The relatively easier part of this plan would be the termination of the whaling treaty. Indeed, under present circumstances, it can be argued that there are grounds for challenging the continued existence of the ICRW. The Vienna Convention on the Law of Treaties,[15] which does not directly apply to the ICRW but which has been said to generally reflect customary international law,[16] allows treaty parties to invoke (among other reasons) changed circumstances and material breach as grounds for termination of a treaty.[17] It might, then, be contended that the historical decline in the practice of whaling, taken together with changed attitudes favoring preservation of whales, should combine to cause the termination of the treaty. Or, on the other side of the coin, it might be charged by whaling state parties that other parties, through their delegations' votes in the IWC to establish a Southern Ocean sanctuary, have committed material breaches of the ICRW's limited grant of authority to the IWC, thereby allowing the whaling states to invoke these breaches as grounds for termination of the treaty.

Unfortunately for those who would prefer a preservation regime, getting rid of the ICRW and its commission today would almost certainly merely replace it with customary international law, which, on the high seas, almost certainly still allows freedom to hunt whales as part of the freedom-to-fish regime[18] and, in two-hundred-mile exclusive economic zones, grants coastal states sovereign rights to exploit living resources, including whales.[19] Presumably for this reason, anti-whaling interests have been careful not to advance the arguments for the ICRW's termination but have instead worked hard to manipulate the treaty to anti-whaling ends.

One treaty-termination doctrine of international law would, if applicable, achieve the goals sought by whale preservationists. According to the Vienna Convention on the Law of Treaties, if a treaty, during its tenure, were to come into conflict with a newly emerged peremptory norm of international law *(jus cogens)*, the treaty would terminate.[20] Thus, should a peremptory norm forbidding commercial whaling emerge, the ICRW would terminate because its clear purpose is to allow commercial whaling, and it would be replaced by the new peremptory norm. In my view, that norm is now in the emerging stage but has not yet arrived.

Of course, another lawful means by which commercial whaling could be made illegal under international law would be the successful negotiation and implementation of a new global international agreement forbidding whaling to replace the ICRW. This would, however, succeed only if all whaling states were to become parties to the new treaty, an unlikely scenario today.

AMENDING THE ICRW

One of the ways to eliminate the dissonance between the ICRW's clearly expressed objectives and the recent actions of the IWC would be, of course, to amend the treaty's objectives to remove any reference to "development," "utilization," "whaling industry," and so forth, so as to allow, even mandate, the "conservation" and preservation of whales for their own sake. Although the ICRW provides a procedure for amending its Schedule of regulations for whaling, the treaty makes it clear that these amendments must comply with its objectives and purposes, "provide for the conservation, development, and optimum utilization of the whales resources," "be based on scientific findings," and "take into consideration the interests of the consumers of whale products and the whaling industry" (Article V). The convention has no express provisions for formal amendment of its objectives and purposes. However, the Vienna Convention on the Law of Treaties, here again certainly reflecting customary law, contemplates amendment of multilateral treaties (Article 40) but further states that, "[u]nless the treaty otherwise provides," a party that does not consent to the amendment is not bound by it and the original agreement remains in effect for that party and the other parties to the treaty.[21] Thus the current anti-whaling majority in the ICRW could not, by majority vote, impose any formal amendment of the treaty's purposes on the minority of state parties who disagree with the amendment.

With formal amendment a very unlikely scenario for getting rid of commercial whaling, then, is it possible that the ICRW could be amended by implication or by new custom? This possibility raises questions similar to, but not the same as, those concerning a new *jus cogens*. An analogy could be drawn to the 1958 Convention on the High Seas, which by its terms ensures freedom of fishing seaward of coastal state territorial seas.[22] Yet by the mid-1980s, long before the 1982 UN Convention on the Law of the Sea and its two-hundred-mile exclusive economic zones came into force in 1994, virtually all the parties to the High Seas Convention had claimed and acquiesced in claims to exclusive coastal state authority over fishing far beyond their territorial seas. Indeed, the United States remains a party to the 1958 High

Seas Convention, but no one would seriously contend that its exclusive economic zone, extending out to two hundred miles,[23] is today in violation of
that treaty's guarantee of freedom to fish beyond its twelve-mile territorial
sea. Why not? It might reasonably be contended that the 1958 convention
has been amended by the development of customary international law ap-
proving extra-territorial-sea exclusive economic zones or exclusive fishing
zones.[24] Perhaps this is an appropriate explanation for a treaty that by its
terms mainly purported to restate or codify custom, which the High Seas
Convention claimed to do.[25] Such a treaty should arguably not be interpreted to prevent the parties from participating in developing new customary law and thereby changing the treaty's terms. It might even be argued
that this was not beyond the contemplation of the treaty's drafters. The
ICRW, however, was clearly established not to codify custom but instead to
counter the then-current and still largely prevailing customary norm of unrestricted freedom to hunt whales on the high seas, but did so for the expressed purpose of conserving the whale resources for the benefit of the
whaling industry. Thus, an evolutionary change in custom favoring protection of whales from whalers, if such a change is indeed on the horizon,
should not as easily be seen as automatically amending the treaty.

A similar analysis of the fate of the High Seas Convention's freedom-to-
fish provision would instead characterize the change as an implied amendment of the treaty.[26] That is, by their actions in claiming and acquiescing in
claims to broad exclusive coastal state authority over fishing beyond the territorial sea, the parties to the High Seas Convention implicitly agreed
among themselves that the treaty was amended to allow such claims out to
two hundred miles. While this might indeed explain the apparent change in
the High Seas Convention, it does not help us resolve the dilemma of the
IWC because some of the ICRW's parties have been actively resisting any
changes in the convention that would make it an agreement solely to protect and preserve whales, and treaty parties should not be bound by implicit
amendments to which they object any more than they should be bound to
express amendments to which they do not consent.

Clearly, then, amendment of the ICRW to conform its objectives to the
actions of the IWC is presently not a means of avoiding the dissonance. This
pretty much leaves the approach employed—not surprisingly—by the anti-
whaling group: interpretation (or reinterpretation) of the ICRW's original
provisions.

INTERPRETING THE WHALING CONVENTION

The Vienna Convention's provisions on treaty interpretation are also considered a reflection of customary international law.[27] These rules place primary emphasis, in interpretation disputes, on good faith and the ordinary meaning of a treaty's terms in their context and in light of its object and purpose. Recourse to the preparatory work leading up to a treaty's conclusion, for example, is reserved for situations in which the treaty's language is ambiguous or leads to absurd applications.[28] While subsequent practice of a treaty's parties must be taken into account in the interpretation process, the Vienna Convention makes it clear that such practice is relevant only if it "establishes the agreement of the parties" regarding the treaty's interpretation.[29]

The Vienna Convention's strong emphasis on the ordinary meaning of a treaty's language is not helpful to those who wish to qualify the ICRW as a whale preservation agreement. Analysis of the whaling convention's language setting forth its purposes and objectives, especially when viewed in light of its context, reveals that it is among the clearest wording of any international agreement.[30] It has required some linguistic gymnastics from those analysts who have attempted to manipulate this wording to justify the IWC's actions in purporting to establish a Southern Ocean sanctuary and an indefinite moratorium on commercial whaling. These exercises have usually seemed to focus on the word "conservation" in the statements of the convention's purposes while ignoring the other principal part of the formula that very clearly establishes whale harvests as the ultimate objective, as if the IWC has been given an option to choose between them.[31] The whole tenor of the treaty, however, makes it abundantly clear that these two primary objectives are inseparable, joined as they are, like the ends of a barbell, by that bothersome conjunction "and," not the multiple-choice disjunction "or." To appreciate the error of viewing the two linked goals as separable, imagine some future IWC mandating the "development" or taking of sperm whales to extinction (because, for example, they are ruthlessly attacking the new pleasure submersibles), thus assertedly complying with the convention's development goal while ignoring the conservation goal. This hypothetical IWC action would surely be considered ultra vires, being no more compatible with the convention's purposes than one that pursues the conservation goal to the exclusion of the development goal. Indeed, the ICRW's frequent references to its purposes and objectives, taken collectively, easily support a plain-meaning reading that identifies the conservation goal as a *means* of achieving the more important sustainable-whaling *end*.

The subsequent practice of the majority of the ICRW's parties, through their votes in the IWC, would at first glance seem to support a reinterpretation of the treaty's objectives in favor of preservation over development, but, as already noted,[32] the Vienna Convention, surely reflecting customary law, requires that such practice amount to an agreement of the treaty's parties. This certainly refers to "authentic" interpretation, or interpretation by agreement of all the parties, which, of course, is not the case with the ICRW. Nevertheless, all is not necessarily lost.

CREATIVE CONSTRUCTION: THE ICRW AS A CONSTITUTION

While there are undoubtedly many shades of differences among treaties, based primarily on their intended functions (see, e.g., the suggestions above concerning the 1958 High Seas Convention[33]), at least three broad divisions of types can be drawn: (1) bilateral agreements, especially short-term and for a limited purpose, that emulate private contracts between individuals; (2) "law-making" conventions, usually multilateral, that attempt to create something like legislation; and (3) constituent instruments or documents that establish organizations or institutions, which themselves are expected to take lawmaking or regulatory actions over an indefinite future existence. Although the Vienna Convention makes no distinction in its interpretation rules among these three treaty types—and in fact explicitly states that its uniform rules apply to "any treaty which is the constituent instrument of an international organization"[34]—it seems very appropriate to do so, perhaps by an emphasis on "context." For example, interpretation of bilateral "contracts" between states might more easily make early reference to the parties' negotiations (preparatory work) and subsequent practice in performance of the agreement to determine the true intent of the two parties, while stronger emphasis solely on the words of the treaty itself might make more sense for a carefully drafted multilateral treaty that attempts to create a legislative regime for guidance of its many parties (this seems to be the Vienna Convention's approach for all treaties). And, most pertinently for the present discussion, a treaty (or part of a treaty) that, like a constitution or charter, creates a new long-term international organization with regulatory powers should be interpreted, like a charter or constitution, in a flexible evolutionary manner that allows it to function over time through changing circumstances and attitudes not present at its creation and even unforeseen by its creators. Indeed, the International Court of Justice (ICJ) has recently indicated that it approves of this interpretive approach for a treaty-constitution. In its response to a request from the World Health Organization (W H O) for an advisory opinion on the legality of the use of nuclear weapons, the court was faced with the issue of whether the W H O's constitution authorized its assembly to make such a request. Although the court ruled that the World Health Assembly had exceeded its powers, the court nevertheless confirmed the special challenge of interpreting such constituent treaties as the W H O constitution:

> 19. In order to delineate the field of activity or the area of competence of an international organization, one must refer to the relevant rules of the organization and, in the first place, to its constitution. From a formal standpoint, the constituent instruments of international organizations are multilateral treaties, to which the well-established rules of treaty interpretation apply. . . . But the constituent instruments of international organizations are also treaties of a particular type; their object is to create new subjects of law endowed with a certain autonomy, to which parties entrust the task of realizing common goals. Such treaties can raise specific problems of interpretation owing, *inter alia*, to their character which is conventional and at the same time institutional; the very nature of the organization created, the objectives which have been assigned to it by its founders, the imperatives associated with the effective performance of its functions, as well as its own practice, are all elements which may deserve special attention when the time comes to interpret these constituent treaties.[35]

There are undoubtedly many examples of illustrative interpretations of this type of document, in both international and domestic contexts. Let me refer to just a couple of famous ones that are at least roughly analogous: the U.S. Constitution and the UN Charter.

While the U.S. Constitution is not, of course, an international agreement, it is the oldest written true constitution now extant, and its interpretation over the generations of its existence provides a primary model of the properly flexible reading of a constituent document. A famous example is the many interpretations and reinterpretations of the language establishing the American federation's basic structure.[36] In creating the federation, the framers of the Constitution clearly intended to place severe restrictions on the powers of the central (or federal) government, enumerating only a few federal powers and leaving nearly all domestic governance to the several states of the new union.[37] This intent was underscored early by the Tenth Amendment to the Constitution. Nevertheless, over the decades and centuries, the U.S. Supreme Court has approved substantial expansions of the central authority until today federal laws and regulations pervade nearly every aspect of American life. This expansion of federal governmental authority has sometimes been assisted by constitutional amendments,[38] but it has often been based in the Court's interpretation of the founders' document and justified by evolutionary changes encountered in the nation's remarkable journey through history. These circumstances have ranged from such cataclysmic events as the Civil War and the Great Depression to the failures of the U.S. states to protect basic civil rights and the increased importance of the nation's foreign involvements. The Court's expansive interpretations have included application of some parts of the Bill of Rights to states via the Fourteenth Amendment and in particular a very broad reading of the power granted by the Constitution to Congress to govern commerce between the states.[39] More recently, the Supreme Court has placed national controls on the ability of states to regulate birth control and abortions by finding a federally protected right of privacy in the "penumbras" and "emanations" of the Constitution's Bill of Rights.[40]

Many of these interpretations of the Constitution have been criticized by advocates of "states' rights," who argue, among other points, that the Court has strayed far beyond the original intent of the document's framers to limit the central government's powers.[41] Supporters of the Court's admittedly expansive interpretation argue that the Constitution is a special type of document, one that creates and sets in motion a governing authority that, to succeed over the expected long tenure of its operation, needs to be applied flexibly to new circumstances not foreseen by the drafters and new attitudes that perhaps they would not even have shared.[42] While quite recently the Supreme Court has taken some steps that might indicate a move toward limiting the expanding universe of central authority,[43] it is undoubtedly true today that the original design for the American federation contemplated far more restricted federal powers. Nevertheless, I believe that the founders would not disapprove of the current balance. Assisted by the Supreme Court, the Constitution's ability to flex and survive the many strains and blows of the last two centuries is an impressive testament to the wisdom (and, perhaps, luck) of the founders. The words of their document, even while they were stretched and molded, have always remained in place to serve as the lighted beacons to guide the Court through the nation's dark times, and, despite the criticism, the federation is still very much intact and working to achieve the basic objectives of those who brought it into existence: to establish a government incorporating a complex system of horizontal and vertical checks and balances that apportions governance sensibly and guarantees freedom from the tyranny of concentrated power.[44]

The Supreme Court's reshaping of the federation through interpretation and construction of the nation's constituent instrument thus provides no strong precedent for those who are now attempting to manipulate the ICRW's explicit grant of powers to the IWC. While governmental actions and the Court's rulings have, over time, heavily adjusted the balances among the multiple centers of governance authority, all still operate in the same overlapping arenas assigned to them by the document's drafters. The states have not disappeared or been relegated to insignificant roles. All elements of the original clockworks construction are still in place and still fulfill their original purposes. None of these purposes has been ignored in the adjustment processes.

This adherence to primary objectives also marks the otherwise sometimes-questioned interpretations of the United Nations Charter,[45] the most important constituent instrument in the history of international agreements. Like the ICRW, the Charter is of course a treaty, but one that, like the ICRW but far more so, creates an international organization designed by its founders to exercise the powers granted to it through foreseeable and unforeseeable changing circumstances and attitudes over time. Indeed, the two treaties—the Charter and the ICRW—have been in existence for nearly the same length of time and thus almost entirely overlap the same significant period of world history. Some of the highlights in the interpretive history of the Charter might therefore help illuminate the permissible bounds for interpretation of those special treaties that bring into existence active international organizations, including in particular the ICRW.

Perhaps the evolution in the interpretation of the Security Council's powers provides an appropriate example. Early in the existence of the organization it was wisely decided, basically by the practice of the council members, that the rather clear Charter language requiring the "affirmative vote" of all five permanent members of the council on important decisions should be read to allow council action despite the abstention or absence of one or more of the permanent members.[46] Only the casting of a negative vote, or veto, prevents the council from proceeding on matters of importance. The members realized that progress toward the Charter's major objective of international peace would be effectively hamstrung without this "reinterpretation" of the Security Council's ability to act. As it turned out, of course, even the veto power of the permanent members of the council made it extremely difficult for the body to take significant action during the many years of the Cold War, but a strict interpretation of the "affirmative vote" requirement might well have threatened the continued existence of the council or of the UN itself.

The relative ineffectiveness of the Security Council during the Cold War did, moreover, provide the asserted justification for other, unilateral reinterpretations of the Charter. As is well known, Article 2(4) prohibits members from resorting to military force against another state, subject to the self-defense exception found in Article 51. Nevertheless, over the years of the Charter, many states have unilaterally engaged in military intervention in the territory of other states, with varying justifications. One of the justifications that eventually won wide support was that the intervention was necessary to prevent or stop gross violations of basic human rights—so-called humanitarian intervention. The articulation of this excuse for military intervention also often referred to one of the Charter's basic objectives—promotion and protection of human rights[47]—and the ineffectiveness of the Security Council, because of the permanent-member veto power, in carrying through on this objective.[48] Thus the pursuit of a primary Charter purpose was used as a basis for a reinterpretation of the document's otherwise rather clear language. It is important to note also that, in the frequent statements of justification for this flexible interpretation of the Charter, its other major objective—"to save succeeding generations from the scourge of war"[49]—entered the interpretation to place strict limitations of purpose and proportionality on the use of military force to protect fundamental human rights.

Even the Security Council's authority to intervene in a UN member's internal affairs to counter human rights violations has been called into question. Article 2(7) of the Charter seems to prohibit such intervention in general, with an exception for enforcement actions of the Security Council under Chapter VII. Yet in Chapter VII, Article 39 requires, as a condition to such enforcement action, that the Security Council first make a determination that there is a breach, or at least a threat of breach, of the "international" peace. On several occasions over the years, however, the Security

Council has authorized intervention in the internal governance of members' states when evidence of human rights violations has been brought to its attention. Again, this expansive interpretation of the Charter's language establishing the limits on the council's powers has often been justified because it is in pursuit of one of the most important reasons for the creation of the United Nations, the objective of protecting and promoting basic human rights and fundamental freedoms. When this interpretation is challenged, however, it might be asserted that the Security Council has the authority to determine the limits of its own authority—an assertion similar to that made by those who defend the recent questionable actions of the IWC, because there is no dispute-settlement mechanism in the ICRW and no equivalent to judicial review long taken for granted with respect to the U.S. Constitution.[50] Again, in its response to the World Health Organization's request for an advisory opinion on the legality of nuclear weapons, the International Court of Justice, quoting from its 1962 *Certain Expenses* case, conceded that international organizations must, "in the first place, at least," determine their own jurisdiction.[51] But, in its answer to the argument that the relevant WHO resolution had been adopted by the required majority and must therefore be presumptively valid, the court said, "The mere fact that a majority of States, in voting on a resolution, have complied with all the relevant rules of form cannot in itself suffice to remedy any fundamental defects, such as acting *ultra vires*, with which the resolution might be afflicted" (paragraph 29).

These few examples of arguable interpretations of two well-known constituent instruments—the U.S. Constitution and the UN Charter—merely hint at the myriad of such examples concerning national constitutions and international-organization charters worldwide. Collectively, they do establish the basic point that international agreements that serve as charters or constitutions for regulating bodies can and should be interpreted flexibly as their work and its context evolve over time. Yet, of course, there must be limits to such flexibility, and the limits must be found in the objectives and purposes of the organizations that the agreements create. Otherwise, they become anything the majority or politically powerful members want them to become, a clearly unacceptable philosophy that history has shown in a multitude of governance situations leads to oppression. In the few instances I related concerning the "reinterpretations" of the U.S. Constitution and the UN Charter, the accepted interpretations are those that make adjustments based on implications consistent with the clear objectives and purposes of the framers of the constituent instruments. There are, of course, always dissents, but the "reinterpretations" that ignore or counter the instrument's objectives and purposes are not acceptable adjustments but are instead violations.

The IWC is certainly the sort of international organization or body established to make regulatory decisions over time in changing circumstances, and thus the ICRW, or at least that part that creates the IWC and provides that body's powers, is the type of treaty the Vienna Convention and the ICJ refer to as a "constituent instrument" or constitution. And it follows that the ICRW's provisions setting out the powers of the IWC can and should be interpreted flexibly in an evolutionary manner. Indeed, those who defend the IWC's recent actions have made this point. But even they must recognize that there are limits on the flexibility of such interpretations. At a minimum, it should be obvious that any action that runs counter to the ICRW's clearly stated purposes for creation of the IWC can and should be considered ultra vires.[52] Moreover, any IWC action that purports to operate in an arena beyond its clearly circumscribed authority would not take effect. To raise an extreme hypothetical example: suppose the IWC, by a majority vote, purported to establish a regime regulating immigration into the United States and binding on the United States. The U.S. reaction would certainly be to ignore the IWC action, with no thought to any supposed requirement that it file a formal objection under the convention's objection procedure or that its position would need to be tested in some dispute-settlement forum. And the United States would be right. The IWC action would simply be beyond the powers of the body and thus of no effect at all.

Similarly, the IWC action that purports to establish a permanent Southern Ocean sanctuary in the face of very convincing evidence that sustainable harvests of minke whales can be taken there not only exceeds the powers of the IWC, as William Burke has amply demonstrated,[53] but the action runs counter to the unambiguously stated purpose of the ICRW and the IWC to conserve whales for the benefit of the whaling industry and its customers. The IWC's action is ultra vires and thus should be of no effect at all. A good case can therefore be made for the proposition that the Southern Ocean sanctuary simply does not exist. It can and perhaps should be ignored. Similarly, a permanent moratorium on commercial whaling would certainly be ultra vires. Although the present IWC moratorium purports to be temporary and not inconsistent with the objective of increasing whale populations for eventual sustainable harvest, there are reports that one or more delegations to the commission intend—almost surely in bad faith and thus viola-

tive of the most basic of state obligations, *pacta sunt servanda*, and the obligation to interpret treaties in good faith[54]—to maintain the moratorium indefinitely as a means of preserving whales for their own sake.

Furthermore, it is not a convincing response to assert that the IWC is the
final arbiter of the limits of its own powers.[55] It is true that the ICRW contains no system for settling disputes concerning its interpretation. On the
other hand, it does not explicitly grant a self-interpretation power to the
IWC. Do these omissions justify an inference that the IWC, by majority vote,
is entitled to be the judge in its majority's own cause? As noted above, the
International Court of Justice has certainly more than hinted, in its refusal
to admit a WHO request for an advisory opinion on the legality of nuclear
weapons, that an international organization's action not authorized by the
purposes for its establishment, even if taken by the procedurally requisite
majority of its members, is not a valid interpretation or reinterpretation of
its powers.[56] Instead, such a reinterpretation is arguably at most an express
or implicit agreement among the states voting for it to amend the treaty
among themselves. Dissenting states are, of course, not bound by the
amending reinterpretation.

WHALES, THE IWC, AND THE RULE OF LAW

As one who admits to an admiration of cetaceans, I am not particularly
bothered by IWC action that saves whales. I have recordings of the songs of
humpback whales that I have listened to with fascination and awe for
decades. I have watched in distress as forty-one sperm whales lay dead or
dying on an Oregon beach. I actively support the attempt to return Keiko,
the "Free Willy" killer whale, to the open sea. Nevertheless, as one who believes that adherence to the rule of law is absolutely essential to the future
well-being of our planet and all of its creatures, including humans and
whales, I am bothered by what seem to me to be extra-legal manipulations
of an important and very visible treaty.

Is there such a thing as the rule of law in the international community?
This is of course a long- and much-debated topic, even among those whose
specialty is international law and relations. My own answer is that, even if
international law is not really law, nearly every nation-state conducts its international relations as if it is, which is about as much as such an abstrac-
tion as "law" should need to validate its existence. That is, every state cites
international law to justify its own conduct and to critique the conduct of

other states. All states hire lawyers to argue their legal positions and send law-trained diplomats to negotiate international agreements to impose something like legislation on themselves and others. States more than occasionally (but not often enough) submit disputes to international tribunals to be decided on the basis of a commonly recognized system of customary and conventional law. Whatever "international law" is, whether it matches our own notion of "law" or not, it is certainly important in the conduct of international relations for all states, because every state has an abiding interest in a long-term system of order within which to conduct its affairs with other members of the community of states and other subjects of international law.[57]

Certainly this system of order is less than perfect. In my view, its major imperfection is the absence of a compulsory-jurisdiction court or court system. This gap often allows states to avoid even clearly stated legal obligations by articulating theories of legal justification that remain untestable in a law-applying tribunal. Whenever this is done unscrupulously and in bad faith, patently flaunting the absence of a neutral legal testing ground, it encourages other states to do likewise and adds to the degradation of a system of order that is much needed in today's complex world society. For cynics and "realists," this does not matter, because they are apparently convinced that any recognition of a so-called system of international law is false play-acting anyway. For those of us who know better, who realize that nation-states abide by and depend on law in thousands of actions every day—even for their very existence as abstract units of the community of recognized states—the rule of law in the Global Village is as essential as law, primitive or sophisticated, has ever been for human interrelationships throughout the millennia of human existence.

Nevertheless, international law is still primitive and tenuous, a historically unprecedented experiment in global cooperation, and, as such, it needs the support that all successful legal systems require to remain viable: largely voluntary compliance in good faith. Every clear violation or obvious avoidance of an accepted norm, especially in high-profile situations, adds to the degradation of the system by condoning a culture of self-interested noncompliance. Eventually the experiment must fail and the rule of law will be replaced by an even more primitive feudal system in which might, indeed, does make right. The most important players in the process of support or denigration of law are, then, the most powerful members of the international community. If law is to rule, these states must almost always try to resist the impulse to violate the law just because their power and influence allows them, in the current state of affairs, to get away with it.

It is therefore unfortunate that the United States has been a leader in the I W C coup that is ignoring the clear mandates of the I C R W. Without U.S. support and economic power, the anti-whaling states that have packed the I W C would probably not have been nearly as successful as they have been in getting away with their violation of the limitations of the I W C's authority. Whenever the United States, as the most important member of today's international community, flouts the law or transparently manipulates treaty language—as it has sometimes been accused of doing in other contexts, such as the Reagan administration's "reinterpretation" of the U.S.-U S S R antiballistic missile treaty to justify its Strategic Defense Initiative[58]—its action condones similar lawlessness by others and adds to the degradation of international law in general.

Like most special-interest lobby groups, those who nearly single-mindedly promote protection and preservation of whales and are therefore vociferously opposed to all whaling (or at least commercial whaling) often fail to see or be concerned about the larger picture. If they did and were, they should know that promotion of whale preservation by packing the I W C and manipulating its mandate by overinterpretation of the I C R W's objectives can lead to counterproductive legal backlashes, both within the I C R W and in a much larger arena. The I C R W, for instance, allows each state party complete discretion to define and conduct "scientific research" whaling despite any attempt of the I W C to otherwise impose restrictions on whale harvests.[59] Anti-whaling members of the I W C have in recent years chastised whaling state members for allegedly abusing this discretionary power.[60] Whether these charges are true or not (and they are contested), it hardly lies in the mouths of those who have reinterpreted the I W C's authority beyond its clear bounds to condemn fellow I C R W parties for applying similar expansive interpretations to powers conferred on them in the same convention. Those who disregard inconvenient legal strictures should not be surprised when others follow their example.

Moreover, many of the same special-interest lobbyists who seek to protect whales are also participants in the movement—so far, astoundingly successful[61]—to weave a global web of conventional international law to protect and preserve the marine and global environment in general. It is hardly in the interest of the nongovernmental organizations and the states they influence to ignore international law in one arena while insisting on observance in others. The rule of law is much needed in the campaign to promote protection of the global environment, as well as in myriad other campaigns to restore and preserve public order in the international community.

Saving the whales might indeed be a noble objective, but those who reject or contort law in the attempt to achieve this goal are, in my view, aligned with all others who, for their own special causes, are also willing to burn the (global) village to save it. Of course, the village will not be saved; it will be ruined.

So: enough preaching. What to do? If we truly want to save the whales (and much else), we must, I think, resist the desire for instant gratification. The probable fact is that, although anti-whaling sentiment is growing, the majority of people on the planet do not presently agree that it is a moral outrage to allow regulated sustainable harvests of whales. Nevertheless, I do believe that such a view is on the horizon. By the time we reach that horizon, international law will certainly have created a peremptory norm, or *jus cogens*, making all whaling illegal and, under the accepted rules of treaty law, thereby cause the termination of the ICRW or any other then-existing treaty that purports to allow or regulate the hunting of whales. In the meantime, whales can be lawfully protected from us humans only by conventional replacement or formal amendment of the ICRW. It is not helpful to their cause for those concerned about whales to misconstrue the convention while demonizing those who disagree. We should instead work at hastening the arrival of the desired *jus cogens* by choosing the art of persuasion over the craft of manipulation.

That being said, there is no more appropriate forum than the IWC for the practice of such persuasion, and for engendering the state practice that will create the *jus cogens* that will outlaw whaling. The practice must, however, be conducted within the bounds the ICRW sets for the IWC, as broadly interpreted in the flexible, evolutionary manner allowed for constituent instruments. While this interpretive approach should not allow the IWC majority to excise or abandon the convention's clearly stated objectives, it would certainly allow the application of the newly emerged precautionary principle,[62] which in this context should dictate resolving reasonable scientific doubt in favor of protecting whales. However, where there is clear and credible evidence that sustainable harvests can occur, as appears to be the case for minke whales in the Southern Ocean, the anti-whaling majority in the IWC has no legal or good faith option but to allow regulated commercial whaling to occur—which does not mean that expressions of dismay and reluctance would be inappropriate accompaniment for affirmative votes or abstentions.

I am convinced that anti-whaling sentiments, reasonably expressed within the IWC and elsewhere, will eventually win out, and the hunting of whales will one day share history's attic with the now-condemned practices of slavery and human sacrifice. I think any graph of pro-whale sentiment over the last several decades would confirm this general historical trend. The question for those opposed to whaling should be how to hasten the coming of that day. In my view, the current shenanigans within the IWC, not to mention the truly atrocious behavior of some well-known anti-whalers outside the IWC, are terribly counterproductive to this goal. Much more importantly, though, they are abusive of the international rule of law at a time when we all need it badly.

NOTES

1. See, for example, William T. Burke, "Memorandum of Opinion on the Legality of the Designation of the Southern Ocean Sanctuary by the IWC," *Ocean Development and International Law* 27 (1996): 315; William T. Burke, "Legal Aspects of the IWC Decision on the Southern Ocean Sanctuary," *Ocean Development and International Law* 28 (1997): 313.

2. *International Convention for the Regulation of Whaling*, 2 December 1946, 10 U.S.T. 952, 161 U.N.T.S. 72 (hereinafter ICRW).

3. See, for example, "Rio Declaration on Environment and Development of the United Nations Conference on Environment and Development," 14 June 1992, UNCED Doc. A/CONF. 151/5.Rev. 1, reprinted in *International Legal Materials* 31 (1992): 874.

4. See generally, for example, William T. Burke, *The New International Law of Fisheries: UNCLOS 1982 and Beyond* (New York: Oxford University Press, 1994); Albert W. Koers, *International Regulation of Marine Fisheries* (West Byfleet: Fishing News, 1973).

5. See, for example, Patricia Birnie, "Are Twentieth-Century Marine Conservation Conventions Adaptable to Twenty-first-Century Goals and Principles?" *International Journal of Marine and Coastal Law* 12 (1997): 488, 491, 509.

6. See, for example, Patricia Birnie, *International Regulation of Whaling: From Conservation of Whaling to Conservation of Whales and Regulation of Whale-Watch-*

ing, 2 vols. (Dobbs Ferry, N.Y.: Oceana, 1985); Anthony D'Amato and Sudhir K. Chopra, "Whales: Their Emerging Right to Life," *American Journal of International Law* 85, no. 1 (January 1991): 21, 30–40.

7. See, for example, the discussions of the Revised Management Procedure in Burke, "Memorandum of Opinion," and in John A. Knauss, "The International Whaling Commission: Its Past and Possible Future," *Ocean Development and International Law* 28, no. 1 (1997): 79.

8. See, for example, D'Amato and Chopra, "Whales," 21–23, and sources footnoted there.

9. Ibid.

10. Again, in historical terms. Actually, very recent trends, perhaps spurred on by the IWC actions of the hard-line anti-whalers, might show an upturn in the number of countries, both within and outside the IWC, and people actively involved in harvesting cetaceans.

11. See, for example, Burke, "Legal Aspects"; Knauss, "International Whaling Commission," 82.

12. See Chairman's Report, *Proceedings of the 34th Meeting of the IWC,* 1983, 21, which reports the IWC's decision to reduce to zero the commercial catch limits of all stocks of whales beginning in 1986, subject to later assessments. All subsequent "assessments" have led to a continuation of the zero-limit moratorium to the present date.

13. See, for example, the Web site of the High North Alliance, in particular its national and EU policies page (http://www.highnorth.no/cont-nat.htm), for, inter alia, statements of positions attributed to some member states of the IWC.

14. Burke, note 1 above.

15. Vienna Convention on the Law of Treaties, 23 May 1969, UN Doc. A/CONF. 39/27 (hereinafter Vienna Convention).

16. Article 4, Vienna Convention. The Vienna Convention entered into force for its parties on 27 January 1990. By the terms of Article 4, it does not apply retroactively to treaties, such as the whaling convention, concluded prior to that date. The International Court of Justice, however, has indicated that it views at least some of the Vienna Convention as reflective of customary international law, for example, Fisheries Jurisdiction Case *United Kingdom v. Iceland,* ICJ 3 (1973), and "Advisory Opinion on Namibia," ICJ 16 (1971). Publicists tend to consider most of the Vienna Convention as a codification of customary principles concerning the law of treaties. See, for example, Peter Malanczuk, *Akehurst's Modern Introduction to International Law,* 7th ed. (New York: Routledge, 1997), 130.

17. Articles 60, 62, Vienna Convention.

18. See Articles 87, 116, "Freedom of Fishing on the High Seas," *United Nations Convention on the Law of the Sea,* 10 December 1982, UN Doc. A/CONF. 62/122, reprinted *International Legal Materials* 21 (1982): 1261 (hereinafter UNCLOS); Article 2(2), "Freedom of Fishing on the High Seas," *Convention on the High Seas,* 29 April 1958, 13 U.S.T. 2312, 450 U.N.T.S. 82 (hereinafter High Seas Convention).

19. See Articles 55–75, UNCLOS.

20. Article 65, Vienna Convention: "If a new peremptory norm of general international law emerges, any existing treaty which is in conflict with that norm becomes void and terminates." Article 53 defines a peremptory norm (*jus cogens*) as "a norm accepted and recognized by the international community of States as a whole as a norm from which no derogation is permitted and which can be modified only by a subsequent norm of general international law having the same character."

21. Article 40(4), Vienna Convention.

22. Article 2(2), High Seas Convention.

23. President, Proclamation, "The U.S. Exclusive Economic Zone, Proclamation 5030." *Weekly Compilation of Presidential Documents* 19 (1983): 383, reprinted, *International Law Materials* 22 (1983): 464. Congress subsequently implemented the proclamation for U.S. fisheries management purposes, giving the United States exclusive fisheries management authority out to two hundred miles. The current codification can be found in "Magnuson-Stevens Fishery Conservation and Management Act," U.S.C.A. 16 secs. 1802(11) and 1811 (West Supplement, 1998).

24. See Malanczuk, *Akehurst's Modern Introduction,* 367.

25. See Preamble, High Seas Convention: "The States Parties to this Convention, Desiring to codify the rules of international law relating to the high seas. . . ."

26. See Malanczuk, *Akehurst's Modern Introduction,* 367.

27. Articles 31–33, Vienna Convention

28. Article 32, Vienna Convention.

29. Article 31(3)(b), Vienna Convention.

30. Burke's analysis clearly leads inevitably to this conclusion. See note 1 above.

31. Birnie makes perhaps the best case for a flexible, evolutionary approach to interpreting the whaling convention, an approach that is perfectly appropriate, indeed demanded by the nature of the convention. Birnie, "Twentieth-Century Marine Conservation." However, as I will explain, with all respect for Birnie and her expertise, I remain unconvinced by her argument. I am also unpersuaded by a similar argument in D'Amato and Chopra, "Whales" 32–35.

32. See note 29 above and accompanying text.

33. See note 25 above and accompanying text.

34. Article 5, Vienna Convention.

35. International Court of Justice, "Legality of the Use by a State of Nuclear Weapons in Armed Conflict," Advisory Opinion, 8 July 1996, General List No. 93. The full text of the opinion can be found on the World Wide Web at http://www.law.cornell.edu/icj/icj2/anw-003f.htm.

36. See generally, for example, John E. Nowak and Ronald D. Rotunda, *Constitutional Law*, 5th ed. (St. Paul, Minn.: West Publishing, 1995).

37. Ibid., 118.

38. Especially the so-called Civil War amendments, the Thirteenth, Fourteenth, and Fifteenth.

39. U.S. Constitution, art. 1, sec. 8, cl. 3. See Nowak and Rotunda, *Constitutional Law*, chap. 8.

40. *Griswold v. Connecticut*, 381 U.S. 479, 484 (1965); *Roe v. Wade*, 410 U.S. 113 (1973).

41. This and other issues of constitutional interpretation have of course been a hotbed of scholarly debate for generations. A smattering of recent entries into the hugely voluminous literature on the topic includes Christopher Wolfe, *How to Read the Constitution: Originalism, Constitutional Interpretation, and Judicial Power* (1996); Walter F. Murphy, James E. Fleming, and Sotirios A. Barber, *American Constitutional Interpretation*, 2d ed. (New York: Basic Books, 1995); Symposium, "Interpretive Methodologies: Perspectives on Constitutional Theory," *Hastings Constitutional Law Quarterly* 24 (1997): 281–664; Symposium, "Originalism, Democracy, and the Constitution," *Harvard of Law and Public Policy* 19 (1996): 237–531.

42. Ibid.

43. See, for example, *United States v. Lopez*, 514 U.S. 549 (1995). (Congress does not have power under the Commerce Clause to forbid possession of firearms in school zones).

44. I briefly analyzed the history of American federalism in Jon L. Jacobson, "Governance of the U.S. Exclusive Economic Zone: A Challenge to the American Federation," in *The International Implications of Extended Maritime Jurisdiction in the Pacific*, ed. John Craven, Jan Schneider, and Carol Simpson (Honolulu: Law of the Sea Institute, 1989): 329, 333–36.

45. United Nations Charter, 26 June 1945, 59 Stat. 1031, T.S. No. 993.

46. The Charter provides in paragraph 3 of Article 27 that decisions of the Security Council on nonprocedural matters require "an affirmative vote of nine members including the concurring votes of the permanent members." Despite the clarity of this language, the practice of the Security Council since the Korean conflict has established that an abstention by or absence of a permanent member does not count as a veto. See, for example, Bailey, *The Procedure of the U.N. Security Council*, 2d. ed. (New York: Oxford University Press, 1988): 224–25.

47. Article 1(3), UN Charter.

48. See generally Louis Henkin et al., *International Law: Cases and Materials*, 3d. ed. (St. Paul, Minn.: West Publishing, 1993): 929–35, and the numerous sources cited therein.

49. Preamble, UN Charter.

50. See Patricia Birnie, "Opinion on the Legality of the Designation of the Southern Ocean Whale Sanctuary by the International Whaling Commission," *Proceedings of the 47th Meeting of the iwc*, 1995 (iwc/47/41), available on the World Wide Web at http://www.highnorth.no/op-on-th.htm; compare: Rosalyn Higgins, *The Development of International Law through the Political Organs of the United Nations* (New York: Oxford University Press, 1963), 66 n. 27 (each UN organ to determine its own competence); but compare *Libya v. United States* (Provisional Measures), [cite] (dissenting opinion of Judge Weeramantry).

51. icj, "Legality of the Use by a State of Nuclear Weapons in Armed Conflict," par. 29.

52. See generally C. F. Amerasinghe, *Principles of the Institutional Law of International Organizations* (New York: Cambridge University Press, 1996): 163–87, together with the ICJ's subsequent advisory opinion in response to an ultra vires who request for an advisory opinion on the legality of nuclear weapons.

53. Burke, "Memorandum of Opinion," and Burke, "Legal Aspects."

54. Article 26, Vienna Convention: "Every treaty in force is binding upon the parties and must be performed by them in good faith." See also Article 31(1): "A treaty shall be interpreted in good faith in accordance with the ordinary meaning to be given to the terms of the treaty in their context and in the light of its object and purpose." Although the Vienna Convention on the Law of Treaties does not apply retroactively to the whaling convention, these articles certainly are restatements of long-standing customary international law binding on all iwc parties.

55. See Birnie, "Opinion on the Legality."

56. ICJ, "Legality of the Use," par. 29.

57. One of the best cases ever made for this proposition is Louis Henkin, *How Nations Behave*, 2d ed. (New York: Columbia University Press, 1979); President, Proclamation, "The U.S. Exclusive Economic Zone, Proclamation 5030." *Weekly Compilation of Presidential Documents* 19 (1983): 383

58. See the debate between Reagan administration officials and international law scholars, "Strategic Arms Limitation: Treaty Obligations and the Strategic Defense Initiative," *American Society of International Law Proceedings* 73 (1986): 73–95.

59. Article VIII, Whaling Convention.

60. See Ray Gambell, "The International Whaling Commission Today," in *Whal-*

ing in the North Atlantic: Economic and Political Perspectives, ed. G. Petursdottir (Reykjavĩk: Fisheries Research Institute, University of Iceland Press, 1997).

61. The variety and quantity of the recent and growing international-community recognition—in both "hard" and "soft" law expressions—of support for protection of the human environment, biodiversity, and preservation of threatened species, transgenerational equity, and the like, should be vastly impressive to even the most hardened cynic with any sense of history. As an illustration of this ongoing development, see the collected documents and other materials, nearly all of which are post-1972, in the thick and weighty supplement to Lakshman D. Guruswamy, Geoffrey W. R. Palmer, and Burns Weston, *International Environmental Law and World Order* (St. Paul, Minn.: West Publishing, 1994).

62. See, for example, Principle 18 of the Rio Declaration on Environment and Development for a general statement of the "precautionary approach." The most detailed "hard law" implementation of this principle to date is probably that found in the 1995 Agreement for the Implementation of the Provisions of the United Nations Convention on the Law of the Sea of 10 December 1982 Relating to the Conservation and Management of Straddling Fish Stocks and Highly Migratory Fish Stocks, UN. Doc. A/CONF. 164/37, 8 September 1995.

3 / Science and the IWC

WILLIAM ARON

Upon reading the preamble of the International Convention for the Regulation of Whaling, a professional resource manager, ignorant of the history of the International Whaling Commission (IWC), would almost certainly conclude that the organization was conservation-oriented (conservation being defined as the sustainable use of living resources) with a strong commitment to scientific understanding. Unfortunately, since its inception in 1946 the commission has proved anything but conservationist, and the role of science in its decision making has been generally ineffective.

This chapter will provide views based in part on the literature, including the commission reports, and on firsthand experience as a member of the IWC Scientific Committee and as an assistant to the U.S. commissioner, and as commissioner for one meeting (in Australia, 1977). It is also strongly dependent on my experiences of administering and developing programs to provide science advice to resource managers. For those interested in both the management of whales and whaling, the good starting points remain: Allen,[1] Scarff,[2] Schevill,[3] Small,[4] and Tonnessen and Johnsen,[5] with some important

I have had many constructive comments and discussions with friends and colleagues during the preparation of this chapter. Special thanks must be given to Jeff Breiwick, Ed Mitchell, Howard Braham, Doug Demaster, Ray Gambell, Sally Mizroch, and Dan Goodman. They, however, should not be blamed for shortcomings or opinions contained in the chapter. It is clear that there is almost nothing that can be said about whales and whaling that will generate consensus.

information in Young.[6] Tonnessen and Johnson is particularly important in detailing the pre-IWC management of whaling. Some of the flavor and controversy within the Scientific Committee can be derived from the paper by Tore Schweder[7] and its strong rebuttal by Sidney Holt.[8] These last two documents, while harsher than the general tone of the Scientific Committee (at least during the decade of my attendance), provide a sense of the difficulties that occur in dealing with uncertain data and strong personal philosophies, which in some cases are influenced by national positions.

The prime purpose of this chapter is to provide insight on how science can contribute to management, and what the relationship should be between the scientific community and those charged with making decisions on the utilization of whales. To do this, however, we need to look at the history and show the interaction between science and management in three periods: (1) 1946–1960, (2) 1961–1971, and (3) 1972–present.

THE BEGINNING: 1946–1960

The period from the start of the commission until 1960 was somewhat organizational, as could be expected in any newly established entity, and was characterized by polite and perhaps overly considerate judgments by the scientists involved with the commission work. The dozen or so scientists (see Schweder or the commission reports) who participated in the Scientific Committee during this period did not attend all of the meetings (mean attendance was seven, again see Schweder) and were primarily well-known whale biologists with strong specialties ranging from physiology and natural history to taxonomy. Remington Kellogg, who served on the committee and was also the U.S. commissioner, was a distinguished cetacean systematist and director of the Museum of Natural History of the Smithsonian Institution. No scientist from the rapidly evolving field of population dynamics and estimation was a member. The outlook of the group is well demonstrated by their recommendations for research in the second report of the commission (1951). The twelve items listed (including food studies, whale body proportions for possible stock identification, swimming and breathing behavior, embryo and pregnancy studies, and the frequencies of sighting single whales and schools of various numbers) are biologically important and would provide considerable insight into the biology of whales, but would provide little, if any, value for population estimation.

The group, most likely because of the ethos of the time, was often timid in providing advice. For example, even with the limited data, it was clear to the Scientific Committee that the fin whale was showing signs of depletion.[9] They noted that the previous season's large catch of about 26,000 fins (increased from the 1950–51 season catch of 17, 474) occurred because fin whales were targeted owing to the difficulty of catching blue whales. They wanted to recommend a catch of no more than 19,000 animals but believed "a cut of this magnitude would scarcely be acceptable for the season 1955–56," so instead they urged an immediate reduction to 25,000 fins and some reduction in the allowable blue whale unit (B W U, with one B W U equal to six seis, two fins, or two and a half humpbacks). The take of fin whales remained at 25,000 for the rest of the decade and the B W U was kept at 15,000.

Much has been made of the fact that even this very modest reduction was not endorsed by E. J. Slijper (from the Netherlands, who is viewed by many as an apologist for the whaling industry), thus allowing the commission some scientific reason to continue full-scale whaling. However, the view of the strong majority of the Scientific Committee to lower the catches is clear in the commission reports, thus making it likely that business would have continued as usual even in the absence of the Slijper dissent. This period of the commission and, in particular, the role of the Dutch scientists is discussed by Schweder.[10]

There was a concern by the scientists at this time (as reflected in private conversations) that a failure by the Scientific Committee to recommend "reasonable" quotas would result in the commission completely ignoring the Scientific Committee's advice. (In 1972, at my first Scientific Committee meeting, after a session in which I was fairly vocal about the data demonstrating the need to lower quotas, a more experienced colleague pulled me aside and told me that, even though he agreed, the proposed reductions were too severe and the commission would simply ignore the advice if it were proffered.) The commission's use of the B W U as the management tool makes it clear that regulations were designed more to control oil production and for the convenience of the whaling fleets (by permitting them to shift targets in the absence of the most desired species) than to conserve whales. I could not locate any serious challenge by the Scientific Committee to the dangers of the B W U for management during this initial period. The B W U was not eliminated until 1972, but by then the serious damage had been done.

The Scientific Committee members were, as they are today, servants of the commission and in some cases reflected national positions. The record of overharvesting, with minimal-to-no penalties by the commission (only expressions of regret), made it unlikely that strong dissent would result in positive change. It is clear that the commission recognized but discounted the views of the Scientific Committee. For example, in 1955, when proposals to reduce the BWU quota from 15,500 to 14,500 over two seasons failed to be implemented because of objections (from the United States among others), the commission noted "that the bulk of scientific opinion in the commission was in favor of still greater reduction."[11] Likewise, while it was recognized that the population of blue whales was seriously depleted, Iceland objected to a prohibition on the taking of blue whales in the North Atlantic, clearly not accepting the advice of the Scientific Committee but providing no evidence to dispute the soundness of that advice.[12] The decade ended with the Scientific Committee concluding that they "would like again to emphasize that they are seriously concerned about the present condition of the stocks of whales."

THE COMMITTEE OF THREE: 1961–1971

The second phase of science and the IWC began with the appointment of the Committee of Three (Douglas Chapman, K. Radway Allen, and Sidney Holt) in 1960.[13] The three were distinguished population specialists, each with considerable experience in providing advice to management authorities. None had prior experience with whale issues. Their charge by the commission was to address the question of the Antarctic catch limit so it could be brought into line with scientific findings, and to recommend conservation measures that would increase sustainable yield. The resolution creating the committee was specific in not requiring their report to pass through the Scientific Committee.

The Committee of Three held its first meeting in Rome in April–May 1961, simultaneous with the Scientific Committee, meeting both separately and as observers at the Scientific Committee meeting. This was done because, in examining the terms of reference of their charter, the committee recognized its dependence on the Scientific Committee for data and data handling and sought to assure the closest possible liaison between the two groups. The initial report demonstrated the requirement for changes in the way the IWC conducted its scientific operations, laying out the need for collation and tabulation of existing data to determine what, if any, additional information was required.[14] The committee indicated the need for financial assistance to handle the computations needed for its analyses, and it urged a substantial increase in the whale-marking operations to make them useful for stock assessments.

The Committee of Three, with the addition of John Gulland (who made important calculations and analyses for the final and the supplementary report), continued its work in Seattle during December 1962. The findings of that session suggested that immediate action by the commission was critical for the protection of whale stocks, particularly the blue, humpback, and fin whales.[15] Despite incomplete analysis, the committee stated that "the general conclusions, both quantitative and qualitative have become clear, and point to the need for action so drastic and of such urgency that the Committee of Three think it essential that the commission should be given an immediate interim report so that national delegations should have ample opportunity to consider the implications of the proposals before the 1963 meeting of the Commission." The interim report clearly states that sustainable yields of both blue and humpback whales could be obtained only by a complete cessation of catching these species for a considerable number of years and that to permit a rebuilding of the fin whales would require a catch limit of less than nine thousand. The interim report also called for the commission to abandon the B W U for management purposes.

The final report of the Committee of Three was issued in time for the 1963 meeting of the commission.[16] It was the first rigorous quantitative study on the population dynamics of Antarctic whales, and it confirmed with even greater emphasis the findings in the second interim report. The committee recommended a complete cessation of catching blue and humpback whales (Group IV and V) and a quota on fin whales of seven thousand or less. They repeated, with emphasis, the need to eliminate the B W U, warning the commission that its application would almost certainly involve overexploitation of stocks that were already below their optimum level.

The Scientific Committee generally concurred with the findings of the Committee of Three but recommended even more conservative steps.[17] They recommended a total cessation of all taking of humpbacks as well as blue whales for a period of fifty years and a cessation of fin whale taking for eight years, and then twenty thousand per year thereafter, or a take of three to four thousand for eleven to thirteen years before resuming a take of twenty thousand per year.

The commission's response to the recommendations of the scientists was only partial acceptance.[18] Use of the B W U was continued "as the only practical method that could be administered." The taking of both humpback and blue whales was largely stopped, but the nine thousand B W U quota did allow for the taking of about fourteen thousand fin whales, as well as other species. At this meeting, the commission also asked that the Committee of Four (the old Committee of Three with the addition of Gulland) continue its work on the Antarctic for another year. The Committee of Four reported that the data from the 1963–64 season confirmed their prior predictions.[19]

With the cessation of taking of blue and humpback whales and the difficulty of finding fin whales, the final chapter of this period was highlighted by the switch to sei whales with a take of more than twenty thousand in the 1964–65 season (nearly two and a half times greater than the previous season), followed by a rapid decline of this species, closing with a take of about 5,400 in 1971–72. This increase in the take of sei whales also reflected their newly discovered value as a source of whale meat and whale by-products.[20]

Tables 3.1 and 3.2[21] show two important features: (1) the shift from blue to fin to sei whales as each species was depleted in turn; and (2) the fact that the whaling nations exceeded the B W U quotas about half of the time during the period the I W C managed whales through the B W U.

This period (1961–1971) was marked by changes in the Scientific Committee, both in terms of an increase in attendance of qualified scientists and a great strengthening of its quantitative competence. Both Chapman and Allen (each of whom later became chair of the Scientific Committee) became members of the Scientific Committee, with Holt and Gulland also continuing their participation as observers and advisors. Member nations added scientists with strong mathematical skills. The Scientific Committee's work was completed during the week prior to the commission meetings and recommendations were generally by consensus.

The Antarctic fin whale stocks, however, remained an item of contention. At the final Scientific Committee meeting of this period in 1971, all members of the Scientific Committee, apart from Japan and the U S S R, accepted a best estimate of sustainable yield at 2,200 fins, and believed that there was no convincing evidence that the population was increasing.[22] The Japanese and Soviet scientists believed that these estimates were too low and that the fin whale population was increasing. The Japanese suggested an available yield of 4,600 fins. The actual catch for the 1971–72 season was 2,683 fin whales. The B W U quota was not reached; in fact, it had been reached only once since the 1962–63 season.

STOCKHOLM, MORATORIUM, AND SOUTHERN OCEAN
SANCTUARY: 1972—PRESENT

A radical change for the commission and the Scientific Committee began in
1972. The twenty-fourth meeting of the commission took place in London
in June 1972 immediately following the United Nations Stockholm Conference on the Human Environment.[23] That conference was highlighted by a
nearly unanimous vote (fifty-three to zero, with three abstentions by whaling nations) on a resolution calling for a ten-year moratorium on all commercial whaling operations.

The moratorium issue was on the agenda for the Scientific Committee.
By consensus the Committee agreed that a blanket moratorium could not
be justified scientifically. It challenged the strength of the world interest in
whales demonstrated by the Stockholm meeting and recommended an expanded whale research program in place of a blanket moratorium. The Scientific Committee's views on a blanket moratorium were repeated at their
next meeting and were cited by the commission in voting down the moratorium proposal. This was one of few times that whaling nations enthusias-
tically supported a Scientific Committee view. For comparison, the Scientific Committee suggested in 1972 that the maximum sustainable yield
(M S Y) for Antarctic minke whales was 5,000, but it recognized that the data
were scanty and the number could be higher. It spelled out the need for a
cautious approach in taking the minke at its 1973 meeting in light of the absence of good data (with a controversy that ranged from an M S Y value of
five thousand by Douglas Chapman to more than twelve thousand by Seji
Ohsumi). Minke whale catches in the two whaling seasons following these
meetings were 5,745 and 7,713. In the arguments regarding the minke whale
recommendations it was expressly stated by one of the Japanese scientists
that they could not accept the low number urged by most of the Scientific
Committee members because that number was too low to meet their industry's need. As Colin Clark later demonstrated, the tension between those
pressing for conservative actions and those representing whaling interests
was exacerbated by the low productivity of whale stocks. If whale stocks
were rigorously and conservatively managed, investors would realize very
low profits.[24]

Although the blanket moratorium was defeated in the commission and
was not tenable on its scientific merits, pressure for a moratorium continued on a political level, both inside and outside of the commission. Activists

TABLE 3.1 Catches of Whales in the Antarctic, 1919–20 to 1976–77

Year	Blue	Fin	Humpback	Sei-Bryde's	Minke	Sperm	Others	Total
1919–20	1,874	3,213	261	71	—	8	14	5,441
1920–21	2,617	5,491	260	36	—	31	13	8,448
1921–22	4,416	2,492	9	103	—	3	—	7,023
1922–23	5,683	3,677	517	10	—	23	—	9,910
1923–24	3,732	3,035	233	193	—	66	12	7,271
1924–25	5,703	4,366	359	1	—	59	—	10,488
1925–26	4,697	8,916	364	195	—	37	10	14,219
1926–27	6,545	5,102	189	778	—	39	12	12,665
1927–28	8,334	4,459	23	883	—	72	4	13,775
1928–29	12,734	6,689	48	808	—	62	—	20,341
1929–30	17,898	11,614	853	216	—	73	1	30,655
1930–31	29,410	10,017	576	145	—	51	2	40,201
1931–32	6,488	2,871	184	16	—	13	—	9,572
1932–33	18,891	5,168	159	2	—	107	—	24,327
1933–34	17,349	7,200	872	—	—	666	—	26,087
1934–35	16,500	12,500	1,965	266	—	577	—	31,808
1935–36	17,731	9,697	3,162	2	—	399	—	30,991
1936–37	14,304	14,381	4,477	490	—	926	1	34,579
1937–38	14,923	28,009	2,079	161	—	867	—	46,039
1938–39	14,081	20,784	883	22	—	2,585	1	38,356
1939–40	11,480	18,694	2	81	—	1,938	705	32,900
1940–41	4,943	7,831	2,675	110	—	804	—	16,363
1941–42	59	1,189	16	52	—	109	—	1,425
1942–43	125	776	—	73	—	24	—	998
1943–44	339	1,158	4	197	—	101	—	1,799
1944–45	1,042	1,666	60	78	—	45	—	2,891
1945–46	3,606	9,185	238	85	—	273	—	13,387
1946–47	9,192	14,547	29	393	—	1,431	1	25,593
1947–48	6,908	21,141	26	621	—	2,622	—	31,318
1948–49	7,625	19,123	31	578	—	4,078	—	31,435

Year	Blue	Fin	Humpback	Sei-Bryde's	Minke	Sperm	Others	Total
1949–50	6,168	18,061	2,117	101	—	2,570	—	20,017
1950–51	6,966	17,474	1,630	367	—	4,742	—	31,179
1951–52	5,124	20,520	1,546	32	—	5,344	—	32,566
1952–53	3,866	21,197	954	123	—	2,185	—	28,325
1953–54	2,684	24,986	594	251	—	2,700	—	31,215
1954–55	2,163	25,878	493	146	—	5,708	—	34,388
1955–56	1,611	25,289	1,432	276	—	6,881	—	35,489
1956–57	1,505	25,700	679	712	—	4,345	—	32,941
1957–58	1,684	25,222	396	2,385	—	6,310	—	35,997
1958–59	1,191	25,837	2,394	1,402	—	5,437	—	36,261
1959–60	1,230	26,415	1,338	3,234	—	4,138	—	36,355
1960–61	1,740	27,374	718	4,310	—	4,666	2	38,810
1961–62	1,118	27,099	309	5,196	—	4,829	1	38,552
1962–63	947	18,668	270	5,503	—	4,771	—	30,159
1963–64	112	14,422	2	8,695	—	6,711	—	29,942
1964–65	20	7,811	—	20,380	—	4,352	—	32,563
1965–66	1	2,536	1	17,587	—	4,555	—	24,680
1966–67	4	2,893	—	12,638	—	4,960	—	20,495
1967–68	—	2,155	—	10,357	—	2,568	—	15,080
1968–69	—	3,020	—	5,776	—	2,682	—	11,478
1969–70	—	3,002	—	5,867	—	3,090	—	11,959
1970–71	—	2,890	—	6,153	—	3,055	—	12,098
1971–72	—	2,683	3	5,456	3,021	3,366	—	14,529
1972–73	7	1,761	5	3,864	5,745	4,203	—	15,585
1973–74	—	1,288	—	4,392	7,713	4,927	—	18,320
1974–75	—	979	—	3,859	7,000	4,162	—	16,000
1975–76	—	206	—	1,821	6,034	2,829	—	10,890
1976–77	—	—	—	1,858	7,900	2,002	—	11,760
Total	307,370	640,357	35,435	139,127	37,413	136,177	779	1,296,658

SOURCE: Sally A. Mizroch, "The Development of Balaenopterid Whaling in the Antarctic," *Cetus* 5, no. 2 (1984).

TABLE 3.2 Catch Limit and Actual Catch
(in BWUS) in the Antarctic

Year	Catch Limit	Catch
1946–47	16,000	15,338
1947–48	16,000	16,364
1948–49	16,000	16,007
1949–50	16,000	16,059
1950–51	16,000	16,413
1951–52	16,000	16,006
1952–53	16,000	14,855
1953–54	15,500	15,439
1954–55	15,500	15,300
1955–56	15,500	14,874
1956–57	14,500	14,745
1957–58	14,500	14,850
1958–59	15,000	15,301
1959–60	15,000	15,512
1960–61	—	16,433
1961–62	—	15,253
1962–63	15,000	11,306
1963–64	10,000	8,429
1964–65	—	6,986
1965–66	4,500	4,089
1966–67	3,500	3,511
1967–68	3,200	2,804
1968–69	3,200	2,469
1969–70	2,700	2,477
1970–71	2,700	2,469
1971–72	2,300	2,252

SOURCE: Sally A. Mizroch, "The Development
of Balaenopterid Whaling in the Antarctic," *Cetus*,
5, no. 2 (1984).

pressed for boycotts, particularly of Japanese exports, and it was clear that some compromise solution had to be found. Australia proposed a plan (the New Management Procedure, or NMP) for limited moratoriums, which would ban whaling on species that had been reduced below a level that would produce 90 percent of the MSY but would allow whaling on stocks above that level, with restrictions that would assure long-term safety and sustainability of the stocks. The NMP was adopted in 1974 and implemented for the 1975–76 whaling season.[25]

An important feature of the NMP was the commission's new dependence on Scientific Committee advice in its quota determinations. The Committee was empowered to provide advice independent of the needs of the whaling industry. Participation in Committee meetings had grown, both in terms of countries represented and in numbers of scientists. No scientist or group of scientists could dominate the meetings or unduly influence the final results. The Scientific Committee also reflected a strong balance between highly competent, field-experienced biologists with those skilled as population dynamicists.

The Scientific Committee continued to expand both in numbers and skill during this period. This growth is demonstrated by even a cursory examination of the working papers used during the first two phases of the Scientific Committee's existence and those of recent times. Apart from the final report of the Committee of Three there is little during the early periods of the Scientific Committee to compare to the quality of the scientific documents provided to and by the Scientific Committee for their deliberations in the recent period.

In trying to implement the terms of the NMP, it became obvious that the procedure was flawed, particularly because of the paucity of critical data and serious uncertainties about the classification scheme. While it provided a scientific rationale that allowed selective moratoriums to be implemented, it did not, in the view of some, provide assurance that the harvested stocks were being taken at catch levels that would ensure their long-term sustainability.

The scientific uncertainties in implementing the NMP coincided with a period of rapid growth in the membership of the commission, almost exclusively by nations committed to a ban on commercial whaling. The original fourteen-member commission swelled to thirty-three members by 1981 (thirty-nine by 1996). The domination of the commission by members opposed to commercial whaling allowed the uncertainties of the Scientific

Committee to be translated into a ban on commercial whaling.[26] The Scientific Committee provided no consensus view either in support of or in opposition to a blanket moratorium, a change from the previous decade, when a blanket moratorium was treated as an equivalent of the BWU and was found unsuitable as a management tool. To a degree, this result was caused by an expansion of the Scientific Committee to about fifty members (with additional representation of original IWC nations and new members). The strong sentiment against commercial whaling was reflected in the Scientific Committee as in the commission. The Scientific Committee was to grow to more than one hundred members by the next decade.

The tradition of the Scientific Committee report is to blur disagreements between its members in the absence of consensus, rather than to vote to show the degree of support for or opposition to a view. The Scientific Committee report simply stated that some members took one view and other members disagreed with that view. Attendees at the 1979 Scientific Committee meeting indicated to the author that only a few Committee members supported a blanket moratorium, while a large majority felt that controlled whaling under the NMP did safeguard whale stocks for future generations at very small risks. No flavor of this division can be seen in the report![27]

The blanket moratorium of 1982 went into effect during the 1985–86 whaling season. The Scientific Committee's unsuccessful attempt to revise the NMP, with the commission's assistance, led to the recognition that a totally different approach was required to develop catch limits. In 1985, the Scientific Committee committed itself to a comprehensive assessment of whale stocks, which included the examination of alternative management approaches.[28] This commitment led to the development of the Revised Management Procedure (RMP).

In 1992 the Scientific Committee unanimously recommended adoption of the draft specification of the RMP they had developed, noting that they had some specific details to complete before the RMP could replace the NMP.[29] The commission passed the RMP by resolution. A good review of the problems of the NMP and the development of the RMP may be found in Young. Chapter 2 of this work, primarily authored by Douglas Butterworth, provides considerable detail on the science of the RMP.[30]

The Scientific Committee completed its work on the RMP in 1993 and once again unanimously passed it on to the commission.[31] Acceptance by the commission would clearly be a first step in allowing the resumption of commercial whaling. The commission failed to adopt the completed R M P, leading to the resignation of the Scientific Committee chair, Phillip Hammond, who concluded that he could no longer justify to himself "being the organizer of and the spokesman for a Committee which is held in such disregard by the body to which it is responsible."[32]

The R M P is the scientific element of an overall plan. Key elements include (1) being stock-specific rather than species-specific, (2) requiring regular systematic surveys to determine abundance if commercial whaling is to continue, (3) incorporating uncertainty in a risk-averse manner, and (4) attempting to make the quota-setting process as objective as possible. The Revised Management Scheme (R M S) includes necessary elements of enforcement, observer coverage, data collection, fiscal arrangements, and so forth[33] No comprehensive R M S has been approved by the commission at this time. It should be noted that at both the 1995 and 1996 commission meetings resolutions were passed that advanced the establishment of the R M S.[34] The 1995 resolution appears to accept the Scientific Committee's R M P by its incorporation of the Scientific Committee annex that described the plan.

During the 1997 meeting, discussions on the R M S (still unpublished chair's report) continued, but its ultimate approval remains doubtful, particularly in light of many nations' objections to commercial whaling. At this meeting Australia articulated its opposition to the R M S.

THE USES OF UNCERTAINTY

The history of the I W C has seen a continuous improvement in the Scientific Committee and in the advice it provides. In its early period as a whalers' organization, the Scientific Committee was largely ignored to meet the short-term economic goals of the industry. The potential of long-term benefits that could be realized by a soundly managed sustainable resource (as called for in the convention) was ignored in the pursuit of immediate profits. It is possible some nations found comfort in the scientific uncertainties and the lack of consensus regarding the status of whale stocks, but this optimistic outlook was clearly proved foolish by the sequential collapse of each of the economically important whale stocks. Even though the signs were clear prior to the appointment of the Committee of Three, the recommendations to reduce catches or face disaster were essentially ignored. It is evident that catches were finally reduced largely because of the unavailability of the whales rather than because of concern about conservation.

For a short period following the UN meeting in Stockholm and the passage of the N M P, the Scientific Committee's advice was reasonably implemented, use of the B W U was dropped, selective moratoriums to protect and restore overharvested species were put in place, and quotas were established at levels that were unlikely to cause reductions of the harvested stocks. The well-documented uncertainties of the N M P surfaced early in its application and coincided with a rapid increase in I W C members having strong anti-whaling sentiments. The reverse of the early history of the I W C then occurred. The Scientific Committee's advice was largely ignored by a commission committed to a ban on commercial whaling, perhaps taking comfort that scientific uncertainties justified a blanket moratorium.

A severe blow to the role of the Scientific Committee was the failure of the commission to adopt its carefully and unanimously supported R M P, thus prompting the resignation of its chair. Such an adoption, of course, would have marked the first step in the process to establish quotas for the resumption of commercial whaling. This step is anathema to most of the I W C members, despite their membership in an organization created expressly for the regulation of whaling, not its prevention. As noted above, the R M P was finally accepted by the commission in 1995.

In light of the general failure of the commission to heed the advice of its Scientific Committee, is there a need for significant improvements for the Scientific Committee to be effective in providing advice to the commission? Should changes be made that would control either personal or national agendas of Scientific Committee scientists and thus allow the generation of bias-free, completely creditable Scientific Committee reports? The problem, however, is not the Scientific Committee but the commission itself and a membership that has from its beginning ignored the terms of the convention it signed. The use of uncertainty is a comfortable excuse to do what you wish—whether it is the overharvesting of whales or the prevention of commercial taking. It is clearly easier to use uncertainty than to admit that for economic, moral, cultural, or political reasons whaling is repugnant or essential for human survival. For the Scientific Committee to fully function and be useful to the commission, the commission must either commit itself to the convention or renegotiate its terms and then adhere to the new terms.

The Scientific Committee, despite its difficult history, has been remarkably successful as the prime source of scientific documents on whales and whaling. Their documents, once they appear in the reports of the commission, are rigorously peer-reviewed, and they are broadly available to the community. Certainly, beginning with the Committee of Three's final report, the published documents of the Scientific Committee can form the basis for management decisions. The difficulty occurs, however, in the working documents used during the Scientific Committee meeting. A massive collection of material is generally presented to the Scientific Committee members shortly before the meetings and in some cases while the meetings are in progress. In the past, management recommendations were based upon urgent and less-than-critical review of such papers. Under the R M P this kind of action is unlikely to occur, but it is critical that advice be based upon documents that have been adequately peer-reviewed and are broadly accessible well in advance of commission action, hence the necessity for establishing deadlines that allow for both review and availability.

Can the Scientific Committee and its operation be improved? Of course, but I contend that the changes would be small. I believe the Scientific Committee membership has become unwieldy. Reducing attendance is politically difficult and cannot be done by the Scientific Committee itself. The commission should address this issue and limit attendance. In part, this change could occur by having specialists who may address specific agenda items while limiting their active participation to those items.

Consensus agreement on critical issues is desirable but not essential. Regardless of the depth and breadth of a research program on any living marine resource, but especially on the whales of the open ocean, there will always be uncertainty. This uncertainty will, in many instances, produce disagreements. The disagreements should be emphasized to the managers of the resource both in terms of revealing the data gaps that must be closed to reduce or eliminate the uncertainty and the sources of the disagreement and the rationale behind the different views. The use of identified minority reports in recent Scientific Committee documents is an important aid to the managers and also reduces the likelihood of unjustified interventions.

To prevent personal bias or national agendas from influencing scientific conclusions is not simple but depends on the goodwill and integrity of the participants and the full exposure of the documents and the database used in the decision-making process. While this may be imperfect, scientific knowledge and growth have been based on the above premises. Both willful distortions and unintentional errors are almost always detected within reasonable periods of time, particularly in arenas of considerable controversy, in the presence of often highly polarized, competent interest groups.

In the management of living natural resources there will always be uncertainty. Frequently the impacts of normal variability outweigh human influences. Estimating the changes generated by anthropogenic activities, including possible climate change, pollution, and the consequences for the ecosystem of either selective harvesting or protection, is exceedingly difficult, if not impossible. The problem is rendered all the more difficult for oceanic animals with great geographic ranges, particularly since most member nations are unwilling to provide adequate research budgets to develop necessary data and analyses.

The International Whaling Commission and its use of scientific advice is an anomaly in the world of resource management agencies. In every fisheries management agency I know, scientists are pleased when allocations of fish and shellfish are made that are consistent with scientific advice. Sadly, there have been too many cases in which the advice is ignored and stocks are badly overfished. I am unaware of any resource commission charged with regulating harvests that has failed to establish quotas given scientific advice that provides a strong rationale for a safe harvest.

The Scientific Committee has been remarkably successful in generating good advice to the commission. The failure of the commission to largely heed this advice is the commission's responsibility. Improvements in the future, as noted above and especially elsewhere in this volume, depend on changes made by the commission to commit to a clearly stated mission. Despite past frustrations and the frequent absence of guidance from the commission, I believe the Scientific Committee, without major change in either structure or operating procedures, can function effectively and provide a sound basis for managing whales and for resolving the conflicts that inevitably occur. A good start for the commission would be full implementation of the RMS.

Readers patient enough to reach this point may have noted the absence of any discussion of aboriginal whaling. This omission is largely due to the fact that such whaling is treated very differently from commercial whaling. Aboriginal whaling is based primarily on the needs of the community taking the whales. The needs discussion and decision are outside the Scientific Committee's mission. The Scientific Committee is consulted, however, to ensure that any taking is conservative and is generally well below the replacement yield of the stock in question. Gray and bowhead whales have been taken to meet aboriginal needs. In the face of the conservative harvests allowed for these species, the gray whale has increased sufficiently to be removed from the Endangered Species List, and the bowhead population has also enjoyed significant increases.

The Scientific Committee took the initial steps that forced the regulation of the bowhead hunt. The failure of the United States to take the steps recommended by the Scientific Committee in 1976 to reduce the expanding hunt and limit the numbers of struck but lost whales[35] prompted the Scientific Committee to recommend a zero harvest at their next meeting in 1977.[36] This recommendation was implemented by the commission and created significant problems both for the U.S. government and its North Slope aboriginal people. At a special meeting of the commission in 1977 the moratorium was lifted and rigid quotas were established to permit the limited takes believed to be essential for the cultural survival of the North Slope communities.[37]

NOTES

1. K. Radway Allen, *Conservation and Management of Whales* (Seattle: University of Washington Press, 1980).

2. James E. Scarff, "The International Management of Whales, Dolphins, and Porpoises: An Interdisciplinary Assessment." *Ecology Law Quarterly* 6, no. 2–3 (1977): 323–638.

3. William E. Schevill, ed., *The Whale Problem: A Status Report* (Cambridge: Harvard University Press, 1974).

4. George L. Small, *The Blue Whale* (New York: Columbia University Press, 1971).

5. Johan Nicolay Tønnessen and Arne Odd Johnsen, *The History of Modern Whaling* (Berkley and Los Angeles: University of California Press, 1982).

6. Nina M. Young, ed., *Examining the Components of a Revised Management Scheme* (Washington, D.C.: Center for Marine Conservation, 1993).

7. Tore Schweder, *Intransigence, Incompetence, or Political Expediency? Dutch Scientists in the International Whaling Commission in the 1950s: Injection of Uncertainty*, SC/44/O 13 (Cambridge, England: IWC, 1992).

8. Sidney Holt, "Incompetence or Political Expediency," International Whaling Commission, 1992 (SC92/SCWP3), unpublished.

9. *Proceedings of the 6th Meeting of the IWC*, 1955.

10. Schweder, *Intransigence*.

11. *Proceedings of the 7th Meeting of the IWC*, 1956.

12. *Proceedings of the 8th Meeting of the IWC, 1957.*

13. *Proceedings of the 12th Meeting of the IWC, 1961.*

14. *Proceedings of the 13th Meeting of the IWC, 1962.*

15. *Proceedings of the 14th Meeting of the IWC, 1964.*

16. Ibid.

17. Ibid.

18. *Proceedings of the 15th Meeting of the IWC, 1965.*

19. Ibid.

20. John A. Gulland, *The Management of Marine Fisheries* (Seattle: University of Washington Press, 1974).

21. From Sally A. Mizroch, "The Development of Balaenopterid Whaling in the Antarctic," *Cetus* 5, no. 2 (1984): 6–10. This data set, derived from official statistics submitted by governments, includes the falsified records submitted by the USSR, which substantially underreported its catch. Despite the underreporting, it remains the best data set available until the figures are corrected. Even with the cheating the point the table makes remains unchanged.

22. *Proceedings of the 22d Meeting of the IWC, 1972.*

23. *Proceedings of the 24th Meeting of the IWC, 1974.*

24. Colin W. Clark, "Economic Aspects of Renewable Resource Exploitation as Applied to Marine Mammals" *Food and Agricultural Organization, Fisheries Series* 3, no. 5 (1981).

25. *Proceedings of the 26th Meeting of the IWC, 1976.*

26. *Proceedings of the 33d Meeting of the IWC, 1983.*

27. *Proceedings of the 30th Meeting of the IWC, 1980.*

28. *Proceedings of the 36th Meeting of the IWC, 1986.*

29. *Proceedings of the 43d Meeting of the IWC, 1993.*

30. Young, *Revised Management Scheme.*

31. *Proceedings of the 44th Meeting of the IWC, 1994.*

32. "Resignation of the Chairman of the Scientific Committee," Circular Communication to Commissioners, Contracting Governments, and Members of the Scientific Committee (1 June 1993).

33. Young, *Revised Management Scheme.*

34. *Proceedings of the 46th Meeting of the IWC, 1996; Proceedings of the 47th Meeting of the IWC, 1997.*

35. *Proceedings of the 27th Meeting of the IWC, 1977.*

36. *Proceedings of the 28th Meeting of the IWC, 1978.*

37. Ibid.

4 / Is Money the Root of the Problem?

Cultural Conflicts in the IWC

MILTON M. R. FREEMAN

The international management of whaling is currently characterized by conflict rather than cooperation. One cause of this conflict appears to be linked to the progressive delocalization of management responsibility, with a result that actors now very far removed from the resource users' environment, lives, and needs have assumed an ever-increasing role in the management process. The result has been an increasing tendency toward desensitizing and politicizing the management process without any increase in the sustainability of the resource user–resource stock interdependency, a relationship that is now recognized as a critically important conservation principle.[1]

One significant consequence of moving management decision making progressively further from a local, national, or regional locus is the marked detachment of decision makers from the actual circumstances faced by resource users. Such detachment, in the case of the International Whaling Commission, manifests itself as cultural insensitivity (or indifference) toward certain others' legitimate concerns and expectations as parties to a treaty, and a growing distrust by resource users of the sincerity, legitimacy, and usefulness of the existing management process. Under conditions in which the majority of participants appear unwilling and unable to accept compromise and move forward, it is unlikely that equitable and effective management decisions will be made. If these conditions persist, it is only a matter of time before the regime disintegrates or is replaced.

In this chapter I will illustrate some important cultural factors contributing to the prevailing conflict and lack of goodwill negotiation within the International Whaling Commission (iwc). It is hoped that the understanding gained from this chapter will improve the reader's cultural awareness and sensitivity and will help decision makers who seek workable solutions and an end to the stalemate that exists in the iwc. This chapter will also serve to warn whaling nations that are not presently members of the iwc what problems they will face if they become members.

Although it is a lack of political will on the part of the majority of members that has resulted in the current impasse, one barrier to rational and workable solutions at the iwc is the confusion and misunderstanding about some of the critical issues in dispute. Such misunderstandings, in turn, contribute to the absence of political will required to change the deadlocked situation, for there is little doubt that in iwc debates, and in the information available to the general public and politicians, there is considerable ignorance of the issues germane to a considered understanding of contemporary whaling.

These misunderstandings result in large part from ongoing and effective campaigns by whale-protection organizations seeking to end the killing of whales by any means. Such campaigns frequently characterize whales as collectively endangered, rare, or severely depleted and under massive threat from whalers and from a deteriorating ocean environment; they claim that whales are uniquely "special" in respect to their social life, mental abilities, and relationship to humans. Nearly all such assertions have only a limited basis in science,[2] yet they are frequently repeated by politicians in the anti-whaling nations. One scientific authority explains the situation in these terms:

> ... in recent years the authors of one popular book after another have *started* from the premise (sometimes explicit, sometimes implied) that the cetaceans represent a higher order of social evolution and, by implication, intelligence. . . . Some of these accounts have received wide publicity in magazines and newspapers and on television, to the extent that complex dolphin sociology and high cetacean intelligence have joined motherhood and apple pie in the public mythology. The realities . . . although interesting are considerably more mundane.[3]

None of the information contained in this chapter is new; it has all been presented and discussed in various iwc subcommittees, working groups, and commission plenaries for more than a decade, and it is well substantiated in the scientific literature referred to in the reports that constitute part of the official record of IWC meetings. What is lacking at the IWC is not access to accepted scientific findings but rather the courage to be open-minded and honest in the discharge of legal obligations required of signatories to an international treaty.[4]

It would be naive not to recognize that for the majority of participants in IWC discussions, the current nonresolution of the whaling problem is the desired outcome. However, this chapter is addressed to those who question the wisdom of perpetuating the current stalemate at the IWC. As many national legislatures seek to curb government expenditures, the annual expense of sending delegates to distant meetings that accomplish nothing may be seriously scrutinized. The annual meetings of the IWC, in addition to wasting public funds that might be better used dealing with environmental problems, are arguably illegal (see chapter 1, "A New Whaling Agreement and International Law," by William T. Burke, in this volume).

This chapter has three other implicit conservation objectives, namely, that in furthering the goal of whale management, decision makers do the following:

1. Take legitimate human needs fully into account.
2. Recognize that cultural differences exist and that minorities (and not only the majority) have rights, especially when seeking to base actions on ethics;
3. Use scientific data properly in making management decisions.

If these objectives can be advanced in the process of managing whaling activities, then to a certain extent another of the purposes of this book will have been addressed: namely, contributing insights applicable to other international management regimes and thereby improving the prospects for better global governance. Indeed, it might be argued that if an issue as emotionally and politically charged as whaling can be dealt with fairly, rationally, and effectively, then other far less emotive issues will surely seem less intractable in intergovernmental discourse.

RECOGNIZING HUMAN NEEDS AND THE QUESTION OF SUBSISTENCE

The 1946 International Convention for the Regulation of Whaling (ICRW) and the International Whaling Commission, which it established, recognize the need for some human groups to kill whales. Thus, aboriginal whalers are permitted to continue whaling, even if the whale stocks in question are considered to be seriously depleted (as has been the case in Alaska, where, in 1977 and years following, the annual bowhead whale kill was considered by some to be far in excess of what the then-known population could sustain). However, in the case of "nonaboriginal" whalers, the IWC majority denies that a similar need to kill whales exists, even where they (like "aboriginal" whalers) engage in subsistence whaling. Indeed, in the case of nonaboriginal whalers, whaling is considered unacceptable even if the whale stocks in question are not endangered, and are even abundant, and the catch is sustainable according to the IWC's own Scientific Committee. One reason for this paradox derives from confusion over the use and meaning of the critical term *subsistence*.

In the IWC, the term *subsistence* is used only in association with the word *aboriginal* (as in "aboriginal subsistence whaling"). Therefore, because the IWC majority believe subsistence does not exist unless practiced by aboriginal peoples, by definition, nonaboriginal people do not engage in subsistence. Yet in the real world, many nonaboriginal whalers engage in subsistence.

The most commonly understood meaning of the term *subsistence* in everyday English is "the barest means in terms of food, clothing, and shelter needed to sustain life" or "the irreducible minimum (as of food and shelter) necessary to support life."[5] Therefore, with this commonplace understanding of the term, subsistence whalers are poor, they experience problems in obtaining regular or sufficient supplies of food or other necessities of life, they necessarily consume all the food they produce and thus, lacking any tradable surplus, will not engage in commercial buying and selling of edible whale products. Thus, to the person unfamiliar with social scientists' understanding of the term, subsistence is found among people who are economically disadvantaged and experience problems of survival.

However, anthropologists' understanding of the term *subsistence* is based upon empirical study of contemporary human societies, all of which today engage in monetized trading. Many of the societies that in the past engaged in subsistence continue to do so today, and indeed, the use of money is essential to maintain subsistence values and practices in today's monetized world,[6] as it has been, throughout the world, for several centuries.[7] Thus, the current scientific understanding of subsistence has progressed far beyond its earlier vernacular usage.

Indeed, for many years, anthropological orthodoxy considered hunter-gatherers (the classical subsistence practitioners) to constitute "the original affluent society,"[8] which clearly negates the popular image of people living at a mere survival level. Affluence implies, if not material abundance, then at least a lack of want that allows considerable leisure time. Where, to the layperson, subsistence is mostly directed to material and economic circumstances, to anthropologists, subsistence relates to various nonmaterial social relationships and cultural norms. Thus, for example, subsistence has been defined as "*a set of culturally established responsibilities, rights, and obligations that affect every man, woman, and child each day.*"[9]

Clearly, there is a marked difference between the understandings of the lay public and most IWC decision makers on the one hand and the social science specialists who study subsistence throughout the world on the other hand. The notion that whale hunters practicing subsistence are necessarily poor and operate outside of a monetized economy has been contradicted by scientific reports submitted at IWC meetings for many years.[10]

What has been documented and reported to the IWC is that equipment and supplies used by subsistence whalers are purchased with money and that cash or in-kind (but purchased) contributions of whaling equipment and supplies may be provided to whalers by others who benefit from the products of the hunt. In the culture of these whaling societies, such contributions, having social as well as economic value, will be reciprocated in the future with a share of meat, blubber, or other products.

Within IWC discussions however, the principal basis for distinguishing aboriginal subsistence whaling from commercial whaling is a belief that aboriginal whalers do not engage in monetized economic exchange, or stated another way, that commoditization of the whale is not part of aboriginal subsistence whaling. Again, evidence to the contrary has been well documented at the IWC.[11]

In Chukotka (in northeast Russia), whaling categorized by the IWC as aboriginal subsistence whaling from the early 1950s until the end of the Soviet era (ca. 1990) was a state-run enterprise carried out by nonaboriginal wage-employed nonlocal whalers. The product of this whaling supported the cash economy of the coastal villages, which was based upon wage earnings derived from fur farming. Immediately following the Soviet era, the villages were required to pay for the whales delivered by this nonlocal whaling enterprise, so after the collapse of the local and regional economy, this form of aboriginal subsistence whaling quickly ended. Since 1993, the whale hunt has been resumed by local hunters, with money from the sale of the meat in some villages paid to the aboriginal hunters who land the whales; the carcass is distributed in the villages to local consumers and to local commercial fur farming operations.[12]

In Greenland, whaling operations (also classified as aboriginal subsistence whaling by the IWC) have involved wage employment and cash sales of edible whale products since the mid-eighteenth century.[13] Today, boats used in coastal whaling include modern commercial fishing vessels that use a harpoon cannon. In this form of aboriginal subsistence whaling, the whale products provide part of the revenues earned by the boat owners and are used to pay crew wages, bank loans, and other expenses involved in owning and operating a multipurpose commercial fishing/shrimping/hunting vessel. A portion of each whale taken in Greenland is likely to be sold by the hunter (whether the hunting takes place from a kayak or from a large steel vessel) either directly to consumers or to local processing plants that package and then distribute the products nationally to retail stores.[14] Wages for plant workers and retail store employees and the profits of these corporations are generated through purchases of edible whale products in the supermarkets and community businesses of Greenland.

Only the Alaskan whalers, among IWC-designated aboriginal subsistence whalers, have surrendered their right to engage freely and openly in a modern market economy (ironically, through the insistence of the U.S. Department of Commerce!). This is not to say that Alaskan whaling is functionally separated from a fully commercialized economy, nor is it legally required to be separated (see below), for it remains thoroughly integrated into the national and global capitalist system. However, as with all societies engaged in subsistence wherever they are found, Alaskan whalers participate in a so-called mixed economy, one that has totally replaced the nonmonetized economic system that existed before contact with traders, whalers, and missionaries occurred in the nineteenth century. As has been observed with respect to Alaskan subsistence, *"the one most important characteristic . . . is that subsistence is now integrated with the cash economy in the lives of all Alaskan Natives."*[15] Furthermore, commercial exchange of subsistence products occurs in Alaska, including the whaling districts.[16] This buying and selling of marine mammal products is fully consistent with U.S. domestic legislation, for under Subsection 1371[b] of the Marine Mammal Protection Act natives living in coastal communities are allowed to engage in such within-state commercial activity.

Indeed, Alaskan aboriginal whaling could not operate without substantial monetary transfers taking place from the market economy: maintaining a bowhead whaling management system that satisfies IWC demands costs these subsistence hunters more than $1 million each year for salaries, consultant fees, travel, meetings, research, and administrative expenses, with three-quarters of the cost contributed by the local Inupiat municipality.[17] Furthermore, the capital invested in bowhead whaling equipment by Inupiat whalers themselves was estimated at about $10,000 per crew in 1980,[18] which represents an estimated current investment by Alaskan bowhead whalers of about $2.6 million for whaling equipment and supplies, all of which are purchased with money.

THE MONEY FETISH AT THE IWC

Why, then, does the IWC insist that only whaling by nonaboriginals is "commercial," and why does the United States forbid the buying and selling of whale meat in its own aboriginal societies when it accepts such transactions in other aboriginal whaling societies? Why has it become so important to the United States and other anti-whaling nations at the IWC to berate nonaboriginal whalers for the use of money within their whaling operations, when these same IWC nations tolerate the selling of whale meat by (some) aboriginal whalers?

Despite the absence of consistency in these actions, there may be two factors influencing such behavior. First, there are the incessant claims made by anti-whaling campaigners that commercial whaling is unsustainable. Such assertions are based upon faulty theoretical arguments developed in the 1970s, when concern about the unsustainable exploitation of whales was increasing. These theoretical analyses (most prominently by Colin Clark) suggested that commercial whaling could not be undertaken in a sustainable manner due to the low reproductive potential of whales.[19] By today's standards of analysis[20] these overly simplistic explanations (e.g., ignoring stock size and market forces) may appear both naive and misleading, but at the time they appeared to explain in mathematical—and hence apparently rigorously scientific—terms an observable phenomenon, namely that the industrial hunting of large whales was inevitably leading to the destruction of the resource.

There are also important cultural reasons for a distaste concerning the killing of mammals[21] and especially concerning the commercial exploitation of wildlife, which is associated in western societies with a deeply rooted ambivalence toward money matters. Aristotle saw the exchange of commodities outside of household-based exchange as being unnatural and dangerous, and the aphorism (from 1 Timothy 6:10) that "the love of money is the root of all evil" is widely acknowledged in Christendom. Early this century, both Karl Marx and Georg Simmel commented on the corrupting influence of money; according to Marx, the "noble savage," living in balance with nature, engaged only in altruistic, nonmonetized barter relations with neighbors, until becoming corrupted and greedy after contact with agents of capitalism.[22]

Such early ambivalences continue to be embraced by many environmental organizations, which believe or espouse the view that "big business" (and industry-supported governments) are the villains responsible for environmental decline. Certainly, there are many examples of commercial overexploitation of natural resources (historic whaling being a good example). Therefore, by uncritical generalizing, it is argued that unless all commercial interest is removed from resource use (and from whaling in particular), there is near certainty that these natural resources cannot be used in a sustainable manner. This money fetish dominates the simplistic model followed by many anti-whaling advocates who appear to believe the following:

1. Aboriginal subsistence whalers don't sell whale meat for *cash*—only commercial whalers sell whale meat for *cash*.
2. Aboriginal subsistence whalers consume all they produce within their communities—commercial whalers make *profits* selling the surplus widely outside of their own communities.
3. Aboriginal subsistence whalers still use *inexpensive*, handmade, traditional tools—commercial whalers use *expensive* manufactured technology.
4. Aboriginal consumers of whale meat are *poor*—those consuming commercially derived whale meat are *rich*.
5. Aboriginal subsistence whalers need whales to *survive*—but commercial whalers kill whales to feed *greed*.

In sum, aboriginal whalers and their family members can be tolerated (because of their poverty, lack of opportunities, and isolation from the national mainstream), but nonaboriginal whalers and their customers are intolerable (because of their relatively privileged position in their national mainstream).

As this is an easily understood analysis of the situation promoted at the IWC, the majority of delegates apparently have no need to consider the matter in any depth nor to seriously address the claims of hardship made by coastal people who are negatively affected by the continued ban on whaling. However, the commercial whaling issue is retained on the IWC annual meeting agenda as it provides a welcome opportunity for anti-whaling nations to express moral outrage over continued whaling by nonaboriginal people. These annual speeches satisfy the needs of domestic political constituents who are well represented in the NGO corps observing this annual event.[23]

However, when considering questions about the sustainability of resource exploitation, whether commercial or noncommercial, a critical issue involves the *scale* of operation. There are significant differences between a large-scale (industrial) activity and a small-scale (artisanal) activity, although both may use money in their operations and both may exploit the same biological resource (be it trees, fish, or whales). Such differences in scale of operations will certainly have policy and management implications, particularly in relation to the size of the resource (whale) stock.

Indeed, ignoring the size of the stock is one principal flaw in Colin Clark's economic analysis, for he did not distinguish between whale fisheries (requiring the same number of whales to be economically viable) hunting from small stocks of whales and those hunting from large stocks. Thus, for example, a whale fishery requiring about five hundred whales annually in order to be economically viable will in a relatively few years exhaust a stock of whales numbering only a few thousand animals. However, that same whaling operation, taking the same number of whales from a stock numbering, say, one hundred thousand will almost certainly be sustainable. In both cases the stocks have the same low reproductive rate.

A second issue to be considered is the nature of money as a generalized currency. In many nonaboriginal societies money as we know it has been in use for centuries, in contrast to some aboriginal societies where the introduction of money may be relatively recent. For example, money was likely introduced into Alaska only a little more than a century ago, whereas in Japan metallic coins were in widespread use as early as the beginning of the twelfth century.[24] The Marxian belief that the introduction of money into societies has a disruptive influence may well be true in the early years of its appearance, as noted by many anthropologists working in such situations.[25] However, such initially disruptive influences are likely to be corrected in time. Such a normalization process has been well illustrated in Japan, for example, where on the many annual occasions when gifts are exchanged, money is now the prescribed gift item. Indeed, in Japan, gifts offered to the gods must always be money (*osaisen*), and money is the required gift offered at such solemn events as funerals. However, this was not always the case; in medieval times, rice was the prescribed gift at funerals, for rice is cosmologically always pure, whereas money was considered to be equivocal—it could be either pure or impure. However, there are ways by which impure money has been rendered pure, for example, through acts of consecration that have involved associating money in various ways either with deities or with rice—which is itself a deity equivalent.[26]

In contrast to these views, money remains decidedly equivocal in Christian societies, where money is closely associated with such evils as avarice, betting, bribery, and usury. Thus, in such societies, discussions about personal money matters are often approached with unease or avoided altogether. Attitudes toward arguing over the price of a purchase, whether in a street market or in an established business, vary according to culture, being commonplace and enjoyable in some societies, and rare and utterly distasteful in certain other societies. So considerations of monetized transactions must always involve an awareness of the appropriate cultural context in which they take place.

All this suggests that evaluating commercial transactions in another society, using only the norms of a different society, may lead to dangerously mistaken conclusions. As anthropologists have observed, "*Not only is it entirely illegitimate to conflate money with capitalist relations and market values, but the extent to which either money or the capitalist market ushers in the world of moral confusion is culturally extremely variable.*"[27]

However, a failure to acknowledge cultural variability is common during IWC discussions about commerciality in Japanese community-based whaling, which since its origins in the seventeenth century has always been a commercial activity (albeit one with enduring ties to subsistence). Indeed, in medieval times, community-based whaling operations were the largest industrial enterprises in Japan, each employing several hundred workers and involving huge financial transfers each season.[28]

PROBLEMS ASSOCIATED WITH THE EXTENDED PAUSE IN COMMERCIAL WHALING

With the IWC-imposed pause in commercial whaling commencing in 1988, Japanese small-type whalers lost access to minke whales, a species that for decades had provided greater than three-quarters of the quantity and value of their annual catch. Moreover, for two of the four Japanese coastal whaling communities, the loss of minke whaling was more than an economic loss, for minke whales were central to these communities' subsistence and identity.[29] For one whaling community and the surrounding area in particular, minke whales also featured prominently in residents' social, ceremonial, symbolic, and religious life.[30] Indeed, in that particular community, a total of thirty-three culturally significant events were identified that customarily include the consumption of minke whale meat.[31]

Social scientists have long recognized the centrality of food in establishing and maintaining the social order in households and communities, as Mary Douglas has observed: "food is the medium through which a system of relationships within the family is expressed."[32] Indeed, more than half a century ago, the fundamental sociocultural importance of food was known to social scientists:

> In the literature covering the sociology of diet there is a growing tendency to consider food as an indicator of cultural values and social processes. In all societies, whether folk or urban, attitudes to food tend to become implicated in the social structure—food is both object and subject of the social structure.[33]

More recently, the importance of gastronomy (or food culture) has been recognized as extending far beyond its role in structuring social arrangements. Today, the ideological, symbolic, and moral meanings implicit in customary foodways are also well recognized:

> Food choices implied shared meanings among members of a culture . . . and may serve as a locus and indicator of social roles and ideology. . . . meals in different cultures are constructed according to sets of identifiable rules which reflect a variety of ideological, symbolic, social or other concerns in a culture.[34]

> At any given time the pervading cultural environment provides moral standards affecting every kind of resource. Food is inevitably brought into the moral perspective . . . many of the important questions about food habits are moral and social.[35]

Given the relationship between gastronomy and the established social and moral order of society, social change theorists advise against coercive efforts by the state to change food habits. For example,

Nowadays there is widespread concern about imposing alien foods or introducing even small changes too hastily. The local foodsystem needs to be understood and appreciated in the context of its relationships with other family institutions, and the interlocking of the family with the larger social institutions of the community.[36]

At its annual meetings the IWC has been made abundantly aware of the extensive scientific literature dealing with the socially damaging effects of forcibly disrupting customary foodways in the traditional whaling communities. Despite providing and explaining this information over many years, there is no indication of a willingness by the IWC majority to take account of the damaging effects the whaling ban has had upon customary foodways within certain coastal whaling communities despite the explicit requirement contained in the IWC's binding convention (the ICRW) to take the whale consumers' concerns fully into account in decision making.

Within the IWC, despite the fact that whale consumption has, inter alia, social, symbolic, moral, esthetic, ritual, religious, and nutrition/health significance, the anti-whaling majority has focused attention overwhelmingly upon one single element, namely that of monetized exchange that occurs in an economic system that clearly contains both monetized and nonmonetized exchanges. Thus, since the IWC Scientific Committee has concluded that the northwest Pacific stock of minke whales can sustain a harvest, the main objection in the IWC to nonaboriginal minke whaling remains its commercial nature. At the 1993 IWC annual meeting, it was proposed that if Japan could remove the commercial elements from its community-based whaling then some countries would agree to allow what was termed an "interim relief allocation" (i.e., a small sustainable quota). This proposal was made in order to alleviate the acknowledged distress caused by the continuing ban (zero quota) imposed on minke whaling.

Therefore, in what has subsequently proved to be a futile attempt to comply with the demands of the anti-whaling majority in the IWC, Japan introduced a management proposal in 1994 that sought to remove commerciality (i.e., monetized exchanges) from the long-practiced commercial coastal minke whale hunt.[37] During the 1994, 1995, 1996, and 1997 IWC meetings, a series of Japanese good faith proposals were debated and successively modified.[38] However, the only achievement was an unequivocal demonstration that this proposal can never succeed, because those opposed to whaling will never agree just how little the involvement with cash must be before whaling meets their idea of what constitutes a "noncommercial" activity. It might be added here that the IWC has no definition of the word *commercial* and appears to have no interest in trying to discuss or adopt such a definition.

The unsuccessful Japanese proposals have attempted to remove commerciality in minke whaling by transferring ownership of landed whales from the boat owner to the local municipality. In this proposed scheme, the municipal authorities would reimburse the boat owner for strictly determined expenses incurred in landing the whale but would ensure that no personal financial gain resulted from such whaling activity. The municipality itself would arrange for distribution of the whale in a manner that would not generate a profit, by dispensing coupons to townspeople in exchange for a levy that would merely cover the expenses incurred by the management scheme and reimbursement of production costs but would produce no additional revenue for the township.

This proposal in its successive revisions was still deemed unacceptable by the IWC majority, who objected to the failure to totally eliminate the use of money in the distribution (i.e., exacting a levy in exchange for coupons) and to the distribution of whale meat to townspeople who operate boardinghouses for tourists or who own restaurants where whale meat might be served. The objection to using whale meat in boardinghouses, inns, and restaurants reflects the lack of cultural sensitivity among opponents of whaling, for Japan had been continually urged in the IWC to replace whale hunting with whale-based tourism. In turn, Japan has continually pointed out that one of the most important reasons Japanese tourists might have to visit a whaling town is to savor the local culinary specialties.[39] In communities where minke whale is used as the key ingredient in large number of socially and culturally important food events, these events are frequently held in restaurants—a practical necessity given the numbers of invited guests and the small size of Japanese homes. In this latter regard, Japan had further proposed (without success) to control the prices of meals served in local restaurants and boardinghouses in order to eliminate profit derived from those items.[40]

All of these successive efforts to conduct constructive dialogue within the IWC (including organizing a 1997 intersessional IWC meeting to seek a resolution of the impasse on this particular issue) have resulted in a total lack of progress toward finding an interim, or ad hoc, solution to the problem.[41] This failure stems from a lack of appreciation of the similarity of the subsistence needs of small traditional whaling communities in Japan to those existing in other nations where such needs have indeed been acknowledged and where catch quotas are successfully negotiated.

One principal reason for failure to achieve a resolution in the Japanese case is the failure of delegates to appreciate that Japanese society has for centuries viewed economic matters as being fundamentally and explicitly social and moral rather than commercial in nature.[42] The similarity, indeed the degree of functional equivalence, existing between subsistence as practiced in Inuit and Japanese community-based whaling societies is striking and has been argued repeatedly at the IWC.[43] Because of this documented similarity, little objective justification exists for the IWC majority's insistence that Japan remove traditional economic institutions from its whaling culture, for to do so would seriously weaken, if not destroy, underlying social institutions and cultural beliefs and practices whose continuation is the primary purpose of seeking a reinstatement of limited minke whaling.

QUESTIONABLE ANALYSIS AND DEMEANING STEREOTYPES

There is another current reason for critically examining whale protectionists' fear of commoditization of natural resources. Today, conservationists (in contrast to those calling for total protection or preservation of resources) point to the importance of allocating economic value to resources in order to provide incentives for their sustainable use: "the fundamental flaws of the preservationist approach have prompted growing international recognition of the need to integrate economic development and environmental protection . . . [and recognition] that conservation and development are essential parts of one indispensable process."[44] This "growing international recognition" has been made explicit in reports of the UN World Commission on Environment and Development (1987), the IUCN/UNEP/UNEF World Conservation Strategy (1991), Agenda 21/UN Conference on Environment and Development (1992), and the Convention on Biological Diversity (1992).

The environmental literature now speaks both to the conservation benefits of sustainable use of living resources and to the problems and dangers to conservation posed by the activities of protectionists: for example, "the preservationist mentality often presents a serious threat to successful conservation,"[45] "sustainability is rarely achieved through external policing and prohibition,"[46] "listing species to achieve legal protection [from international trade] is not working,"47 and "the government's listing of a species . . . is often the species' obituary notice."48

Thus, what is being advocated by some of the more progressive conservation organizations and thinkers is quite the opposite of what one sees at the IWC. For example, in the Sustainable Use Program of the World Conservation Union (IUCN), prevailing thinking is that "governments and conservation organisations need to begin to promote using wild species rather than stopping such use . . . conservation organisations [need] to think of themselves as marketing agents."49

However, such thinking has yet to be adopted by the 1970-style antiwhaling campaigners, who find that the earlier rhetoric (namely, that any and all commercial use is bad) remains effective in raising money for their protectionist causes. A similar mind-set continues to dominate at the IWC, where the protection of whales from nonaboriginal use remains the primary political objective. Indeed, not content with just that level of protectionism, the actions of some countries (notably Australia and New Zealand) signal their intent to extend their preservationist actions to further restrict aboriginal whaling.

In such a deteriorating climate for rational debate, it has become necessary for a country such as the United States (which does support whaling by aboriginal peoples) to insist on maintaining an unambiguous categorical distinction between aboriginal and nonaboriginal whalers. However, attempting to define the terms *subsistence* or *commercial*—terms used by the IWC to regulate whaling—is clearly both difficult and contentious. Therefore what is preferable is something simple in the extreme, namely, a simple categorical distinction that asserts that commercial whalers sell whale meat for cash, but aboriginal subsistence whalers do not. As it is an undisputed article of faith among whale protectionists that killing whales to make money is an outrage, this simple categorization ensures that the focus of hostility at the IWC remains directed against nonaboriginal whalers.

Such ambiguity and duplicity, made possible by ignoring the known scientific evidence of a large number of characteristics common to both aboriginal and nonaboriginal community-based whalers, causes no discernible problem for the majority of IWC delegations. This majority places importance on simplifying complex issues (thus encouraging dangerous stereotyping) and ignoring whatever scientific advice might challenge the political correctness of decisions. Although consensual agreement may be obtained in some matters involving aboriginal whaling, consensus is virtually never achieved when nonaboriginal whaling is being discussed, for these issues can be decided by the numerical strength of the anti-whaling bloc vote.

Despite the categorical difference that has been constructed for political reasons to distinguish between aboriginal and nonaboriginal community-based whalers, what commonalities exist among these "smallholders"? Both are involved, inter alia, in subsistence (i.e., they sustain community traditions and identity by engaging in a range of whaling-associated activities that are highly valued in their own cultural communities); both operate family- or community-based whaling operations; both exist within an economy that involves subsistence and commercial activities; and both contribute significantly, through their whaling, to the economic viability and social vitality of their communities. It should be noted that the IWC definition of aboriginal subsistence whaling to a great extent encompasses most of what occurs in nonaboriginal community-based whaling, for it is defined as

> whaling, for purposes of local aboriginal consumption carried out by or on behalf of aboriginal, indigenous, or native peoples who share strong community, familial, social, and cultural ties related to a continuing traditional dependence on whaling and on the use of whales.
>
> Local consumption means the traditional uses of whale products by local, aboriginal, indigenous, or native communities in meeting their nutritional, subsistence, and cultural requirements. The term includes trade in items that are by-products of subsistence catches.
>
> Subsistence catches are catches of whales by aboriginal subsistence whaling operations.[50]

A fundamental difference exists between the earlier historic period of excessive industrial whaling and contemporary or recent community-based whaling. The large-scale factory ship operations of the past were sustained by a global demand for whale oil, a commodity that was used in a large number of different industries throughout the world. While whales remained plentiful, this industry thrived, but once the effects of overexploitation became apparent, chemists successfully discovered new and abundant (and therefore cheap and reliable) sources of raw materials. The result, during the 1960s, was the beginning of the end of large-scale industrial whaling in those countries where whaling existed only to provide cheap raw materials. Today, no comparable market demand or economic incentive exists to stimulate a return to such damaging industrial operations. In addition, the nations that formerly consumed and traded whale oil have now closed their borders to imports of all whale products.[51]

The failure of the IWC to institute sustainable whaling during the first decades of its existence was due to a number of factors, including an imperfect management system and the lack of political will to improve the system. The management deficiencies were only partly a result of the imperfect knowledge of whale biology and whale numbers at that time. Other flaws were inherent in the management system itself; for example, the "blue whale unit," which measured catch in units of whale oil (irrespective of species or stock of whale being taken), the absence of an international observer scheme (and hence the opportunity to cheat), and the absence of a satisfactory dispute-settlement mechanism (that allowed nations to dissociate themselves from IWC decisions if their national interests were being disadvantaged).

In those early post–World War II days, when the whaling industry was supplying a global market for oil, the IWC was in fact an oil cartel, concerned primarily with maintaining the world price of whale oil by regulating its supply. However, the situation that encouraged such excesses has not existed for several decades. Thus, during the 1970s, the industrial whaling nations left the whaling business (e.g., Australia, Canada, France, Germany, Netherlands, South Africa, the United Kingdom, and the United States), and by the 1980s, with the exception of pelagic whaling by Japan and the Soviet Union, the few whale fisheries that remained were small scale, operating out of relatively remote communities in a dozen or so countries.

Whaling continues to be of interest to those societies with a longstanding dietary, cultural, or socioeconomic dependence upon community-based whaling: these societies are classified by the IWC as either "aboriginal subsistence whalers" or "commercial whalers." Whaling continues to be practiced, or is subject to growing interest, in several nations that operate outside of IWC jurisdiction. At the present time, it has been estimated that about 97 percent of whaling in the world operates entirely outside of IWC jurisdiction,[52] a situation that is unlikely to be addressed unless current IWC attitudes and practices change, and unless the IWC begins to exhibit greater sensitivity to cultural diversity. If the tendencies apparent in the IWC persist, the minute proportion of whaling regulated by the IWC can only decrease, making that body less relevant and, arguably, even less credible in the future.

Improving the credibility of the IWC will depend upon a majority of its members recognizing that, despite the noneconomic importance of whaling in their own nations, the commission is still the operational instrument of an international *whaling* convention charged with the orderly development of *whaling*. This international convention, and therefore its original purpose, remain unchanged. The credibility of the IWC may improve if members accept that, although resource management is an imperfect science, management decisions will be more equitable and justifiable if based upon science-supported (rather than emotional and narrowly political) decisions. In his essay *Environmentalism for the Twenty-first Century*, Ronald Bailey reminds his readers that "there is no perfect solution to any [environmental] problem; tradeoffs have to be made. The good cannot be held hostage to the perfect."[53]

However, even prudent management decisions may require courage, especially in politicians from industrialized countries in which sectors of the public may noisily oppose the killing of attractive animals. However, the noise reflects the squeamishness of urban populations who have "little direct experience with animal husbandry or large wild animals [and who] have achieved a level of affluence permitting them to press for the cessation of activities that are no longer important in their own societies, but which are otherwise still essential among other peoples . . . about whom they may know very little."[54] When the squeamishness and the noise are largely based upon ignorance of the facts,[55] politicians need to exert leadership, for the issue involves not only the international regulation of whaling but also the human rights of people who have only politicians to look to when their communities, cultures, and livelihoods are threatened. It was, after all, politicians from these nations who pledged their countries to uphold the UN Covenant on Economic, Social, and Cultural Rights (1966), wherein is stated the following:

> All people may for their own ends freely dispose of their natural wealth and resources . . . based upon the principle of mutual benefit, and international law. In no case may a people be deprived of their means of subsistence.

Armed with easily obtained facts, the exercise of political leadership in the United States should not be difficult to justify; when provided with information, the majority of citizens are perfectly capable of making up their own minds. In support of that conclusion, a recent public opinion poll has indicated that 71 percent of U.S. citizens supported the hunting of nonendangered minke whales, while only 18 percent opposed such hunts.[56]

If championing human rights holds only limited appeal, then certainly the international regulation of whaling itself requires leadership at this time. Indeed, the issue may be larger than that of whaling; as one distinguished legal scholar has observed,

> The development of effective regimes in the area of international environmental law and resources depends upon trust—in other nations and in science. Japan's dilemma [in the IWC], unresolved, may reinforce the wariness of other nations to enter into international regimes. At worst, the moral of the IWC's history could be this: Will any nation that signs a global environmental or resource convention find itself ensnared in a regime that appears to discard its original premises and to pay little heed to its own scientific advisors?[57]

NOTES

1. For example, National Research Council, *Proceedings of the Conference on Common Property Resource Management* (Washington, D.C.: National Academy Press, 1986); Fikret Berkes et al., "The Benefits of the Commons," *Nature* 340 (13 July 1989): 91–93; Elinor Ostrom, *Governing the Commons: The Evolution of Institutions for Collective Action* (Cambridge: Cambridge University Press, 1990); IUCN/UNEP/ WWF, *Caring for the Earth: A Strategy for Sustainable Living* (Gland, Switzerland: IUCN, 1991); Oran R. Young et al., "Subsistence, Sustainability, and Sea Mammals: Reconstructing the International Whaling Regime, *Ocean and Coastal Management* 23 (1994): 117–27.

2. Milton M. R. Freeman, "A Commentary on Political Issues with Regard to Contemporary Whaling," *North Atlantic Studies* 2, no. 1–2 (1990): 106–16; Milton M. R. Freeman, "Science and Trans-Science in the Whaling Debate" in *Elephants and Whales: Resource for Whom?* ed. Milton M. R. Freeman and Urs P. Kreuter (Basel, Switzerland: Gordon and Breach, 1994), 143–58.

3. David E. Gaskin, *The Ecology of Whales and Dolphins* (London: Heinemann, 1982), 115. Also see M. M. Bryden and Peter Corkeron, "Intelligence," in *Whales, Dolphins, and Porpoises,* ed. Richard Harrison and M. M. Bryden (New York: Facts on File, 1988), 160–65.

4. See chapter 1, "A New Whaling Agreement and International Law," by William T. Burke, in this volume.

5. *Webster's New World Collegiate Dictionary,* 4th ed. (New York: Macmillan, 1999); *Merriam-Webster's Third New International Dictionary* (Springfield, Mass.: Merriam-Webster, 1993).

6. Milton M. R. Freeman, "The International Whaling Commission, Small-Type Whaling, and Coming to Terms with Subsistence," *Human Organization* 52, no. 3 (1993): 243–51.

7. Nicholas Peterson, "Cash, Commoditisation, and Changing Foragers," in *Cash, Commoditisation and Changing Foragers,* ed. Nicholas Peterson and Toshio Matsuyama, Senri Ethnological Studies, no. 30 (Osaka: National Museum of Ethnology, 1991), 1–16; Daniel Styles, "The Hunter-Gatherer 'Revisionist' Debate," *Anthropology Today* 8 (1992): 13–17.

8. Marshall D. Sahlins, *Stone Age Economics* (New York: Aldine, 1972).

9. George Wenzel, *Animal Rights, Human Rights: Ecology, Economy, and Ideology in the Canadian Arctic* (Toronto: University of Toronto Press, 1991), 60.

10. G. P. Donovan, ed., *Aboriginal/Subsistence Whaling (with Special Reference to the Alaska and Greenland Fisheries),* Reports of the International Whaling Commission, special issue 4 (Cambridge, England: IWC, 1982), 38–39; Finn O. Kapel and Robert Petersen, "Subsistence Hunting—the Greenland Case," in *Aboriginal/Subsistence Whaling (with Special Reference to the Alaska and Greenland Fisheries),* ed. G. P. Donovan, Reports of the International Whaling Commission, special issue 4 (Cambridge, England: IWC, 1982), 51, 69–71; Tomoya Akimichi et al., *Small-Type Coastal Whaling in Japan: Report of an International Workshop* (IWC/40/23) (Cambridge, England: IWC, 1988); Robert Petersen, *Traditional and Present Distribution Channels in Subsistence Hunting in Greenland* (TC/41/22). (Cambridge, England: IWC, 1989); Richard A. Caulfield, *Qeqertarsuarmi Arfanniarneq: Greenlandic Inuit Whaling in Qeqertarsuaq Kommune* (TC/43/AS4) (Cambridge, England: IWC, 1991); Richard A. Caulfield, *Whaling and Sustainability in Greenland* (IWC46/AS1) (Cambridge, England: IWC, 1994); Brian Moeran et al., *Similarities and Diversity in Coastal Whaling Operations: A Comparison of Small-Scale Whaling in Greenland, Iceland, Japan, and Norway* (IWC/44/SEST6) (Cambridge, England: IWC, 1992); Japan, *A Critical Evaluation of the Relationship between Cash Economies and Subsistence Activities* (IWC/44/ SEST5) (Cambridge, England: IWC 1992); Japan, *Action Plan for Japanese Community-Based Whaling (CBW): Distribution and Consumption of Whale Products* (IWC/ 46/Rev. 2) (Cambridge, England: IWC, 1994).

11. For example, Marc G. Stevenson, Andrew Madsen, and Elaine Maloney, *The Anthropology of Community-Based Whaling in Greenland: A Collection of Papers Sub-*

mitted to the International Whaling Commission, Occasional Paper 42 (Edmonton, Alberta: Canadian Circumpolar Institute, 1997).

12. Milton M. R. Freeman et al., *Inuit, Whaling, and Sustainability* (Walnut Creek, Calif.: Altamira Press, 1998), 84–88.

13. Finn Lynge, *Arctic Wars, Animal Rights, Endangered Peoples,* trans. Marianne Stenbaek (Hanover, N.H.: University Press of New England, 1992), 47; Ole Marquardt and Richard A. Caulfield, "Development of West Greenlandic Markets for Country Foods since the Eighteenth Century," *Arctic* 49, no. 2 (1996): 107–19.

14. Richard A. Caulfield, *Greenlanders, Whales, and Whaling: Sustainability and Self-Determination in the Arctic* (Hanover, N.H.: University Press of New England, 1997), 92, 103–9.

15. Steve J. Langdon, "Alaska Native Subsistence: Current Regulatory Regimes and Issues," *Alaska Native Review Commission Reports* (Anchorage: Alaska Native Review Commission, 1984), 5.

16. Ibid., 8; Rosita Worl, "The North Slope Alaskan Complex," *Senri Ethnological Studies* 4 (1980): 314.

17. Alaska Eskimo Whaling Commission (AEWC), *Tri-Annual Alaska Eskimo Whaling Captains Convention* (Barrow: AEWC, 13–15 February 1995).

18. Donovan, *Aboriginal/Subsistence Whaling,* 39.

19. For example, Colin W. Clark, "A Delayed Recruitment Model of Population Dynamics, with an Application to Baleen Whale Populations," *Journal of Mathematical Biology* 3 (1976): 381–91.

20. Jon Conrad and Trond Bjørndal, "On the Resumption of Commercial Whaling: The Case of the Minke Whale in the Northeast Atlantic," *Arctic* 46, no. 2 (1993): 164–71.

21. Lynge, *Arctic Wars;* Milton M. R. Freeman, "Issues Affecting Subsistence Security in Arctic Societies," *Arctic Anthropology* 34, no. 1 (1997): 7–17.

22. From Jonathan Parry and Maurice Bloch, eds., *Money and the Morality of Exchange,* (Cambridge: Cambridge University Press, 1989), 1–7.

23. For a perceptive analysis of the different role players at the IWC, see Masami Iwasaki-Goodman, "Social and Cultural Change in Ayukawa-hama (Ayukawa Shore Community)," (Ph.D. diss., University of Alberta–Edmonton, 1994).

24. Edwin O. Reischauer and Albert M. Craig, *Japan: Tradition and Transformation* (Boston: Houghton Mifflin, 1978), 63.

25. For example, Paul Bohannan, "The Impact of Money on an African Subsistence Economy," *Journal of Economic History* 19, no. 4 (1959): 491–503.

26. Emiko Ohnuki-Tierney, *Rice as Self: Japanese Identities through Time* (Princeton, N.J.: Princeton University Press, 1993), 71–73.

27. Parry and Bloch, *Morality of Exchange,* 18; see also Caulfield, *Greenlanders,* 112–14, 142–44, 168–69.

28. Arne Kalland, *The Spread of Whaling Culture in Japan* (TC/41/STW3) (Cambridge, England: IWC, 1989); T. Takahashi et al., *Japanese Whaling Culture: Continuities and Diversities* (IWC/41/STW2) (Cambridge, England: IWC, 1989).

29. Michael Ashkenazi and Jeanne Jacob, *Summary of Whalemeat as a Component of the Changing Japanese Diet in Hokkaido* (IWC/44/SEST2) (Cambridge, England: IWC, 1992); Lenore Manderson and Haruko Akatsu, "Whale Meat in the Diet of Ayukawa Villagers," *Ecology of Food and Nutrition* 30 (1993): 207–20.

30. Akimichi et al., *Small-Type Coastal Whaling;* Theodore C. Bestor, *Socio-Economic Implications of a Zero Catch Limit on Distribution Channels and Related Activities in Hokkaido and Miyagi Prefectures* (IWC/41/SE1) (Cambridge, England: IWC, 1989); Stephen R. Braund, Milton M. R. Freeman, and Masami Iwasaki, *Contemporary Sociocultural Characteristics of Japanese Small-Type Whaling* (IWC/41/STW1) (Cambridge: IWC, 1989); Lenore Manderson and Helen Hardacre, "Small-Type Coastal Whaling in Ayukawa." Draft Report of Research: December 1988–January 1989 (IWC/41/SE3) (Cambridge, England: IWC, 1989); Masami Iwasaki-Goodman and Milton M. R. Freeman, *Social and Cultural Significance of Whaling in Contemporary Japan: A Case Study of Small-Type Coastal Whaling* (IWC/46/SEST3) (Cambridge, England: IWC, 1994).

31. S. R. Braund et al., *Quantification of Local Need for Whale Meat for the Ayukawa-Based Minke Whale Fishery* (TC/42/SEST8) (Cambridge, England: IWC, 1990).

32. Mary Douglas, *In the Active Voice* (London: Routledge and Kegan Paul, 1982), 86.

33. J. W. Bennett, "Food and Social Status in a Rural Society," *American Sociological Review* 8 (1943): 561.

34. Michael Ashkenazi, "From Tachi Soba to Naori: Cultural Implications of the Japanese Meal," *Social Science Information* 30, no. 2 (1991): 287–88.

35. Mary Douglas, ed. *Food in the Social Order: Studies of Food and Festivities in Three American Communities* (New York: Russell Sage Foundation, 1984), 11.

36. Ibid., 12.

37. Japan, *Action Plan for Japanese Community-Based Whaling (CBW): Distribution and Consumption of Whale Products* (IWC/46/Rev. 2) (Cambridge, England: IWC, 1994).

38. Japan, *Action Plan for Japanese Community-Based Whaling* (IWC/47/SEST1) (Cambridge, England: IWC, 1995); Japan, *Action Plan for Japanese Community-Based Whaling (CWB): (Revised)* (IWC/47/46) (Cambridge, England: IWC, 1995).

39. Jo Stewart-Smith, *In the Shadow of Fujisan: Japan and Its Wildlife* (Harmondsworth, England: Viking/Penguin, 1987), 175; Akimichi et al., *Small-Type Coastal Whaling,* 100–103; Japan, *Report to the Working Group on Socio-Economic Implications of a Zero Catch Limit* (IWC/41/21) (Cambridge, England: IWC, 1989); Nelson H. H. Graburn, "Whaling Towns and Tourism: Possibilities for Development of Tourism in the Former [*sic*] Whaling Towns: Taiji, Wada, Ayukawa," Report submitted by the government of Japan, *Proceedings of the 42d Meeting of the IWC,* 1990.

40. Japan, *Action Plan for Japanese Community-Based Whaling (CWB): (Revised)* (IWC/47/46) (Cambridge, England: IWC, 1995).

41. Arne Kalland, "Some Reflections after the Sendai 'Workshop'" *Isana* 16 (1997): 11–15.

42. Bestor, "*Zero Catch Limit.*

43. Akimichi et al., *Small-Type Coastal Whaling,* 79–84; Japan, *A Critical Evaluation of the Relationship between Cash Economies and Subsistence Activities* (IWC/44/SEST5) (Cambridge, England: IWC 1992); Japan, "*Commercial*" vs. "*Subsistence,*" "*Aboriginal*" vs. "*Nonaboriginal,*" and the Concept of Sustainable Development in the Context of Japanese Coastal Fisheries Management (IWC/46/SEST1) (Cambridge, England: IWC, 1994); Moeran et al., *Coastal Whaling Operations.*

44. World Wide Fund for Nature (WWF), *Integrating Economic Development with Conservation* (Gland, Switzerland: WWF, 1993), 5.

45. WWF, *Economic Development,* 2.

46. K. Makombe, ed. *Sharing the Land, Wildlife, People, and Development in Africa,* IUCN/ROSA Environmental Issues Series, no. 1 (Harare, Zimbabwe: IUCN, 1994), 8.

47. Stephen R. Edwards, "Conserving Biodiversity: Resources for Our Future," in *The True State of the Planet,* ed. Ronald Bailey (New York: Free Press, 1995), 254.

48. R. Littell, *Endangered and Other Protected Species: Federal Laws and Regulations* (Washington, D.C.: Bureau of National Affairs, 1992), 30.

49. Edwards, "Conserving Biodiversity," 257.

50. Report of the Special Working Group of the Technical Committee Concerning Management Principles and Development of Guidelines about Whaling for Subsistence by Aborigines, IWC Document IWC/40/5.

51. Freeman, "Science and Trans-science," 147–49.

52. Editorial, *INWR Digest* 12, no. 1 (1997).

53. Ronald Bailey, "Prologue: Environmentalism for the Twenty-first Century," in *The True State of the Planet,* ed. Ronald Bailey (New York: Free Press, 1995), 5.

54. James R. McGoodwin, *Culturally-Based Conflicts in the Use of Living Resources and Suggestions for Resolving or Mitigating Such Conflicts,* International Con-

ference of Sustainable Contribution of Fisheries to Food Security, Kyoto, 4–9 December 1995 (UN Report KC/FI/95/TECH/9), 22.

55. Milton M. R. Freeman and S. R. Kellert, "International Attitudes to Whales, Whaling, and the Use of Whale Products: A Six-Country Survey," in *Elephants and Whales: Resources for Whom?* ed. Milton M. R. Freeman and Urs P. Kreuter (Basel, Switzerland: Gordon and Breach, 1994).

56. Responsive Management, *American Attitudes to Whales and Whaling* (Harrisburg, Va.: Responsive Management, 1997).

57. Christopher D. Stone, "Legal and Moral Issues in the Taking of Minke Whales," in *Report: International Legal Workshop, Sixth Annual Whaling Symposium,* ed. Robert L. Friedheim (Tokyo: Institute of Cetacean Research, 1996), xix.

5 / Food Security, Food Hegemony, and Charismatic Animals

RUSSEL LAWRENCE BARSH

The recent refusal of the International Whaling Commission (IWC) to recognize "artisanal" whalers, together with tighter IWC restrictions on "indigenous" whaling, may threaten the food supply and cultural survival of peoples who have relied on whales for food for millennia. The right to decide what is acceptable as food has been captured by a small number of wealthy Western states and Western NGOs through an exercise of their political power. This is not new. It is a continuation of a strategy of cultural hegemony that powerful states have pursued since the Roman Empire. This chapter traces the history of Western hegemonic control of food production and food consumption as a context for criticizing the present international whaling regime and its recent application to the Makah tribe of the Pacific Northwest United States. While peoples such as the Makah have relatively little physical power, they have a growing body of international law on their side, in particular new conventional law on the rights of indigenous peoples and similar small, traditional communities. There are modest grounds for optimism that these peoples will continue to gain standing and influence in international forums.

FOOD SECURITY AS A RIGHT

Managing the harvesting of marine mammals and other wildlife often confounds logically independent objectives: the perpetuation of individual species and the promotion of species that humans find desirable because they are tasty, attractive, or otherwise useful or interesting. In a world affected by human activity for millennia, it is impossible to determine scientifically the appropriate sizes of wildlife populations. A minimum number of bowhead whales consistent with reproductive continuity can be established empirically, but how many more whales is enough—and how many is too much? These policy questions are necessarily answered from a human normative perspective. Humans do not agree on the reasons why whales are good and therefore cannot agree how many whales there should be.

The public discourse over the acceptability of killing and eating marine mammals has been couched in moral and legal terms, but the real issue is relative power. Privileged societies have acquired the power to determine what the world eats and to impose their own symbolic and aesthetic food taboos on others. Placed in proper historical context, contemporary efforts to abolish whaling and sealing are exposed as the flip side of Western European domination of world food supplies.[1]

Over the past five centuries, the European food system (itself a mixture of indigenous and exotic species) has expanded at the expense of other agricultural and nonagricultural food systems, and often at the sacrifice of entire ecosystems. Peripheral societies have had to change what they produce and what they eat, at a loss of diversity and nutritional quality in their diet.[2] Aggressive mass marketing of nutritionally poor European foods such as soft drinks, and aggressive mass marketing of the idea that particular animal species such as seals are too cute to kill, are arguably simply the latest twists in the development of European food hegemony.

Moral indignation, rather than conservation, has driven the anti-harvesting campaigns of the past twenty-five years.[3] Indigenous peoples have responded by complaining about cultural imperialism, pointing out that they have normative systems of their own to ensure sustainability and to curb unnecessary suffering. They have also turned increasingly to international conventional law relating to food security, cultural rights, popular self-determination, and the protection of the lifeways of indigenous, tribal, and traditional societies.

The international effort to rebuild Europe in the wake of the Second World War relied on food warehousing and redistribution programs, which combined the resources of private relief organizations, national aid initiatives, and the United Nations agencies that became UNICEF and the United Nations High Commissioner for Refugees (UNHCR). Public and private investment in the European aid project was largely motivated by fears for the future stability and security of the region. The fact that fascism had thrived on the economic hardships and inequalities of the 1920s and 1930s bolstered the linking of food, peace, and individual human dignity, and the idea of "food security" as a fundamental objective of international cooperation.[4]

The cornerstone of modern international human rights law, adopted by the UN General Assembly in 1948, is the Universal Declaration of Human Rights. Article 25 states that "Everyone has the right to a standard of living adequate for the health and well-being of himself [*sic*] and his family, including food." It took member states another eighteen years to agree on the wording of two complementary, legally binding conventions based upon the Universal Declaration. Preoccupied with the Cold War, Western powers concluded that the whole notion of socioeconomic rights was Communist-inspired. To achieve consensus, the General Assembly split the principles of the Universal Declaration into separate treaties on "civil and political" rights, and "economic, social and cultural" rights. The United States and Japan still insist that the International Covenant on Economic, Social, and Cultural Rights (ICESCR) contains "social goals" rather than legally enforceable rights.

"Recognizing the fundamental right of everyone to be free from hunger," Article 11 of the ICESCR commits states parties "to improve methods of production, conservation and distribution of food" with a view to achieving "the most efficient development and utilization of natural resources" and "to ensure an equitable distribution of world food supplies in relation to need."[5]

The two international human rights covenants also contain a common Article 1, which refers to the right of all peoples to self-determination and adds that "In no case may a people be deprived of its own means of subsistence." There has yet to be an explicit test of this principle in international law, although it has been mentioned in discussions of U.S. efforts to isolate Cuba and Nicaragua, and more recently the food-for-oil deal in Iraq. Its original target was the abuse of economic sanctions such as embargoes, but it was drafted at a time when the food industry was far less globalized and was concentrated in far fewer Western corporations than it is today.

The international consensus on the strategic importance of food security has been implemented mainly through bilateral and multilateral food-aid initiatives, including those administered by UNICEF, UNHCR, and the World Food Program, as well as technical assistance programs aimed at boosting food production and making it more sustainable. As far as a human "right" to food is concerned, however, it remains, like all international human rights norms, little more than a talking point in intergovernmental political positioning at the United Nations. In some circumstances it may be useful tactically to accuse a government of depriving people of food security, but such accusations are not made with consistency, and their impact depends chiefly on differences in power between the accusers and the accused.

Human rights may gain greater influence as a result of greater world economic interdependence. The globalization of trade and communication is two-edged. It will shift power from governments to the private sector, but also from governments to consumers. At first, the shift in power will take place within the wealthy industrialized states of the north and will favor their wealthier and better-informed consumers. Consumers in the industrializing countries of the south will gain power as their numbers and incomes grow. But will consumer demand be significantly more sensitive to human rights issues than government bureaucracies? Or more consistent in their responses?

FOOD HEGEMONY AND IMPERIALISM

Control of food supplies has long been an instrument of empire and was indeed a crucial factor in the emergence of the earliest states in western Asia. Strong, hierarchical organizations were needed to build irrigation canals and central granaries; the institutional capacity to boost cereal production through the control of water supplies was also the capacity to store surplus grain and to distribute the surplus for political ends. Food supply and power were inextricably linked.

The importance of state institutions for storing and transporting food increased as cities grew in size. Whether it was acquired through trade or as tribute, more food had to be transported over greater distances. Managing periodic famines was a significant challenge to the power and legitimacy of early Mediterranean states.[6] The development and defense of extensive trading networks, the annexation of wealthy agrarian societies, and the settlement of colonists in arable but underdeveloped regions were key aims of Greek and Roman foreign policy.[7] Rome not only imported an increasing share of its food but implanted its food production system in Iberia, Britain, and the lower Rhine. Demanding gourmets preferred wine from Bordeaux and knew, as Pliny observed, that the best radishes grew in the cool climate of northern Europe.[8] Roman wealth dominated production, and Roman tastes redefined what Europeans chose to plant and eat.

The second wave of European expansion, more than a millennium later, had the same effects on a far larger scale. European empires acquired maize, potatoes, chili, squash, peppers, cocoa, tomatoes, sweet potatoes, and turkeys from the Americas in the sixteenth century, while exporting the farming of European cereals, peas, cattle, sheep, and pigs to the Americas and Australasia.[9] Many introduced foods were reexported through trade and colonization, such as chilis, potatoes, and tomatoes, which spread quickly throughout south Asia. Global trade and imperialism diversified the cuisines of all countries. In terms of total world food production, however, relatively catholic European tastes have prevailed. Wheat, maize, potatoes, and three native European cereals (barley, oats, and rye) today account for 51 percent of total world food crop production.[10]

The globalization of particular food crops has guided national tastes as well as considerations of economic productivity. Maize became popular in southern Europe but was never a favorite in London, Paris, or Berlin. Maize therefore remained a secondary crop in European colonies, cultivated for the sustenance of colonized peoples rather than the subsistence of laborers in Europe. Wheat gradually surpassed other grains in the European food system because of the growth of industrialized bread making and growing European preferences for manufactured bread over porridges and other cereal products.[11] The British Empire controlled roughly half of the world's trade by 1900, and British demand for bread, beef, and sweets had a decisive impact on world food production.

The earliest stage of English urbanization and industrialization depended on taxes and produce exacted from Scotland and Ireland, but it was soon necessary to shift the burden of food production abroad to allied European kingdoms and recently colonized American territories.[12] After 1750, British territorial acquisitions were used to enlarge and standardize the British food system, based on European cereals and tropical spices.[13] Crops were transplanted from Europe to British colonies, and from one colony to the next. By the 1780s, Great Britain had become a net importer of grain.[14] The United States became Britain's main supplier, followed by the British colonies of Australia, Canada, and India. Britain continued to expand its food system by promoting the cultivation of wheat, even in corners of the empire such as Canada's prairies, where conditions were marginal.[15] Food security was integral to French and German imperial policy as well, leading to the emergence of several competing intercontinental food supply systems in the nineteenth century.[16]

Imports of staple cereals tell only part of the story. Of equal, if not greater, social and economic significance were imports of luxury foods that became a subjective measure of well-being and progress, especially among the working classes. Pepper and sugar shipped from the Levant were the principal spices of Tudor Britain, but by the eighteenth century sugar had become the most valuable British import for domestic consumption, surpassing tea, coffee, wine, and grains.[17] Sugar retained its lead until the 1860s. Establishing an internal source of sugar supply was a major motivation for British colonization of the Caribbean in the seventeenth century, just as internalizing supplies of tea, coffee, cocoa, and copra were motives for subsequent colonization of the western Pacific and East Africa.[18]

Empires not only selectively appropriate other peoples' foods, then, but also decide what foods other peoples may eat and assign foods to status hierarchies.[19] Categorization and ranking of foods may be a direct or indirect consequence of economic policy—for example, when certain foods are promoted as cash crops for export or (like infant formula) promoted as part of public health campaigns. Ranking of foods can also be a demonstration effect: the poor try to eat what they see eaten by the rich. In much of the English-speaking world, the trend toward eating mass-produced "aerated" bread manufactured with bleached wheat flour was primarily a matter of gaining status (and sacrificing nutrition) a century ago.[20]

NEW FORMS OF FOOD HEGEMONY

As long as agricultural output continued to be largely a function of cultivated area and human labor, the sine qua non of food hegemony was the physical acquisition of arable territory. Existing indigenous farming systems could be taxed or appropriated by colonists. Nonagricultural peoples were relocated to free their land for colonization, or else they were managed as a supply of inexpensive farm labor.[21] Indigenous food systems were replaced if they did not satisfy Europeans' tastes.[22] Staple crops were aggressively planted in some colonies simply to feed laborers in other colonies. American grain fed the slaves working on Caribbean sugar plantations in the 1700s,[23] just as East African millet and maize fed the black miners of South Africa a century later.

Until the twentieth century, agricultural profitability was chiefly a function of economies of scale. Entrepreneurs assembled vast plantations, assisted by public policies that ensured them access to cheap land and inexpensive or slave labor. By the 1920s agricultural machinery, manufactured fertilizers and pesticides, and hybrid seed varieties had begun to play a decisive role in agricultural competitiveness. The key to food hegemony was no longer the ownership of more land but the control of technology, which in turn was a function of the ability to mobilize financial capital. Europeans jettisoned most of their overseas territories by the 1960s but continued the use of financial capital to control global food supplies and to market Western food to newly independent countries in the south.

Mass media and scientific advertising,[24] also developed in the 1920s, required considerable outlays of financial capital as well and further strengthened European domination of the food industry. It is true that distinctive packaging and promotional advertising of food brands was well established by 1800.[25] However, the global reach of advertising, and its power to standardize beliefs about "what *is* food" and which foods are associated with high status, was not fully realized until the 1950s.[26] The promotion of infant formula as a substitute for breast-feeding, a food industry initiative that has had severe adverse health impacts in many countries, is one example.[27] The growth of the beef industry is another example of effective promotion and marketing.[28] Mutton and pork were more popular in Europe and North America until the 1870s, when American and Australian beef was promoted as superior in cost and quality. After 1945, beef was successfully promoted globally.

The preeminence of financial capital, technology, and advertising as competitive factors has facilitated the concentration of world food production and distribution in a small number of European and North American transnational corporations. North American transnationals controlled 70 percent of the world's wheat trade in 1979, helped by price supports, transportation subsidies, and state-financed warehousing of surplus stocks.[29] They exert their power by controlling crop varieties, sales of machinery, and the marketing of grain rather than by operating farms. Control of the world beef industry is also concentrated in technology, marketing, and promotional advertising rather than in ranching.[30]

The globalization and concentration of the food industry over the past fifty years has resulted in the paradox of developing countries exporting high-quality foods while facing malnutrition.[31] Vast inequalities of military and economic power mean that rich nations eat from the tables of poor nations, while depriving poor nations of their own locally controlled sources of nutritionally and ecologically appropriate food. Coffee is one of the most striking examples. Nearly all coffee is produced in developing countries, led by Brazil, Colombia, Cote d'Ivoire, India, and Indonesia, but two-thirds of the world's output is consumed in the United States, Japan, and Western Europe.[32]

MARKETING FOOD, MARKETING CULTURE

Empires not only control what is produced but define the *aesthetics* of food production, including the ideal of the landscape. The European idealization of the healthy, bucolic countryside was popularized by the Roman poets Virgil and Horace and has remained virtually unchanged. In this ideal, the boundary between Nature and Culture is blurred in a way that preserves the superiority and dominance of Man. The farm is cheerful and bright with no hint of backbreaking labor or the screams of slaughtered sheep and cattle. The forest beyond the farm is a quiet and shaded retreat, immune from the noise and haste of the city. Of course, the forests of Roman Tuscany and Bordeaux were a reconstructed countryside, cleared of dangerous animals and crossed by maintained roads. In fact, it was an *extension* of the farm: a vast, purely ornamental garden. Culture and Nature were perfectly complementary and conveniently sanitized.

Only a powerful empire could afford such an ideal, which depends on the ability to distance or externalize the processes of production. Horace's Rome had long since logged out the oak forests of Latium and replaced them with olive groves, parks, and those pleasant hobby farms to which Horace and his contemporaries retreated from city life. Real forests were far beyond the Rhine frontier, where few Romans ventured. Between the managed countryside and the wild frontier was the Empire's vast production zone of farms, mines, and barbarian markets for amber, leather, and furs.[33] As Roman power grew and its culture and language spread, the Roman garden expanded and both the production zone and the wild forest frontier retreated farther north.

Bred to Roman sensitivities, the Norman invaders of Britain were distressed to find many of the indigenous inhabitants living, hunting, and woodcutting in oak forests, and what was even worse, using fire to clear away the trees for gardens. Until they were absorbed culturally and biologically by their subjects, the Normans pursued a campaign of removing local people from the forests, tearing down their houses, and replanting trees in clearings, all to create a treed garden for Norman recreation.[34] Deer and swans were among the species that the common folk were not ordinarily permitted to eat, much less hunt.[35] Social rank was associated with the right to appropriate such species and to deny their use as food. Local memories of this policy became entwined in the Robin Hood myth.

The metropolitan powers that currently lead the global crusade against hunting and trapping of wild mammals built their fortunes on the selective *destruction* of wildlife species such as beaver, bison, and plumed birds for industrial and commercial purposes. Today they are the world's largest consumers of meat and (in the form of paper products) trees. The killing of mammals and destruction of forests are acceptable so long as they involve species that the metropoles classify as "food" and forests that are a safe distance from home.

Western opponents of eating wildlife frequently rationalize their own food habits by arguing that raising meat on the farm is more efficient than hunting, produces a more nutritious product, and can even be more humane.[36] This argument permits the citizens of wealthy countries to feel morally superior, without substantially changing their own way of life and food preferences. The deaths of billions of chickens, and tens of millions of cattle, are assigned a much lower moral value than the demise of a single whale or seal.[37] Moreover, Western societies allocate moral worth based upon other species' visual similarity to humans (especially to human infants with large eyes and chubby bodies), to culturally sanctioned pets and toys (soft, furry, cuddly, responsive to humans' attention), and to popular media imagery (e.g., deer, bears). Insects need not apply, nor species still regarded as threats or pests.

Like our Roman forebears, we are still trying to turn Nature into a recreational garden, free from weeds and danger. North Americans greet the news of a bear attack in a national park with fear and demand the destruction of the bear. Bears are supposed to be cuddly; if not, they must be destroyed. Similarly, forests are expected to remain the same forever, like cultivated gardens, rather than burning down periodically to renew their biodiversity.[38] Industrialized countries that have already cut down most of their forests insist that developing nations bear the responsibility and expense of stabilizing those forests that remain. Canada clear-cuts more primary forest acreage than Brazil, but Canadian environmental advocacy is chiefly focused on Amazonia.

Many of the same factors that have favored the rise of transnational corporations, such as fast, inexpensive worldwide communication and travel, have also facilitated the growth of a new category of international actors: nongovernmental organizations (NGOs). Indeed, contemporary NGOs often differ little structurally from transnational corporations. They have a world headquarters that usually is in a major European or North American city, is staffed largely by Western professionals, and has a network of national or regional-level subsidiaries. They frequently raise capital through sales of goods and services rather than donations or subscriptions, and they employ state-of-the-art promotional advertising to compete for public legitimacy and financial support. NGOs struggle for brand-name recognition (the Greenpeace name, the WWF panda logo) and market share. In principle, NGOs are a countervailing force to states and transnational corporations. In practice they may be managed by technocrats whose interests are the same as those of corporate technocrats: personal status and advancement, which depend on the visibility and financial success of the organization.

International NGOs market ideas rather than goods. Since they are geographically, financially, socially, and culturally rooted in Europe, their advocacy tends to reflect European values, priorities, and perspectives. Their shareholders and customers are essentially the same people: mainly Europeans, who have strong beliefs they wish to promote globally. Hence NGOs often use their power as lobbies for European tastes and values, much as transnational corporations use their power to promote tastes for European products. Internationally, NGOs are valuable advocates for human rights and environmental awareness, but the largest NGOs exist, and have significant global power, because Europe retains dominant global power.

FOOD HEGEMONY AND SUSTAINABILITY

Most of the increase in food production over the past century has actually been due to expansion of cultivated areas rather than greater productivity.[39] In North America and south Asia, the area under cultivation doubled, while in Argentina and Australia, the area under cultivation grew by more than twenty times. Very little arable land remains undeveloped in any region of the world. Shifting to European farming methods has meanwhile led to widespread soil degradation. Tillage and cattle grazing on the prairies of Australia, East Africa, and North America resulted in catastrophic erosion and soil loss, such as the American dust bowl of the 1920s. Replacing com-

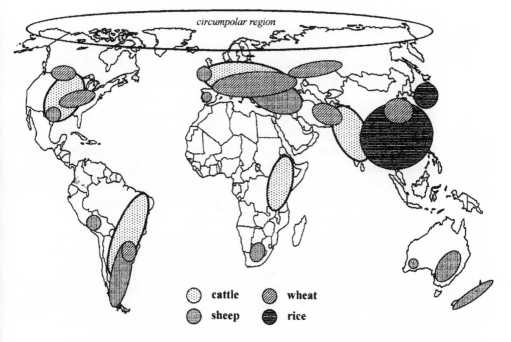

FIG. 5.1 Principal World Agricultural Zones
(over 90 percent of world production)
Map by Barsh based on Burger 1994.

plex intercropped gardens with vast monocultural (single-crop) fields has increased the spread of crop diseases and insects.[40]

For centuries, Europeans compensated for the unsustainability of their agrosystems by the simple expedient of relocating. Europeans cleared most of their own vast forests for farmland between 1100 and 1600 and then began to clear and settle the coastal areas of the Americas and Australia.[41] When the soils of eastern North America played out, European farmers moved inland, just as South American farmers today are moving inland to the Andes and Amazonia.[42] Continued expansion has meant farming increasingly marginal land. Greater mechanical and chemical inputs can be used to offset losses from soil depletion, invasive plants, and insects, but these practices increase costs and put farmers into debt. Fertilizers accelerate soil oxidation, while pesticides and herbicides kill natural pollinators and helpful predators. Mechanical cultivation is also inefficient in terms of its use of energy. Premechanical European farming produced more total energy than it used; contemporary U.S. farms use twice as much energy as they produce.[43]

The expansion of European agrosystems affected not only the soils and ecology of cultivated lands but the ecology of uncultivated areas that were important sources of food, medicine, and fuel and were reservoirs of biodiversity. The deliberate introduction of extremely prolific, competitive species, such as European forage grasses, bees, and pigs, quickly displaced native species far beyond the fences of farms and ranches. European colonization and trade also transported invasive species abroad.[44] Globalization of the food industry, together with the growing trade in ornamental species and exotic pets, pose a continuing threat to ecosystems and global biodiversity.[45] Large-scale production of cash crops for export undermines indigenous systems of food distribution and markets and increases the risk of famine.[46]

Why, then, replace small but sustainable indigenous food systems with extensive monocultural, mechanical agrosystems that are unsustainable in the long term? Such a course is neither ecologically sound nor economically rational. It nevertheless serves the hegemonic interests of European states, their food industries, and their exporters of chemicals, machinery, and agricultural technology.

FOOD HEGEMONY AND HUMAN HEALTH

Food hegemony not only perpetuates historical inequalities of power and wealth but also creates and perpetuates inequalities in human health, disability, and life expectancy. European food systems have not always been successful at replacing the nutritional content of indigenous food systems, which tended to be far more diversified and better adapted to the ecological possibilities of challenging environments such as arid prairies or tundra.[47] Human populations differ in their ability to metabolize certain foods, moreover, and in their minimum requirements for micronutrients.[48] Local cuisines originally reflected differences in population genetics and the risks posed by local parasites and disease.[49]

In the Canadian Arctic, for example, replacing marine mammal meat with store-bought starches, sugars, and saturated vegetable oils—a change hastened by the 1982 European Community ban on seal fur imports—has been implicated in severely elevated rates of diabetes and cardiovascular disease.[50] Inuit and other hunting peoples tend to metabolize animal fats efficiently for energy, while carbohydrates, which are scarce in the traditional food system, are converted to storage fat. By contrast, Europeans and Asians, who have subsisted on cereals for millennia, tend to burn carbohydrates immediately and store fats. For Inuit, a carbohydrate-rich European diet has led to widespread obesity and hypertension. Similar patterns have been reported for indigenous peoples in Australia and Siberia.[51]

Nutritional factors are also suspected in elevated rates of early childhood respiratory infections among indigenous peoples, as well as developmental delays in cognition and perception, which can have lasting consequences for physical mobility and productivity. Among North American Indians and Inuit, declining infant and adult mortality rates are offset by very high rates of chronic illness and disability.[52] Changing diets have not only compromised health but undermined the sustainability of indigenous societies in physical and cultural terms. People afflicted with chronic poor health and physical and cognitive disabilities have difficulty maintaining productive economies or dynamic cultures. Most of their collective energy must be diverted to caregiving and grieving. By keeping subject populations chronically malnourished, food hegemony perpetuates colonial domination.

Food hegemony has also played a role in perpetuating *class* power within countries. Inexpensive food imported from the Americas made it possible for eighteenth-century British workers to increase their per capita food consumption.[53] Imported refined sugar and wheat flour were less nutritious than working-class Britons' customary fare of porridge and milk products, however.[54] Laboring people could afford to buy more calories, but the calories they could afford came with fewer micronutrients. Canning and refrigeration made imported meats more affordable in the 1850s, but by then most workers were living in cities, paying higher rents, and seeing the real value of their wages shrink.[55] While food produced by British colonies made the growth of British industrial power possible, its impact on the health of generations of British workers was modest, even negative.[56]

CONSERVATION AND FOOD HEGEMONY

Indigenous peoples face a triple threat.[57] They are surrounded by settlers whose activities can adversely affect the size and structures of wildlife populations, the quality of water, and the productivity of soils throughout the region. Settlers clear away native vegetation, wittingly and unwittingly introduce invasive plant and animal species, and selectively overharvest native species for food or export.[58] As indigenous food systems disintegrate, traders, missionaries, and aid agencies promote the substitution of inexpensive, less nutritious staple crops and imported foods. The final blow comes from restrictions on whatever customary harvesting activities survive, putatively for the purposes of conservation and animal welfare.

This process is reaching its final stage in Canada's north, where Cree and Dene peoples are embroiled in a controversy over fur trapping. European settlement displaced the Cree northward more than 150 years ago, not only by occupying arable lands but also by intensively harvesting wildlife. Near the rapidly moving frontier, Cree learned to survive by acting as intermediaries in an evolving fur trade, replacing the large wild ungulates in their customary diet with flour and lard. The fur trade accelerated the collapse of beaver and other fur-bearing populations and the further northward movement of the frontier. The Cree were eventually forced into Dene territory, leading to war, a treaty of coexistence, and overutilization of the wildlife in Dene territory, which today must be shared with growing mining and pulpwood industries.[59] The methods of customary Cree and Dene trapping have meanwhile become the targets of global criticism.

A similar process is under way in Amazonia, where indigenous peoples are trapped between the ecological impacts of new settlements and the demands of conservationists. No longer able to cope with wildlife scarcity by periodically relocating their villages, indigenous peoples have compensated with greater firepower.[60] This shift has subjected them to criticism for waste and greed and has served as the justification for increasingly restrictive hunting regulations.

The adverse impact of conservation measures on indigenous peoples is amplified by the nature of those measures—usually quotas based on relatively rigid population targets for each managed species. Ecosystems are inherently volatile. Natural populations can remain in long-term dynamic balance while fluctuating greatly in their absolute numbers and relative proportions. Some of these fluctuations can be forecast with modest accuracy in timing and direction if not in total scale, such as the biennial cycles in migrations of Pacific pink salmon. Most, however, are too chaotic. Population targets are a distortion of natural ecological processes, then, and management interventions based on relatively rigid targets can actually increase the amplitude of the unavoidable fluctuations.[61]

In a traditional wildlife-harvesting society, ecological sustainability and biodiversity tend to be maintained by two means. Harvesting is spread over a large variety of species and fluctuates with the seasonal and annual availability of each species.[62] As a measure of added food security, each community maintains a web of exchange relations and reciprocal resource-access privileges with other groups, often at considerable distances.[63]

Quotas can prevent communities from shifting their hunting patterns flexibly in response to the changing abundance of species and from sharing windfall harvests with one another. This restriction limits their ability to utilize local food stocks fully when they are abundant and may force them to *overutilize* local food stocks in times of scarcity. Quotas therefore reduce the average quantity of wildlife in the diet, *without* improving conservation.

The broad eclecticism of wildlife harvesters' customary diets has important nutritional as well as ecological consequences. Eating from a wide variety of animal and plant species may be necessary to ensure an adequate supply of all necessary micronutrients. A narrow diet places severe pressure on a very small number of species with the result of modifying the structure of the ecosystem and is likely to be deficient in some of the micronutrients necessary for maintaining human health. Hence conservation measures that force indigenous peoples to focus their food-gathering efforts on a smaller menu can have adverse effects on biodiversity and on human health, even if people continue to consume adequate food in caloric terms.

THE WHALING CONTROVERSY

The worldwide recovery of cetacean populations has set the stage for a new and particularly contentious struggle over the right to food. If opponents of marine mammal hunting cannot pretend that stocks are threatened, they are left with the argument that some animals should not be eaten on moral or aesthetic grounds. There are two layers to this argument—first, that the cetaceans (at least, but perhaps not the seals, sea lions, or *sirenia*) are intelligent, possess feelings, and therefore are moral subjects; and second, that the killing and eating of marine mammals is a waste because these creatures have many other important social and aesthetic values and because there are sufficient alternative sources of human nutrition.

The argument from intelligence and feelings, which builds on the works of philosopher Peter Singer, is a slippery slope that implies an I.Q. test for edibility. It cannot be applied without a standard and method of measuring animal consciousness, and a faith that there is a logically defensible threshold on the consciousness curve between amoeba and human. The privileged intellectual status of humans is untenable in the face of a growing body of research on consciousness and cultural diversity among other animals, however.[64] What is emerging is a dynamic view of ecosystems as arenas in which individuals and populations (rather than "species") engage in strategic cooperative as well as competitive games, with changing rules of engagement. Indigenous peoples never doubted the consciousness and willfulness of all creatures, nor their participation in a "cosmic food web" in which *everyone* is destined to become food.[65] While Europeans worry about *not* eating any conscious beings, indigenous peoples have understood that *everything* they eat is conscious and accordingly believe that every creature should be eaten with a measure of respect and regret.

It has often been observed that pigs seem at least as intelligent as dogs, and there have been some experimental results to support such claims.[66] Considerable intelligence and sensitivity has also been attributed to horses, not only by the mad Roman emperor Nero and by Jonathan Swift in *Gulliver's Travels* but also by generations of horse owners and horse breeders. People have had considerably more concrete experience on which to base assessments of porcine and equine consciousness than cetacean intelligence: we have lived together for several millennia in very close quarters. Yet far more Europeans would be appalled by a platter of *muktuk* (whale skin and attached fatty tissue) than by a pork chop or a *boucherie chevaline*. The explanation is cultural. Europeans are familiar with eating hogs and horses; it is part of European culture. Cetaceans are distant and mysterious, and they were never an important part of the European food system.

WHALING AND NUTRITION

One argument that can be made in defense of whaling is that human habitation of some regions is otherwise not feasible. As indicated by figure 5.1, most of the world's cultivated grains and domestic meats are produced in a half dozen tropical and temperate regions. Some coastal regions are not associated with these key conventional food-production zones: the cordilleran eastern Pacific rim, much of Saharan and sub-Saharan Africa, and the boreal forests and tundras of circumpolar North America, Europe, and Asia. The Pacific cordillera and Africa are borderline cases. While their current agricultural output is dwarfed by other regions, and their soils and climates are not ideal for large-scale commercial agriculture, they would be much better candidates for irrigation, mechanization, customized hybrid crop varieties, and other production-boosting technologies than the circumpolar zone. Subarctic agriculture generally requires hothouse systems to boost temperatures for most or all of each growing season.[67] Indoor agriculture involves a substantial investment of materials and fuel, often far exceeding the high cost of importing food.

From an agricultural perspective, the circumpolar region is the best candidate for harvesting wildlife. Primary production is slow and very thinly distributed in the Arctic, and people have used mammals as bioaccumulators of meager and widely dispersed food sources. Terrestrial herbivores such as caribou accumulate the thinly distributed energy of mosses and lichens, while cetaceans and seals accumulate the energy of plankton, fish, and shellfish. Highly migratory cetaceans such as the larger whales are essential to an Arctic hunting strategy because they accumulate food energy over very large areas. Arctic whale hunters are therefore indirectly utilizing photosynthesis in temperate and tropical as well as polar oceans.

Circumpolar communities could conceivably harvest more fish as an alternative to hunting marine mammals—a kind of macrobiotic Arctic diet that targets lower rungs of the food chain. The potential food yield of an expanded Arctic fishery would be modest in the absence of calories imported by migrating cetaceans, however, and it would have a relatively large impact on resident nonhuman predators such as seabirds, seals, and polar bears. Primary production in ice-covered seas is very low, mainly restricted to the ice edge,[68] thus an expanded human fishery would put considerable strain on the entire polar ecosystem. Fishing in the circumpolar ocean is also extremely hazardous. Targeting the largest, fattiest, calorie-rich food packages such as marine mammals makes sense if harvesters must minimize the time they are exposed to the elements. Harvesting more fish might be possible using large steel ships and the most sophisticated fish-finding and harvesting technology, but the expense would be considerable. Who would pay this expense: circumpolar peoples, presumably with revenues derived from exploiting their petroleum and other minerals (and the environmental costs), or the southern peoples who object to whaling and sealing?

It is not entirely clear that the net global environmental impact of Arctic hothouse agriculture or high-technology fishing would be any less than the impact of harvesting marine mammals. Maintaining Arctic food production by other means requires greater energy consumption, chiefly the consumption of petroleum-derived fuels. The macrobiotic strategy, fishing, also deprives marine mammals of more of their prey, albeit in principle this should have somewhat less impact on total marine mammal biomass than whaling and sealing. In economic as well as environmental terms, then, hunting marine mammals appears to be the least costly food-production option for circumpolar human settlements. Two further objections have been raised by opponents of whaling and sealing, however. The circumpolar human population is small, they maintain, and could be supported by importing food, or else relocated to more hospitable climes.

There are considerable technical barriers to transporting food to Arctic communities year round. Roads are few, costly to maintain, and seasonal. Shipping is restricted to two or three ice-free months; air transport is extremely expensive and can be unreliable in winter. The high cost and seasonal nature of shipping mean that imported foods are rarely fresh and must be sold at very high markups. Northern peoples must content themselves with canned fruits and vegetables (very low in vitamins) and with the cheapest canned and processed meats (consisting mainly of fat and carbohydrate fillers). They still pay more for such poor-quality imported food than southerners pay for high-quality fresh and frozen food. As a result, circumpolar residents generally cannot afford an adequate diet unless they continue to hunt.[69]

Even under the best economic conditions, furthermore, imported foods could not replace the micronutrient composition of marine mammal meat. There are several important differences between terrestrial meats, such as beef, and marine mammal meat. Compared to broiled tenderloin steak (much better quality beef than northerners ordinarily can afford), raw whale meat has a somewhat higher ratio of fat to protein, a *much* higher ratio of sodium to potassium (3.5 compared to 0.1), and a *lower* ratio of calcium to magnesium and phosphorus. Whale has a superabundance of vitamin A: about 1,860 milligrams per 100 grams, or nearly twenty times as much as beef. Whale is also saturated with iron in comparison to beef and is a fair source of vitamin C, which frozen or processed beef lacks.

The circumpolar nutrition gap is too large to bridge with vitamin supplements. Vitamins A and C can be replaced at modest cost, but the mineral composition of a marine mammal diet poses special problems. Western diets of beef and grains are rich in potassium and phosphorus, while diets of marine mammals and fish are rich in sodium, magnesium, and iron. The ability of the human body to moderate electrolytes and iron is limited, so differences in dietary intakes have significant implications for the chemistry and tensile strength of bones, heart functioning and blood pressure, and the management of B vitamins.[70] There is growing evidence that circumpolar peoples had adapted metabolically to the high proportions of sodium, magnesium, and iron in their customary diets by slowing the rates at which these minerals are absorbed through the gastrointestinal tract. After switching from "bush food" to store-bought diets, Arctic communities tend to develop an unusual appetite for table salt and bananas, boosting their intake of sodium and phosphorus to compensate for the relatively lower levels of these two nutrients in domestic meats. Recent surveys show that Canadian Inuit are suffering more from heart disease, renal and metabolic disease, iron-deficiency anemia, and osteoporosis since the 1950s, all of which is consistent with changing proportions of dietary minerals.

The Inuit experience suggests not only the futility of importing food as a substitute for whaling and sealing but the serious health risks of relocating circumpolar peoples. Relocation might bring people closer to cheaper sources of fresh domestic meats, fruits, and vegetables, but southern diets that are nutritionally sound for Europeans may not be compatible with northern-adapted metabolisms. Dietary change in situ or as a result of relocation is a gamble with human health.

Nutritional arguments may be weakened somewhat by the growing threat to Arctic ecosystems from industrial pollution. Marine mammals are powerful accumulators of fat-soluble toxic compounds such as PCBs, as well as toxic metals such as mercury, cadmium, and lead, although the potential adverse health effects of these contaminants have yet to manifest themselves in Inuit. "The health risks associated with exposure to contaminants remain though they may be outweighed by the benefits of continued consumption," a Canadian government study concluded.[71] "This poses a confusing public, moral and political dilemma." Arctic peoples may no longer be free to eat the most abundant, accessible, and energy-rich foods in their environment. In principle, it is the responsibility of industrialized nations to reduce the contamination in accordance with Articles 194 and 207 of the United Nations Convention on the Law of the Sea (UNCLOS) and Article 10(c) of the UN Convention on Biological Diversity. As the Tokyo meeting on climate change demonstrated, however, the implementation of environmental treaties remains a hostage to wealthy countries' real priorities: domestic economic growth and employment.

WHALING AND CULTURE

The argument of Arctic nutritional necessity, if successful, may reasonably extend to some nonindigenous societies such as Icelanders and Faeroese, whose homelands also have minimal agricultural capability and whose coastal waters lie along polar cetacean migration routes. Nutritional necessity is less clearly applicable to historical whale-eating peoples who have more satisfactory alternative means of securing adequate food. Food security and self-reliance remain issues for Japan and other crowded Asian nations, which have become net food-importers, but they can afford to eat relatively well without whaling.

A cultural argument for whaling has been advanced successfully by the Makah of Washington State's Pacific coast. The Makah represent an intermediate case between Japan and Inuit. They are citizens of a wealthy, food-exporting country, therefore whaling is not primarily a nutritional issue. The Makah Nation occupies a small tract of forests and beaches that has little economic potential in itself but borders on highly productive ocean fisheries. Aboriginal Makah subsisted on a wide variety of marine mammals, fish, and shellfish. Halibut and whale made the largest contributions to the Makah diet and were also traded widely with neighboring peoples. Whaling was exceedingly hazardous and was undertaken only by members of a secret ceremonial society that placed extraordinary spiritual and physical demands on its initiates. The right to throw the harpoon was acquired by a combination of inheritance and a strenuous four-year ritual spirit quest associated with real or symbolic cannibalism.[72] Harpooners were wealthy and potentially dangerous men, with unsurpassed knowledge of the sea and a personal relationship with the spirits of whales.

Whaling unquestionably played a central role in Makah culture until it was discouraged by U.S. authorities nearly a century ago.[73] By the 1940s, only a few men survived who had gone whaling. Nonetheless, the *idea* persisted that Makah were a strong and special people because they had once been whalers. I encountered that kind of Makah exceptionalism in visits to the community in the 1970s and 1980s when I was researching traditional fishing methods and marine-ecological knowledge. Makah are a small, ethnically Nootkan enclave within the Salish linguistic and cultural region, and they continue to be somewhat isolated, socially as well as geographically. It is not difficult to understand the appeal of whaling traditions to young Makah in the 1990s, as their self-esteem was being slowly eroded by severe unemployment and the onslaught of television and beach resorts.

In 1997, the I W C authorized the Makah to take up to five gray whales per year for subsistence purposes. After concerns over violent protests resulted in several delays and global press coverage, a much-publicized hunt on 17

May 1999 landed a single thirty-foot whale amid a veritable armada of hostile spectators. The meat was distributed like a communion meal throughout the Makah community. One journalist observed that the protracted confrontation with opponents of the whale hunt had strengthened the solidarity and sense of distinctive cultural identity of the Makah.[74] Defiant young Makah wore tee shirts emblazoned with "Pass the Harpoon!" and planned to continue whaling in the fall. Whale hunting has given Makah a sense of unique identity and collective significance.

Even if the Makah attain their IWC quota, they will be able distribute fewer than one hundred pounds of whale meat per year to each member of the community—not enough to make a large nutritional impact. The significance of resuming the hunt is fundamentally *political.* The Makah are exercising the power to be different in a modest but highly visible respect: they eat what they choose to eat. For long-oppressed people, a few pounds of whale meat can signify great freedom.

IMPLICATIONS FOR THE IWC

In 1979, Alaskan Inuit won a "subsistence" exemption from the IWC ban on bowhead whaling by arguing that whaling is not only nutritionally necessary for their physical survival but necessary for the survival of Inuit culture. The IWC has also granted subsistence quotas to Russian Yupiit [Inuit] and Chukchi, Greenlandic Inuit, and Caribbean Bequian whalers.[75]

Arguments based on culture can be problematic because cultures change. The Makah no longer maintain a secret society of whalers that practices ceremonies of purification and spiritual communion with the whales, for example. Institutions that once ensured the technical and ethical integrity of Makah whaling had all disintegrated by the 1940s.[76] No one is suggesting that modern-day harpooners ritually bathe and fast for four years and secure powerful guardians in the spirit world before embarking on their first hunt. Makah "traditions" are being invented to manage contemporary whaling within a contemporary culture that is dominated by television, fast food, mass marketing, and the American dollar. While whaling is still important to Makah identity, it serves different functions and is embedded in a new framework of institutions.

Food is a potent cultural symbol, and food choice is an exercise of individual self-identification and community self-determination. A plausible case can be made for a right to *culturally appropriate* food, distinct from the right to *nutritionally adequate* food, which has been recognized expressly in the International Covenant of Economic, Social, and Cultural Rights. At least one authoritative analysis of the right to food concluded that "cultural acceptability" is as much a part of food adequacy as nutritional quality and food safety.[77] A people should be able to eat what they prefer, or what identifies them with their specific heritage, as long as they utilize their living resources sustainably.

Two problems remain. A people's customary or culturally preferred food may have rights of its own. Eating humans, which served important symbolic functions in many cultures until quite recently, would probably remain unacceptable as an exercise of the right to adequate food. Although no United Nations treaty recognizes nonhuman species as possessing rights, efforts have been under way in New Zealand (thus far unsuccessfully) to extend "human rights" to apes,[78] and national laws in many countries already recognize domestic animals such as dogs and horses as bearers of a right to humane treatment. A conservative Hindu government in India might argue that no human has a right to eat beef. International agencies may eventually have to balance the right to culturally acceptable food with the rights of food species, and if that occurs, the balance will be biased in favor of the food values of the most powerful states. Americans' hamburgers are safe for as long as the United States is a superpower.

A second problem involves the acceptability of *commercial* whaling to satisfy the rights of an unrelated community. Suppose that the IWC were someday to find that certain Japanese prefectures are entitled to continue their historical tradition of eating whales. Could the Makah sell them whale meat? The right to culturally appropriate food is about the meaning of *eating*, after all, rather than the means of production. It should therefore be irrelevant if Japanese consumers simply find it convenient to pay someone else to do the hunting.[79] Suppose, similarly, that the Makah someday find themselves without any harvestable whales, perhaps because of El Niño or the effects of global warming on Pacific currents. Should the Makah be entitled to purchase a few gray whales from Russian or Japanese whalers, in view of the fact that Makah simply lack the capital and technology to pursue whales at long distances?

Culture is the argument that sells, both to the public and to the governments that control IWC decision making. Sentimentalism is as much responsible for the decision to authorize Makah whaling as it is for the decision to prohibit *other* societies from resuming whaling. Policy continues to be driven by cultural imperialism and stereotypes. The ɪwc paid attention when the Makah based their argument on cultural authenticity rather than economic well-being. Having emphasized their cultural authenticity, the Makah were bound to disappoint Western environmentalists who expected them to be unspoiled, vegetarians, or, as University of British Columbia Native education professor Michael Marker has put it, "Yodas of the Deep." The principal argument raised against Makah whaling, as tensions grew, was that they were not culturally authentic *enough*. The anti-whaling campaign accused Makah leaders of cultural fraud.

Nutritionally adequate, culturally appropriate food is arguably a right already recognized by the international community. When peoples such as the Makah assert this right, however, they get caught between hegemonic, Eurocentric notions of what is appropriate to eat as food, and hegemonic, Eurocentric notions of what is culturally authentic.

FORCES OF CHANGE

Indigenous peoples are gaining growing recognition of their right and capacity to manage the ecosystems on which they have traditionally relied.[80] At the same time, measures taken to conserve some of the species of particular interest to indigenous peoples, such as baleen whales, sea otters, and fur seals, have succeeded to a degree that sustainable harvesting is possible. As John Knauss has argued, international management bodies such as the ɪwc could lose credibility as politically neutral, scientific decision makers if they continue to forbid harvesting in the face of these trends.[81]

The United Nations Conference on Environment and Development concluded unanimously in its final declaration that "Indigenous people and their communities, and other local communities, have a vital role in environmental management and development because of their knowledge and traditional practices. States should recognize and duly support their identity, culture and interests and enable their effective participation in the achievement of sustainable development."[82] *Agenda 21*, the global program of action adopted by the conference, called upon all countries, "in full partnership with indigenous people," to take steps "to empower indigenous people," including recognition of the contributions of traditional wildlife harvesting practices to sustainable development. Article 10(c) of the 1992 Convention on Biological Diversity, ratified by a majority of member states, directs governments to "protect and encourage customary use of biological resources in accordance with traditional cultural practices that are compatible with conservation or sustainable use requirements."[83]

New environmental instruments such as the biodiversity convention do not create rights that can be enforced by individuals or their communities. Nevertheless, they outline management principles that may become talking points when governments meet to discuss environmental cooperation and conditions attached to conservation-related financial and technical assistance. International bodies such as the UN Commission on Sustainable Development and the Conference of Parties, which oversees the Convention on Biological Diversity, are opening their doors to representatives of indigenous peoples and similar grassroots communities, moreover, enabling them to lobby government wildlife and conservation ministers more directly and effectively.

The collapse of the Soviet Union has meanwhile opened the way for a new global emphasis on promoting investment and trade and the growth of a powerful new international legal regime centered in the World Trade Organization (WTO). Governments preoccupied with their economic bottom line are more responsive to the decisions of the WTO than the resolutions of multilateral human rights or environmental bodies. With its authority to adjudicate the consistency of national regulations with the principles of international fair trade, and to legitimize countervailing economic sanctions by aggrieved states, the WTO potentially has effective police power. This is cause for optimism and pessimism.

The WTO could be persuaded that any restriction on the harvesting of wildlife for sale must be supported by sound scientific evidence of its conservational necessity. In the 1997 *Import Prohibitions of Certain Shrimp and Shrimp Products* case,[84] a WTO dispute-resolution panel ruled that the United States cannot apply its sea turtle bycatch regulations to imports of Asian shrimp without adequate scientific evidence that Asian sea turtle bycatches are unsustainably high. A similar argument might be applied to European Community bans on furs and marine mammal products such as sealskin or to a future IWC attempt to bar the sale of meat from nonendangered cetacean species. The WTO does not have jurisdiction over wildlife harvesting unless the products are destined for international trade, but whaling communities could provoke test cases and indirectly challenge the IWC by attempting to export token quantities of cetacean products.

Pessimistically, it must be admitted that the WTO best serves the interests of the largest trading countries, such as the United States, which can afford to carry out WTO-sanctioned countervailing measures—and to ignore WTO rulings with which it disagrees. The history of the softwood lumber and winter wheat trade disputes between Canada and the United States has been characterized by repeated Canadian concessions, for example, even after Canada wins favorable trade decisions. Hence, while the new global trade regime is ruthless in requiring scientific proof of a species' endangerment before condoning trade restrictions, it may have little real impact on European or American whaling policy.

States will eventually need to reconcile four international legal regimes that touch on the issue of harvesting marine mammals: trade; human rights; law of the sea (UNCLOS); and conservation, including the Convention on Biological Diversity, the Convention on Trade in Endangered Species (CITES),[85] and the International Convention for the Regulation of Whaling. Must conservation treaties be interpreted in a manner consistent with the human right to adequate and culturally acceptable food? Does fair trade trump conservation and human rights? Multilateral bodies may give inconsistent directives, and governments will find themselves caught between the economic power of the WTO and the public-relations power of human rights and environmental NGOs.

There is little effective coordination of regulatory bodies at the international level. Procedures do not exist to set priorities or promote harmonization, and this situation is likely to continue for some time. Governments prefer the flexibility of having contradictory principles available for use in different situations as their policies and interests dictate. It is rarely the case that states agree on the application of international law to a particular dispute—but it is inevitable that every plausible legal principle will be advanced by *some* government that has an interest in the dispute. Thus the right to adequate food, and the sustainable-use rights of "indigenous and local communities," will likely haunt the agendas of international meetings for many years, undermining the credibility of protectionist arguments based on aesthetics and European values rather than empirical ecology.

The growing complexity and inconsistencies of international legal regimes have contributed to two countervailing trends: regionalization and industry self-regulation. Shared values and interests can be considerably greater at the regional level. Fishery regulation has been pursued at a regional level, for example, in the North Pacific and North Atlantic, while a global treaty on migratory fish stocks remains elusive. Trade liberalization is also proceeding more quickly at the regional level, especially within Europe,

North America, and southeast Asia. There are even regional human rights treaties in Europe, Africa, and the Americas; the Organization of American States is preparing a regional instrument on the rights of indigenous peoples in response to the growing dissatisfaction of Latin American governments with United Nations efforts in this field.

While governments are responding to the ambiguities of international governance by forming regional bodies, industry is responding by promoting certification and other self-regulation schemes that minimize the role of international institutions. Labeling of tuna as "dolphin-safe" is a familiar example. More recently, attempts by some Western governments and environmental organizations to promote a global convention on forest conservation foundered on industry proposals that forest products simply be certified by industry bodies. Consumer goods could be labeled as products of "sustainable forestry," and governments could exclude uncertified products from their markets without violating trade treaties.

If the trends toward regionalization and certification continue, the implications for indigenous peoples who hunt and eat "charismatic mammals" are clear: form regional management organizations such as the Alaska Eskimo Whaling Commission and World Council of Whalers, set regional standards for sustainable management, establish credibility for good science and consistent enforcement, and use distinctive labeling on any products that may enter international commerce. Retaliatory trade actions by governments opposed to hunting could then be challenged as violations of trade law. Barring industry-certified products from the European Community would be clearly exposed as an expression of cultural hegemony, for example, rather than as a legitimate measure for the protection of world ecosystems.

NOTES

1. I have used the terms *West* and *Western European* to refer to the largest global industrial powers, which are chiefly Western European or of Western European origins.

2. Gretel H. Pelto and Pertti J. Pelto, "Diet and Delocalization: Dietary Changes since 1750," *Journal of Interdisciplinary History* 14, no. 2 (1983): 507–28; Ellen Messer, "Anthropological Perspectives on Diet," *Annual Review of Anthropology* 13 (1984): 205–49.

3. James M. Jasper and Dorothy Nelkin, *The Animal Rights Crusade: The Growth of a Moral Protest* (New York: Free Press, 1992), 11–25; George Wenzel, *Animal Rights, Human Rights: Ecology, Economy, and Ideology in the Canadian Arctic* (Toronto: University of Toronto Press, 1991), 156–61; Alan Herscovici, *Second Nature: The Animal-Rights Controversy* (Montreal: CBC Enterprises, 1985), 53–55.

4. Gary Kline, "Food as a Human Right," *Journal of Third World Studies* 10 (1993): 92–107; Thomas J. Marchione, "The Right to Food in the Post–Cold War Era," *Food Policy* 21, no. 1 (1996): 83–102.

5. It could be argued that a right to adequate food is also *implied* in Article 6 of the International Covenant on Civil and Political Rights ("Every human being has the inherent right to life"), originally aimed at curbing executions and political killings.

6. Peter Garnsey, *Famine and Food Supply in the Greco-Roman World: Response to Risk and Crisis* (Cambridge: Cambridge University Press, 1988).

7. Barry Cunliffe, *Greeks, Romans, and Barbarians: Spheres of Interaction* (London: Guild Publishing, 1988).

8. C. Anne Wilson, *Food and Drink in Britain from the Stone Age to the Nineteenth Century* (London: Constable, 1973), 194.

9. Stephen V. Boyden, *Biohistory: The Interplay between Human Society and the Biosphere* (Paris: UNESCO and Parthenon, 1992), 123–25.

10. Noel D. Vietmeyer, "Lesser-Known Plants of Potential Use in Agriculture and Forestry," *Science* 232 (13 June 1986): 1379–84.

11. E. J. T. Collins, "Why Wheat? Choice of Food Grains in the Nineteenth and Twentieth Centuries," *Journal of European Economic History* 22, no. 1 (1993): 7–38.

12. Brindley Thomas, "Feeding England during the Industrial Revolution: A View from the Celtic Fringe," *Agricultural History* 56, no. 1 (1982): 328–42.

13. I continue to use the term *spice* in the sense that it was used in Norman and Tudor Britain to refer to costly imported nonstaple foods, beverages, and flavorings. Wilson, *Food and Drink,* 282.

14. B. R. Mitchell, *Abstract of British Historical Statistics* (Cambridge: Cambridge University Press, 1962).

15. Michael Clayton Wilson and Ineke J. Dijks, "Land of No Quarter: The Palliser Triangle as an Environmental-Cultural Pump," in *The Palliser Triangle: A Region in Space and Time,* ed. R. W. Barendregt, M. C. Wilson, and F. J. Jankunis (Lethbridge: Lethbridge University Press, 1993).

16. Peter J. Hugill, "Structural Changes in the Core Regions of the World Economy, 1830–1945," *Journal of Historical Geography* 14, no. 12 (1988): 111–27.

17. Mitchell, *British Historical Statistics.*

18. J. H. Galloway, *The Sugar Cane Industry: An Historical Geography from Its Origins to 1914* (Cambridge: Cambridge University Press, 1989); Richard B. Sheridan, *Sugar and Slavery: An Economic History of the British West Indies, 1623–1775* (Barbados: Caribbean Universities Press, 1974); H. C. Brookfield, *Colonialism, Development, and Independence: The Case of the Melanesian Islands* (Cambridge: Cambridge University Press, 1972).

19. Sidney W. Mintz, "Zur Beziehung zwischen Ernahrung und Macht," *Jahrbuch für Wirtschaftgeschichte* 1 (1994): 61–72.

20. At the same time, Imperial cities from Rome to London have celebrated their power by embracing the exotic, aping the arts, and eating the foods of their victims. Second-century Romans decorated with Greek vases, drank Rhenish wine, and dabbled in worshiping Isis, an Egyptian goddess. The Victorians adored Turkish fashions and Chinese porcelain, while they consumed Turkish coffee and Chinese tea.

21. Ann McGrath, *"Born in the Cattle": Aborigines in Cattle Country* (London: Allen and Unwin, 1987); David J. Webster, "The Political Economy of Food Production and Nutrition in Southern Africa in Historical Perspective," *Journal of Modern African Studies* 24, no. 3 (1986): 447–63.

22. Claire C. Robertson, "Black, White, and Red All Over: Beans, Women, and Agricultural Imperialism in Twentieth-Century Kenya," *Agricultural History* 71, no. 3 (1997): 259–99.

23. Sheridan, *Sugar and Slavery.*

24. By "scientific" advertising, I mean advertising strategies based on quantitative psychological studies of consumer response.

25. Olive R. Jones, "Commercial Foods, 1740–1820," *Historical Archaeology* 27, no. 2 (1993): 25–41.

26. Leslie Sklair, *Sociology of the Global System* (Baltimore: Johns Hopkins University Press, 1991); United Nations Center on Transnational Corporations, *Transnational Corporations in World Development: Trends and Prospects,* 1988, UN Doc. ST/CTC/89, 222–25.

27. David Kowalewski, "Transnational Corporations and the Third World's Right to Eat: The Caribbean," *Human Rights Quarterly* 3, no. 4 (1981): 45–64; S. Prakash Sethi and James E. Post, "Public Consequences of Private Action: The Marketing of Infant Formula in Less Developed Countries," *California Management Review* 21, no. 4 (1979): 35–48.

28. Marvin Harris, *Good to Eat: Riddles of Food and Culture* (New York: Simon and Schuster, 1985); Jeremy Rifkin, *Beyond Beef: The Rise and Fall of the Cattle Culture* (New York: Dutton, 1992).

29. Raymond F. Hopkins and Donald J. Puchala, *Global Food Interdependence: Challenge to American Foreign Policy* (New York: Columbia University Press, 1980).

30. Rifkin, *Beyond Beef.*

31. Ben Fine, Michael Heasman, and Judith Wright, *Consumption in the Age of Affluence: The World of Food* (London: Routledge, Chapman and Hall, 1996); Michael Redclift and David Goodman, *Refashioning Nature: Food, Ecology, and Culture* (New York: Routledge, 1991); Susan George, *Ill Fares the Land: Essays on Food, Hunger, and Power* (Washington, D.C.: Institute for Policy Studies, 1984); John W. Warnock, *The Politics of Hunger: The Global Food System* (London: Methuen, 1987).

32. J. de Graaf, *The Economics of Coffee* (Wanginingen, Netherlands: Pudoc, 1983).

33. Cunliffe, *Greeks, Romans.*

34. H. E. Hallam, ed. *The Agrarian History of England and Wales,* vol. 2 (Cambridge: Cambridge University Press, 1988), 130–31; Leonard Cantor, *The Changing English Countryside, 1400–1700* (London: Routledge and Kegan Paul, 1987), 96–100.

35. Wilson, *Food and Drink,* 81, 120.

36. Jim Mason, "Brave New Farm," in *In Defence of Animals,* ed. Peter Singer (Oxford: Basil Blackwell, 1985), 89–107.

37. Even Peter Singer (*Animal Liberation: A New Ethics for Our Treatment of Animals* [New York: Avon, 1975]), the most influential advocate of animal rights, draws a moral boundary between animals and plants, which may not be logically defensible, and clearly contradicts indigenous peoples' idea of universal interrelatedness of life forms.

38. Russel L. Barsh, "Chronic Health Effects of Dispossession and Dietary Change: Lessons from North American Hunter-Gatherers," *Medical Anthropology* 18, no. 1 (1998): 1–27.

39. Boyden, *Biohistory,* 130.

40. Carsten Thies and Teja Tscharntke, "Landscape Structure and Biological Control in Agroecosystems," *Science* 285 (1999): 893–95.

41. Michael Williams, *Americans and Their Forests: A Historical Geography* (Cambridge: Cambridge University Press, 1989).

42. Jane L. Collins, "Smallholder Settlement of Tropical South America: The Social Causes of Ecological Disaster," *Human Organization* 45, no. 1 (1986): 1–10.

43. Boyden, *Biohistory,* 133.

44. Alfred W. Crosby, *Ecological Imperialism: The Biological Expansion of Europe, 900–1900* (Cambridge: Cambridge University Press, 1986).

45. Jeffrey A. McNeely, "The Great Reshuffling: How Alien Species Help Feed the Global Economy," in *Proceedings: Norway/UN Conference on Alien Species, The Trondheim Conference on Biodiversity, 1–5 July 1996*, ed. Odd Terje Sandlund, Peter Johan Shei, and Aslaug Viken (Trondheim: Norwegian Institute for Nature Research, 1996), 53–58.

46. Hazel R. Barrett, *The Marketing of Foodstuffs in the Gambia, 1400–1980: A Geographical Analysis* (Aldershot, England: Avebury, 1988), 36–92.

47. Russel L. Barsh, "The Substitution of Cattle for Bison on the Great Plains," in *The Struggle for the Land*, ed. Paul A. Olson (Lincoln: University of Nebraska Press, 1990).

48. Barsh, "Chronic Health Effects."

49. Peter Farb and George Armelagos, *Consuming Passions: The Anthropology of Eating* (Boston: Houghton Mifflin, 1980); Harris, *Good to Eat.*

50. Barsh, "Chronic Health Effects"; Wenzel, *Animal Rights,* 120.

51. E. K. Rousham and M. Gracey, "Persistent Growth Faltering among Aboriginal Infants and Young Children in Northwest Australia: A Retrospective Study from 1969–1993," *Acta Paediatrica* 86, no. 1 (1997): 46–50; William R. Leonard, Anne Keenleyside, and Evgueni Ivakine, "Recent Fertility and Mortality Trends among Aboriginal and Non-aboriginal Populations of Central Siberia," *Human Biology* 69, no. 3 (1997): 403–17.

52. Russel L. Barsh, "Canada's Aboriginal Peoples: Social Integration or Disintegration?" *Canadian Journal of Native Studies* 14, no. 1 (1994): 1–46.

53. Michael Nelson, "Social-Class Trends in British Diet, 1860–1980," in *Food, Diet, and Economic Change Past and Present*, ed. Catherine Geissler and Derek J. Oddy (Leicester: Leicester University Press, 1993), 101–20; Gareth Shaw, "Changes in Consumer Demand and Food Supply in Nineteenth-Century British Cities," *Journal of Historical Geography* 11, no. 3 (1985): 280–96.

54. Carole Shammas, "The Eighteenth-Century English Diet and Economic Change," *Explorations in Economic History* 21, no. 3 (1984): 254–69.

55. Forrest Capie and Richard Perren, "The British Market for Meat, 1850–1914," *Agricultural History* 54, no. 4 (1980): 502–15.

56. Compare Alfredo Castillero-Calvo, "Niveles de vida y cambios de dieta a fines del período colonial en América." *Anuario de Estudios Americanos* 44 (1987): 427–76.

57. I use *indigenous* in this context to refer to groups that continue to derive their subsistence from customary hunting, fishing, pastoral, or horticultural activities in their own territories. Barsh, "Chronic Health Effects."

58. Kent H. Redford and John G. Robinson, "The Game of Choice: Patterns of Indian and Colonist Hunting in the Neotropics," *American Anthropologist* 89, no. 4 (1987): 650–67.

59. René M. J. Lamothe, *It Was Only a Treaty: A Historical View of Treaty 11 According to the Dene of the Mackenzie Valley* (Ottawa: Royal Commission on Aboriginal Peoples, 1993); Robert Jarvenpa and Hetty-Jo Brumbach, "Occupational Status, Ethnicity, and Ecology: Metis Cree Adaptations in a Canadian Trading Frontier," *Human Ecology* 13, no. 3 (1985): 309–29; Shepard Krech III, "The Influence of Disease and the Fur Trade on Arctic Drainage Lowlands Dene, 1800–1850," *Journal of Anthropological Research* 39, no. 2 (1983): 123–46.

60. R. Brian Ferguson, "Game Wars? Ecology and Conflict in Amazonia," *Journal of Anthropological Research* 45, no. 2 (1989): 179–206; James A. Yost and Patricia M. Kelley, "Shotguns, Blowguns, and Spears: The Analysis of Technological Efficiency," in *Adaptive Responses of Native Amazonians,* ed. Raymond B. Hames and William T. Vickers (New York: Academic Press, 1983).

61. Paul D. Spencer, "Optimal Harvesting of Fish Populations with Nonlinear Rates of Predation and Autocorrelated Environmental Variability," *Canadian Journal of Fisheries and Aquatic Sciences* 54, no. 1 (1997): 59–74.

62. Russel L. Barsh, "Indigenous Peoples' Role in Achieving Sustainability," in *Green Globe Yearbook 1992,* (Oxford: Oxford University Press, 1992), 25–34; Philippe Descola, *In the Society of Nature: A Native Ecology in Amazonia* (Cambridge: Cambridge University Press, 1994).

63. Wayne P. Suttles, *Coast Salish Essays* (Seattle and Vancouver: University of Washington Press and Talonbooks, 1987); Rogier Van den Brink, Daniel W. Bromley, and Jean-Paul Chavas, "The Economics of Cain and Abel: Agro-Pastoral Property Rights in the Sahel," *Journal of Development Studies* 31, no. 3 (1995): 373–99; Jacqueline S. Solway, "Foragers, Genuine or Spurious? Situating the Kalahari San in History," *Current Anthropology* 31, no. 2 (1990): 109–22.

64. Marian Stamp Dawkins, *Through Our Eyes Only? The Search for Animal Consciousness* (Oxford: Oxford University Press, 1998); Daniel J. Povinelli, "What Chimpanzees (Might) Know about the Mind," in *Chimpanzee Cultures,* ed. Richard W. Wrangham et al. (Cambridge: Harvard University Press, 1994); A. Whiten et al. "Culture in Chimpanzees," *Nature* 399 (1999): 682–85.

65. Kaj Arhem, "The Cosmic Food Web: Human-Nature Relatedness in the Northwest Amazon," in *Nature and Society: Anthropological Perspectives,* ed. Phillipe Descola and Gisli Palsson (London: Routledge, 1996); Russel L. Barsh, "Taking Indigenous Science Seriously," in *Biodiversity in Canada: Ecology, Ideas, and Action,* ed.

Stephen Bocking (Toronto: Broadview Press, 1999); Gerardo Reichel-Dolmantoff, "Cosmology as Ecological Analysis: A View from the Rain Forest," *Man* 11, no. 3 (1976): 307–18.

66. Dawkins, *Animal Consciousness*, 156.

67. Anna Burger, *The Agriculture of the World* (Aldershot, England: Avebury, 1994), 206–7.

68. W. O. Smith Jr. and D. M. Nelson, "Importance of Ice Edge Phytoplankton Production in the Southern Ocean," *Bioscience* 36 (1986): 251–57.

69. Royal Commission on Seals and the Sealing Industry, *Seals and Sealing in Canada: Report of the Royal Commission* (Ottawa: Minister of Supply and Services, 1986), 2: 248–53.

70. Barsh, "Chronic Health Effects."

71. J. Jensen, K. Adare, and R. Shearer, eds., *Canadian Arctic Contaminants Assessment Report* (Ottawa: Department of Indian Affairs and Northern Development, 1997), 366; compare Health Canada, *State of Knowledge Report on Environmental Contaminants and Human Health in the Great Lakes Basin* (Ottawa: Minister of Public Works and Government Services, 1997), 158–59.

72. Elizabeth Colson, *The Makah Indians: A Study of an Indian Tribe in Modern American Society* (Manchester: Manchester University Press, 1953), 33–34, 175, 251–52; Ruth Kirk, *Hunters of the Whale* (New York: Harcourt, Brace and Jovanovich, 1974); Erna Gunter, "Reminiscences of a Whaler's Wife," *Pacific Northwest Quarterly* 3, no. 1 (1942): 65–69.

73. Colson, *Makah Indians*, 5.

74. M. L. Lyke, "After Whale Hunt, Makah Celebrate New Sense of Unity; Annual Festival Marked by Fewer Protests," *Seattle Post-Intelligencer*, 30 August 1999.

75. Richard Kirk Eichstaedt, "'Save the Whales' v. 'Save the Makah': The Makah and the Struggle for Native Whaling," *Animal Law* 4, no. 1 (1998): 145–71; Milton M. R. Freeman, "The International Whaling Commission, Small-Type Whaling, and Coming to Terms with Subsistence," *Human Organization* 52, no. 3 (1993): 243–51.

76. Colson, *Makah Indians*, 173–77.

77. Asbjørn Eide, *Report on the Right to Adequate Food as a Human Right, Submitted by Mr. Asbjorn Eide, Special Rapporteur*, 1987, UN Doc. E/CN.4/Sub.2/23, 27–29.

78. "NZ Bid to Win Rights for Apes Fails in Parliament," *Nature* 399 (1999): 629

79. In 1954, the Soviet Union forbade traditional Chukchi whaling and instead allowed nonindigenous ships to hunt gray whales with "modern" gear and sell them to Chukchi communities [Milton M. R. Freeman et al., *Inuit, Whaling, and Sustainability* (Walnut Creek, Calif.: AltaMira, 1998), 84–85]. The IWC acquiesced.

80. IUCN (World Conservation Union), World Commission on Protected Areas, and World Wide Fund for Nature, *Principles and Guidelines on Indigenous and Traditional Peoples and Protected Areas* (Gland, Switzerland: IUCN, 1996).

81. John A. Knauss, "The International Whaling Commission: Its Past and Possible Future," *Ocean Development and International Law* 28, no. 1 (1997): 79–87.

82. Principle 22, *Report of the United Nations Conference on Environment and Development (UNCED)*, vol. 1 of *Resolutions Adopted by the Conference*, 1992, UN Doc. A/CONF.151/26/Rev. 1.

83. UN, *Convention on Biological Diversity*, 1994, UN Doc. UNEP/CBD/94/1.

84. Summarized in World Trade Organization, *Annual Report 1998* (WT/DS 58).

85. I do not classify the Convention on International Trade in Endangered Species (CITES) as a trade treaty, although it achieves its purpose—conserving biological diversity—by regulating trade. Unlike the evolving trade-treaty system, CITES *restricts* trade.

PART II
EXPLAINING THE POLITICS
OF THE REGIME

6 / Distorting Global Governance

Membership, Voting, and the IWC

ELIZABETH DeSOMBRE

We are . . . going through our usual game. It is a game in which we know the players very well, we know the rules, and before we start we know the outcome. So it's all very predictable. Now for our part—those of us who have sponsored this one—we know that the minute we put down consumptive use, that the majority countries here are going to oppose it. We know this. But what are we to do? Either we say nothing and we don't have anything in the meeting, or we put it. Good. So that is what we have done. But we know what is going to happen. On the other hand, Mr. Chairman, the people who have proposed the other resolution know very well that the minute they put the word 'cetaceans' we are not going to support that. But they—for what? For their reasons—also put it and put it forward. So all I would say is, with the greatest of respect to our distinguished Commissioner for Chile is that we put these things to the vote, get on with the game, and let us go in to dinner.
—The commissioner from St. Vincent and the Grenadines[1]

There is no question that governance is stalled in the International Whaling Commission. Anti-whaling states fight to keep commercial whaling proscribed, while some whaling states find legal ways around the ban to keep their whaling industries ready for the day when the ban might be lifted. States opposed to whaling couch their concerns in terms of the need to ensure sustainability of stocks, while scientific assessments indicate that some stocks could be sustainably harvested. Representatives in the commission dutifully play their roles in pushing to protect their positions, without much hope of real resolution of the conflicts. Is the current inability to resolve the issue of consumptive uses of whales a simple difference of opinion subject to a democratic process, or is there something more troublesome underlying the difficulties? Are the factors contributing to the stalemate recent arrivals to the politics of whaling, or have they been present all along?

Because of the central principle of sovereignty in the international system, states cannot be forced to join, or stay in, international agreements that they do not want to belong to. The International Whaling Commission (IWC) is like other international environmental agreements that must struggle to provide governance without government; international organizations must convince states to join, agree to, and then implement various provisions to protect a natural resource. The goal of international governance to address resource issues is often to convince states to work together to provide a collective benefit they all value and could not individually ensure.

Two main factors in the history of the International Whaling Commission have contributed to its current difficulties providing this type of governance. The first is the underlying problem of incompatible goals among the actors involved in making and influencing policy with respect to whaling. Some do not accept the premise that whaling is ever acceptable, and others are reluctant to accept any restrictions on their ability to hunt whales. The change from an organization in which all members agreed to "ensure proper and effective conservation and development of whale stocks"[2] to one in which some influential actors argue that whales should never be used consumptively[3] has led to a conflict that no level of scientific certainty about the status of whaling stocks or discussion about the adequacy of management procedures can resolve. This difficulty is discussed elsewhere in this volume and so will be examined here only in the extent to which it interacts with aspects of governance in the IWC to create insurmountable difficulties. Other international agreements, such as the Convention on International Trade in Endangered Species of Wild Fauna and Flora (CITES), have been able to withstand fundamental differences of opinion among their members about the acceptable uses for resources, and other such conflicts are sure to arise in future agreements. It is more important for the purposes of this chapter, therefore, to diagnose the aspects of governance in the IWC that have rendered it incapable of working around these fundamental differences in opinion, or allowed them to arise in the first place.

The second factor contributing to the current difficulties in the IWC is the membership and voting structure of the organization and the ways in which membership has become a tool through which to influence whaling policy. The use of membership as a tool has two components. The first is the process, undertaken primarily but not exclusively by the United States, of bullying states into the agreement, or into changing their policies within the agreement. The second is the effort undertaken by a number of states and nongovernmental organizations (NGOs) to "bribe" states into the agreement, by paying their dues or representing them at meetings, or offering them foreign aid to join or take certain positions. Both strategies have led to an IWC in which membership practices and policies advocated did not represent the real views of interested parties. When states initiate or support certain provisions, not because these provisions are consistent with their own beliefs about how whaling should be conducted but because they have been coerced or bribed into advocating these measures, governance problems already difficult may become insurmountable.

Initial whaling regulations were almost certainly helped by the efforts to bring all whaling states into the agreement. As a common pool resource problem, whale conservation is difficult enough if all relevant actors participate in the agreement; if significant whaling states stayed out, protection of whale stocks would have been harder to ensure. The use of economic threats to convince states to remove objections to whaling quotas, though clearly coercive, can be seen as hegemonic activity in support of a collective action regime.[4] Constant free riding by major whaling states within the organization on regulations to protect a common pool resource would quickly undermine the ability of the organization to govern and (as came dangerously close to happening in this case) not succeed in protecting the resource in question. And at a time when whale stocks were so depleted that nothing short of drastic action would have saved them, it was precisely these participation policies that made the initial moratorium on commercial whaling possible.

But these policies also set the stage for the current problems. States have waged their battles in the IWC through collecting votes and threatening those who do not follow the policies they want, rather than by persuading others of the value of their positions. The creation of the much-needed commercial moratorium happened when stocks had become so dangerously depleted that some states in favor of allowing commercial whaling nevertheless realized that allowing stocks to recover was a necessary part of that process. These states, combined with those who were opposed to whaling on principle and had otherwise managed to win the membership battle, garnered enough votes to create a zero catch limit. The economic threats by the United States were sufficient to get the agreement implemented even by those who did not originally agree. There is little question that the moratorium assisted in the regeneration of some stocks to the point that sustainable hunting might eventually be possible. But the way in which this policy was created now means that global governance is unlikely to be based on anything but bullying and bribing. With a history of overharvesting and noncompliance, and a temporary solution that was created not by persuasion but by coercion, there is little on which to base real negotiation. States must be willing and able to discuss trade-offs and compromises, something for which the recent history of discussion about whaling has not prepared the participating states.

MEMBERSHIP AND DECISION MAKING

The International Convention for the Regulation of Whaling (ICRW) allows accession by any state, whether or not the state is or has ever been engaged in whaling. The representatives of all member states make up the International Whaling Commission, the decision-making body for the regime. Decisions by the commission relating to regulations on whaling require a three-quarters majority vote.[5] Any state may opt out of a decision made this way by lodging an objection to an amendment to the Schedule. Amendments do not take effect until ninety days after the IWC has notified member states of the change. During this period, any state can lodge an objection, indicating that it does not wish to be bound. Although a state can object only during the ninety-day period before the amendment enters into force,[6] this objection stays in place until the state decides to remove it.

BRIBING

Enabled by the open membership policy set out in the ICRW, both sides in the whaling controversy attempted to bring states into the agreement to sway the voting over the proposed moratorium on commercial whaling, either in the process of establishing the moratorium initially or in an attempt to end it. There are several different types of activities that could fall under this category of bribing or buying states into the agreement. The first is offering to pay the state's dues or transportation costs to the annual meetings, or offering to represent it, or help prepare its positions. Presumably, in these cases a state agreeing to join the organization, though it may not have been active on the whaling issue, is not undertaking a radical change in its point of view. States might agree to join for a number of reasons of public relations or international involvement provided dues or representation are covered; they are unlikely to take positions they would otherwise be opposed to unless there is a greater incentive than simply the paying of dues or the preparation of briefing books. These greater incentives represent the second type of bribing that may bring a state into the organization: the offering of foreign aid or other economic incentives if a state joins the agreement. The related issue of making economic threats if states do not join or take certain actions is considered below as "bullying."

Environmental groups have certainly attempted to bring states with no previous history of or interest in whaling into the organization. Although buying states into the agreement is difficult to substantiate, there are a number of states in the agreement that have traditionally not cared about whaling, that joined in the time period when a commercial whaling moratorium was being discussed in the organization. Antigua and Barbuda, Oman, Egypt, and Kenya, for example, are all states that joined in the early 1980s, with no previous history of concern about whaling issues, that voted consistently against commercial whaling.[7] The IWC secretary tells the story of an unnamed member state that simply signed over the check from an environmental organization to pay its dues. Similarly, a representative from a recently joined nonwhaling state showed up late for a meeting and had to ask directions to the NGO section to get his briefing book from the organization that had prepared it for him.[8] A former Greenpeace consultant tells of a plan that added at least six new anti-whaling members from 1978 to 1982 through the paying of annual dues, drafting of membership documents, naming of commissioner to represent these countries, at an annual cost of more than $150,000.[9] Japan's representatives at the forty-seventh annual IWC meeting in 1995 pointed out that "some individuals are listed as government delegates attending the preliminary meetings of working groups and sub-committees but registered as NGO observers in the plenary week."[10]

Japan is alleged to have brought a number of states into the agreement or into supporting particular policies through financial assistance as well. In 1980, in an unsuccessful attempt, Japan targeted the Seychelles, a leading advocate of whale conservation and one of the states often mentioned as potentially brought into the agreement by anti-whaling groups. Japan offered to finance a fishing vessel for the Seychelles for research and training, but threatened to withhold the grant if the country did not withdraw its support of the proposed moratorium. In a letter that later become public, the Japanese ambassador explained that "the fishery industry in Japan strongly oppose the grant in view of your government's attitude at the IWC. . . . if in future your government should change its attitude at the IWC towards Japan, there would be a possibility of my government extending the grant to Seychelles."[11] After this attempt failed, Japan offered $40 million in foreign aid to the Seychelles, which also did not change its position.[12]

It is suggested that Japan used foreign aid a bit more successfully around the same time to ensure votes against the moratorium from Brazil, Chile, and Peru, among others. Brazil in particular was offered a Japanese agricultural investment program on the assumption that "Brazil would reach a sympathetic 'understanding' of Japan's position on whaling."[13] Jamaica and its anti-whaling commissioner left the IWC in 1982 after Japan negotiated to buy 95 percent of that country's crop of Blue Mountain coffee.[14] Similar stories are told about the reasons Costa Rica and the Philippines shifted from voting for to voting against a commercial moratorium.[15]

Japan is also likely to have had success in influencing other Caribbean states on later whaling issues. In 1992, when discussion of the proposed Antarctic whaling sanctuary was heating up, Grenada, St. Kitts and Nevis, Antigua and Barbuda, Dominica, and St. Vincent and the Grenadines all received Japanese funding for new fishing fleets.[16] Dominica, which has received more than $16 million in Japanese aid, rejoined the organization and began voting in favor of commercial whaling the year the aid began.[17] Grenada joined the IWC in 1993,[18] around the time that the Japanese aid arrived. Antigua and Barbuda, St. Kitts and Nevis, and St. Vincent and the Grenadines had all been suspended from voting in the IWC for lack of payment of their dues, but shortly after receipt of Japanese aid they paid their dues and were able to resume voting in the organization.[19] It is also suggested that Japan paid the airfare of these and other Caribbean states to attend the conference to vote against the whale sanctuary.[20] An IWC official explained: "we're not fools. We know the Japanese pay for these people's membership fee, for their hotel bills—even for the limousines they drive around in."[21]

There is no inherent reason that states not engaged in whaling should not have an interest in establishing protection for certain species or in ensuring an international norm of sustainable use of resources. The fact that there are states in the IWC that do not have a direct material stake in whaling regulations is not in and of itself problematic. But to the extent that these states are being used as part of a strategy of bribing them to join or take particular positions within the IWC, the governance of the organization suffers. In working to win a diplomatic battle through the equivalent of stuffing the ballot box, states are furthering the move in the IWC away from an effort to figure out a workable policy and toward an effort by those with different perspectives to outstrategize each other.

Bullying states—primarily the United States, though others have used this strategy to a lesser extent—have attempted to bring states into the IWC, and to influence their policies within the organization, through threats of economic harm. At least three states can be seen to have joined the agreement because of this pressure, and another several have made important changes in whaling policies in response to these threats.

The United States took to protection of cetaceans early, with the Marine Mammal Protection Act in 1972. This act outlawed all commercial whaling for U.S. whalers, even though at the time whaling was legal internationally.[22] The United States began to work internationally to protect whales more vigorously. The Marine Mammal Protection Act created a national Marine Mammal Commission charged with, among other things, suggesting international policies for marine mammal protection. Earlier U.S. legislation, the Pelly amendment to the Fisherman's Protective Act, provided for the United States to restrict fish imports from states that "diminish the effectiveness of an international fishery conservation program."[23] In the legislation, the term "international fishery conservation program" is defined as "any ban, restriction, regulation or other measure in force pursuant to a multilateral agreement to which the United States is a signatory party, the purpose of which is to conserve or protect the living resources of the sea."[24] According to this definition, then, whaling agreements were under the purview of this legislation from the beginning. Under this legislation, the Secretary of Commerce can certify a state that is acting in a manner that diminishes the effectiveness of such an agreement. The president then has the option to prohibit imports of fish products from the certified state.

The Pelly amendment is discretionary: the president is not required to impose sanctions on certified states. To ensure that certification would carry greater weight, Congress passed the Packwood-Magnuson amendment to the Magnuson Fishery Conservation and Management Act in 1979. Certification under the Packwood-Magnuson amendment that a state was "diminishing the effectiveness of the International Convention for the Regulation of Whaling" triggered certification under the Pelly amendment as well. Once a state is certified, the Secretary of State is required to reduce that state's fishing allocation in U.S. waters by at least 50 percent. If by a year later the certification is not removed, no fishing rights are to be granted to the state.[25]

Note that a state can be in full compliance with international law and IWC policies and still be certified under these pieces of legislation. A state that is not a member of an agreement and therefore has no obligations under it, or one that has followed the legal procedures to opt out of certain obligations, can still be considered to be "diminishing the effectiveness" of an agreement. In all cases, the states that the United States has certified under this legislation were acting in ways technically both legal and consistent with IWC policy.

One of the uses the United States made of this legislation was to bring states into the agreement that were whaling but were not members. In 1978 the United States certified Chile, Peru, and South Korea for whaling in excess of IWC quotas, despite the fact that they were not members of the organization, and for their nonparticipation in the organization. All three states began steps to join the organization and were therefore not subject to sanctions.[26] Taiwan was targeted in 1980 for the same reason. It was not actually certified, but was told, through diplomatic channels, that the Pelly amendment process could be applied. In response in part to the threat of certification, it stopped all foreign whaling and then banned whaling altogether.[27]

The second use of these economic threats was to bring about changes in the policies of states within the agreement. Initially these uses occurred largely when states objected to certain amendments to the Schedule and therefore were not bound by them. In 1974 Japan and the Soviet Union were both certified for objecting to the quota on minke whales and taking a higher number than they would have been able to otherwise. Both agreed to accept the quota the following year.[28] In 1980 the United States threatened to certify Spain for its objection to the North Atlantic quota on fin whales, and threatened to certify South Korea for its objection to the ban on using cold harpoons. Spain complied with the quota, and South Korea withdrew its objection to the ban.[29] In the early 1980s, as the moratorium approached, the United States discussed certifying Japan for whaling in excess of the sperm whale quota and did certify the Soviet Union for whaling in excess of the minke whale quota. Both states had lodged objections to these quotas and so were not legally bound by them. Although the Soviet Union was not sanctioned under Pelly, its fishing quota was cut under the Packwood-Magnuson Amendment.[30] A third use of this type of legislation was to influence states conducting scientific whaling, to attempt to limit the amount of whaling done for this reason and make sure it conformed to guidelines set out by the organization. The United States certified or considered certifying Iceland, Japan, and Norway several times each (and South Korea once) for their research activities that involved killing whales.[31]

The most influential use of the U.S. sanctioning legislation, however, was in the creation of the moratorium on commercial whaling. When initially passed in 1982 (to take effect in the 1985–86 whaling season) the major whaling states lodged objections. The Soviet Union, Japan, and Norway all indicated their unwillingness to be bound by the moratorium.

The Soviet Union was certified under the Pelly amendment for its objection; although the United States did not decide to impose the economic restrictions under Pelly, it did cut the Soviet fishing allocation in U.S. waters, under the Packwood-Magnuson Amendment.[32] Shortly thereafter the Soviet Union withdrew its objection to the Southern Hemisphere minke quota and announced that it would cease commercial whaling in 1987.[33] The United States did not actually certify Japan, but the domestic battle over whether to do so went all the way to the U.S. Supreme Court,[34] and it was thus clear to Japan that certification was a serious possibility. Japan agreed to withdraw its objection to the moratorium and cease commercial whaling. The Japanese delegate to the IWC later said that Japan's withdrawal of its objection was "involuntary . . . coerced by a certain nation."[35] The United States did certify Norway for its objection to the moratorium in 1986,[36] and Norway agreed to stop commercial whaling after 1987. It did not, however, remove its objection, which is why it has been able to legally resume commercial whaling. U.S. threats, and imposition (in one case), of economic sanctions can be seen as influencing states to stop commercial whaling once the IWC had agreed on a commercial moratorium.

Those in favor of whaling also used economic threats (sometimes difficult to disentangle from economic promises or the threats to withdraw them) in an attempt to change the positions or behavior of states in the IWC. For example, the former commissioner from Panama claims that he was withdrawn from the delegation, first temporarily in 1978 and then permanently after he again proposed a pelagic whaling ban in 1979, because Japan threatened to cancel its $5 million order for Panamanian sugar if Panama did not withdraw its proposal.[37]

At this point, the battle of economic force, like the battle of numbers, has been dominated by those opposed to whaling. In neither case, however, is the underlying issue of who should have a say in whale conservation addressed.

EFFECTS OF MEMBERSHIP ACTIONS

Without the efforts to bully or bribe states into the IWC, the regulatory process would have taken a different path. Whaling states outside of the agreement would have had little incentive to join. The inability of whaling states within the IWC to agree to catch limits that would actually protect whale stocks led to short-run policies that nearly decimated a number of populations of whales, as discussed elsewhere in this volume. The pressure applied by the United States certainly had an effect initially of gaining adherence by whaling states to IWC regulations on which they would otherwise have chosen to free ride.

Importantly, the efforts by the anti-whaling states to change membership and voting practices are what made the protection of whale stocks possible. As new nonwhaling members joined, votes for zero catch limits for some species (sperm whales in 1981 for instance) became possible.[38] Although some commercial whaling states ceased whaling and were thus more willing to vote in favor of restrictions on whaling, the major new votes in favor of restrictions came from new members. Once the numbers shifted sufficiently, the moratorium on commercial whaling passed.

It is essential to note that, even though at points in the recent history of whaling regulations a number of whale stocks had been so thoroughly overhunted that whalers could not even catch the quotas they agreed to,[39] whaling states vigorously fought the moratorium. Without external pressure after sufficient numbers of state voted in favor of a moratorium, a cessation in commercial whaling would never have come about. There is clearly an argument to be made for the approaches taken to stop the overharvesting of whales.

But these approaches may have had the effect of winning the battle while losing the war. Regulation is hard enough in a commons when all parties believe in the regulations undertaken. There are always incentives for free riding. But when states are bullied into accepting regulations, it may be more likely that they will not accept the underlying premise that conservation is essential. The noncompliance we now know to have been regular practice on the part of the former Soviet Union[40] is one indication of the type of response that is possible when states are faced with the regulation of a common pool resource they have not willingly committed to preserving. Similarly, the increase in Japanese, Norwegian, and Icelandic lethal scientific whaling came as their commercial whaling opportunities decreased; they thus moved from an activity the IWC was allowed to oversee to one that it had no legal rights to regulate.

The energy the whaling states have put into garnering votes, protecting their whaling industries, and trying to overturn the ban mirrors the energy the anti-whaling states have put into protecting the moratorium. Neither side is actually discussing the real efforts that would have to be made to have a sustainable commercial whaling system, and both sides are misrepresenting their positions in official discussions.

Recent developments threaten to expose the posturing being undertaken by the various coalitions in the whaling battles. A proposal to allow whaling within exclusive economic zones with the whale meat used only for local consumption in return for a ban on whaling on the open ocean has been discussed. The interesting aspect of this potential compromise is that it exposes the game playing that states have been doing in representing their negotiating positions. States like the United Kingdom and the United States, which have largely opposed whaling for ethical reasons but have couched it in language of sustainability and compliance, find themselves faced with a plan that would actually meet their stated goals if not their underlying concerns. Conversely, states like Japan and Norway, which have attempted to couch their desire for commercial whaling in other guises, would have to admit that some of their stated goals were not their real interests in order to accept the compromise. Japan's "scientific whaling," which may produce some scientific information, is nevertheless seen by many as an excuse for the gathering of whale meat for human consumption. Japan would need to admit that whale conservation is possible without this type of "research," a position it has staunchly avoided. It remains to be seen whether this compromise will be accepted and the collapse of the institution avoided. But, given that vote buying and political jockeying continue, the outcome does not look promising.

The irony, then, is that in the IWC it is precisely the actions that allowed the organization to function with a higher degree of effectiveness for much of its recent history that are undermining its ability to govern in the long run. Without the membership actions taken, whale stocks would almost certainly have been more seriously depleted, and regulation would be more difficult. But these activities, which may have protected whales in the short run, are now serving to undermine the IWC in the longer run.

IS THE IWC UNIQUE?

To some extent the structural factors in the IWC that made membership coercion possible are not unusual in international organizations. Most international agreements addressing resource issues allow open membership, without consideration of members' use of the resource. The voting structure, not requiring unanimous decisions but including a procedure for opting out, is not rare among regimes that manage particular resource stocks. Even the conflict between conservation and preservation that arose in the IWC is not unheard of.

Universality of participation within environmental regimes is becoming more prominent. There is a logic to it: all potential users of a resource should be involved in its management if regulators want to ensure that changes in actions by states do not undermine the regime. Any state that may later be involved in the regulated environmental activity can undermine the management regime if not participating in the management regime. The relationship of various resources and ecosystems also points in favor of universal membership in regulatory agreements: those who do not hunt whales may be impacted by the number of whales in a given ecosystem, and the ability of whales to thrive may relate to actions taken by others that are not directly related to whaling. The idea that the earth is the common heritage of humankind also mitigates in favor of involvement of those who might not have an interest in hunting whales—should they not have rights in determining the future of this species nonetheless? In short, the open membership policy of the IWC, while it allowed for bullying and bribing states into the agreement, is not unusual in international environmental law, and if anything is becoming less so.

The voting process of the IWC is less common among recent environmental agreements but is representative of some types of resource management regimes, particularly those regulating fisheries. The idea that a supermajority of states, rather than all participants unanimously, can make regulations for the entire group seems problematic in international law but is used elsewhere precisely as it is used in amending the Schedule in the IWC. If negotiations had to begin anew, and meet the concerns of all parties, every time a regulatory decision was to be made, regimes that set annual catch limits (and need to do so in order that harvesting decisions be based on scientific information) would not be able to reach decisions in sufficient time to allow the resources to be used. This type of decision-making process is used in CITES as well, so that each decision to place an endangered species onto an appendix does not require a full-scale diplomatic negotiating process.

The other essential part of this type of regulatory regime is the objections procedure. Many criticize the system under the IWC and analogous regimes that allow states to opt out of obligations.[41] Within this type of regulatory treaty, however, where decisions are not made unanimously, this type of procedure is essential. No state can ever be bound by a regulation with which it does not agree, and states would not accept decisions made by majorities if they did not have another option.

More important, the states that have made use of the objections procedure in the IWC have learned an important lesson from this process: the importance of maintaining their rights to object, and their legal ability to do so, even when they go along with the majority position temporarily. Japan surely regrets removing its objection when pressured to do so by the United States, and in future situations would follow the strategy of preserving its rights to object. Norway, which stopped commercial whaling for a period of time but maintained its legal ability to do so by preserving its objection, made the better strategic calculation. It is unlikely these states would agree to a regulatory system in the future that did not protect their ability to refrain from regulatory options they did not want.

The use of economic threats or bribes in environmental regimes, or in international relations in general, is also not uncommon. The United States has used a variety of trade threats to convince states to change their actions on issues of international environmental regulation.[42] The use of such threats is coming to be seen as more controversial, as international trade bodies rule that they are illegal under international trade law. More efforts are being made to legitimize these types of measure by including them in agreements. But, as Abram Chayes and Antonia Chayes point out, enforcement of treaty norms by international organizations themselves is rare.[43] It may be unrealistic to expect that states with functioning carrots and sticks would refrain from using them in international relations.

The situation in the IWC is thus not unique. The decision-making structure and the attempts to work around it when influence is available can be expected to arise in other issues. As examined elsewhere in this volume, conflicts between consumptive uses of species and the idea that they should never be harvested have also arisen. The IWC exhibits an unusual level of transparency of efforts to influence membership and voting, and that transparency gives us insight into the process. The IWC has also taken most of these not unusual factors to their logical extreme: universal participation may be the norm in environmental agreements, but usually the states involved have a more direct interest in the issue; there may be pressure applied but rarely so frequently or to such a large proportion of the actors involved. What happens with regulation in the IWC then can be seen as giving important lessons for international environmental politics more generally.

We are thus left with a difficult conclusion. Not only are the processes of management in the IWC representative of issues of environmental governance more broadly, but the conclusions suggested by the experience are not optimistic. Governance is stalled because the organization is being run through coercion and bribery, with states that do not represent their own interests. But without such actions, would whaling regulations have ever taken hold? Perhaps the creation of a new form of management, building both on the mistakes of the old system and on the important impacts it had in stopping the nearly unregulated harvesting of whales, could better govern decisions about whaling.

NOTES

1. International Whaling Commission, *45th Annual Meeting*, Kyoto, Japan, 10–14 May 1993, Verbatim Record.

2. Preamble, *International Convention for the Regulation of Whaling*, 2 December 1946, 161 U.N.T.S. 72 (hereinafter ICRW).

3. Although few IWC commissioners would make this statement as such within the organization, the commissioner from the United Kingdom did say that it does not accept the premise that "because something *can* be exploited on a sustainable basis, it *must* be exploited" (IWC/48/OS/UK). Outside official negotiations the commissioner from New Zealand explained that "whales are the most highly developed form of life in the sea . . . and they are too valuable to end up as a steak on a table in Tokyo." Andrew Pollack, "Commission to Save Whales Endangered, Too," *New York Times*, 18 May 1993, C4.

4. A hegemon, a powerful state or group of states, is thought by some as impor-

tant in ensuring international cooperation by bearing a disproportionate responsibility for ensuring that collectively advantageous outcomes occur. See, for example, Charles P. Kindleberger, *The World in Depression, 1929–1939* (Berkeley and Los Angeles: University of California Press, 1973); Robert O. Keohane, *After Hegemony: Cooperation and Discord in the World Political Economy* (Princeton, N.J.: Princeton University Press, 1984).

5. Article III(2), I C R W.

6. If any state does lodge an objection, the entry into force of the amendment is delayed for a further ninety days. If an objection is made during this second period, the time for objections may be extended further. Article V(3), I C R W.

7. Alan Macnow, a consultant for the Japan Whaling Association, argues that these states in particular, along with ten others, were recruited to the organization by anti-whaling groups. "A Whaling Moratorium Opposed by I.W.C.'s Own Scientists," Letter to the Editor, *New York Times*, 29 September 1984, 22.

8. Interview with Ray Gambell, 3 June 1997.

9. Leslie Spencer, Jan Bollwerk, and Richard C. Morais, "The Not So Peaceful World of Greenpeace," *Forbes*, 11 November 1991, 174ff (Lexis/Nexis).

10. Chairman's Report, *Proceedings of the 47th Meeting of the I W C*, 1995, 41.

11. From *South China Morning Post*, as quoted in David Day, *The Whale War* (San Francisco: Sierra Club Books, 1987), 103–4.

12. Day, *Whale War*, 104–5.

13. Ibid., 107.

14. Ibid.

15. Ibid.

16. Paul Brown, "Playing Football with the Whales," *The Guardian*, 1 May 1993, 26. It is interesting to note that Antigua and Barbuda was one of the states originally listed by pro-whaling activists as on the payroll of environmental N G Os.

17. Mark Fineman, "Dominica's Support of Whaling Is No Fluke," *Los Angeles Times*, 9 December 1997, A1.

18. International Whaling Commission, *Annual Report 1993*.

19. International Whaling Commission annual reports; Paul Brown also makes this allegation.

20. Amanda Brown, "Tourism Threat over Whales Vote-Buying Claim," *Press Association Newsfile*, 30 June 1992.

21. Fineman, "Dominica's Support."

22. Public Law 92–522, title II, 16 U.S.C. 1401 ff.

23. Public Law 92–219, 22 U.S.C. 1978.

24. Public Law 92–219, sec. 8(g)(3) and (4).

25. 16 U.S.C. 1821.

26. *Public Papers of the Presidents of the United States: Jimmy Carter* (Washington, D.C.: GPO, 1979), 265.

27. Gene S. Martin Jr. and James W. Brennan, "Enforcing the International Convention for the Regulation of Whaling: The Pelly and Packwood-Magnuson Amendments," *Denver Journal of International Law and Policy* 17, no. 2 (1989): 300. Taiwan's decision is also likely to have been influenced by a 1979 IWC resolution calling on member states to cease importing whale meat from nonmember states.

28. Steve Charnovitz, "Encouraging Environmental Cooperation through the Pelly Amendment," *Journal of Environment and Development* 3, no. 1 (1994): 11.

29. Martin and Brennan, "International Convention," 299.

30. "Whaling Activities of the U.S.S.R," Message to Congress, *Public Papers of the Presidents* (Washington, D.C.: GPO, 1985), 727.

31. See Elizabeth R. DeSombre, *Domestic Sources of International Environmental Policy: Industry, Environmentalists, and U.S. Power* (Cambridge: MIT Press, 2000), 210–12.

32. Martin and Brennan, "International Convention," 306–7.

33. Letter from Secretary of Commerce C. William Verity to President Ronald Reagan, 14 April 1988.

34. *Japan Whaling Association v. American Cetacean Society*, 478 U.S. 221, 105 C. St. 2860 (1986).

35. 1993 IWC Meeting, Verbatim Report, 198. Whether this coercion came from the possibility of certification or from other negotiating strategies the United States pursued vis-à-vis Japan, it is clear that the Japanese felt bullied by the United States into giving up its objection to the moratorium.

36. "Whaling Activities of Norway," Message from the President of the United States, 5 August 1986. (Washington, D.C.: GPO, 1986).

37. Day, *Whale War*, 106.

38. Patricia Birnie, "The Role of Developing Countries in Nudging the International Whaling Commission from Regulating Whaling to Encouraging Nonconsumptive Uses of Whales," *Ecology Law Quarterly* 12 (1985): 962.

39. See Johan Nicolay Tønnessen and Arne Odd Johnsen, *The History of Modern Whaling* (Berkeley and Los Angeles: University of California Press, 1982), 749.

40. David Hearst and Paul Brown, "Soviet Whaling Lies Revealed," *The Guardian*, 12 February 1994, 1; V. A. Zemsky et al. "Report of the Sub-Committee on Southern Hemisphere Baleen Whales, Appendix 3: Soviet Antarctic Pelagic Whaling after WWII: Review of Actual Catch Data," *Report of the International Whaling Commission* 45 (1994): 131–35.

41. See, for example, Valeria Neale Spencer, "Domestic Enforcement of International Law: The International Convention for the Regulation of Whaling," *Colorado Journal of International Environmental Law and Policy* 2 (1991): 109–27.

42. See Elizabeth R. DeSombre, "Baptists and Bootleggers for the Environment: The Origins of United States Unilateral Sanctions," *Journal of Environment and Development* 4, no. 1 (1995): 53–75.

43. Abram Chayes and Antonia Handler Chayes, *The New Sovereignty: Compliance with International Regulatory Agreements* (Cambridge: Harvard University Press, 1995).

7 / Negotiating in the IWC Environment

ROBERT L. FRIEDHEIM

If one wished to succeed in bargaining in the International Whaling Commission (IWC), one would have to bargain smarter, not harder. A great deal of effort has been devoted to the question of the resumption of commercial whaling—perhaps more than could be justified by an "objective" analysis of the issue. The level of effort tells us how important the issue is to the world community. However, the strategy and tactics used by both pro- and anti-whaling states, state coalitions, and nongovernmental organizations (NGOs) has produced stalemate, and unless some of the major players break from their long-held positions, stalemate will be the best that can be hoped for in the foreseeable future.

A recent attempt to break the stalemate has—thus far—not been successful. Michael Canny, current IWC Chairman and Irish commissioner, introduced a proposal that would adopt the Revised Management Scheme, design a global sanctuary for whales, allow closely supervised coastal whaling by traditional whalers, allow no international trade in whale products, and end "scientific" whaling. He even called for an intersessional closed-door "commissioners only" meeting in Antigua on 3–5 February 1998 to consider his plan.[1] Unfortunately, it proved acceptable to no one. While Canny's efforts at providing entrepreneurial leadership are to be applauded, we are back at square one, and more efforts are needed to prevent stalemate from turning into regime collapse.

This is not to say that the major actors are happy with the present stalemate, but they fear they will be worse off if they shift their position or take another approach to the problem. After all, the major players have some of what they want under the present arrangement.[2] Whaling opponents control the IWC's decisions, and they have been able to use their control to impose a worldwide moratorium on commercial whaling and a newer Southern Ocean sanctuary. While rigid defense of these decisions has succeeded in substantially reducing whaling, the anti-whaling majority has not been able to force the pro-whaling minority to actually give up whaling. While they are not well off, proponents of whaling view any initiative that might break the stalemate as risky, since, under present arrangements (using an objection clause of the International Convention on the Regulation of Whaling [ICRW] and a right to capture whales for scientific study[3]) they are able to catch enough whales to keep their hopes and industries alive.

The issue is not ripe for resolution by negotiation. The majority still hopes to impose an essentially legislative outcome on the minority. If they continue on this path, they will succeed either by outlasting the minority or by coercing it. But thus far the minority has been obdurate—they will not concede. While the stalemate may last for some time into the future, it is unlikely to last forever. Fred Ikle pointed out some years ago that bargaining is a process with a threefold choice available to participating decision makers—concede, continue to negotiate, or defect.[4] We have seen no concession, only a willingness to continue to remain—for the moment—in the negotiating arena. But someday the specter of defection or leaving the negotiation will loom large because some of the participating states will believe that they will be better off with no agreement than with an unacceptable agreement. The purpose of this chapter is to show what it would take to negotiate a solution and prevent implosion of the International Whaling Commission.

RIPENESS AS A CONCEPT: ITS APPLICATION TO IWC BARGAINING

"Ripeness" is a negotiating state that many experienced diplomats recognize when it develops but few are in a position to force. It is defined as a "moment of seriousness," a turning point at which all stakeholders recognize that "it is actually possible to arrive at a solution to the problem by a joint decision."[5] Another way of describing it is that the parties realize they will be better off by agreement than disagreement and therefore cooperate in shaping a joint outcome. But ripeness often comes about because of a shift in factors over which the negotiators have little control. Therefore, it is quite difficult to predict when it will develop or what might be done to cause it to develop. Before launching into a description of tactics that might help to shift the discourse into a mode that would encourage ripeness, we warn the reader that it is likely that nothing we recommend will help convince states to shift their fixed positions. In the I W C, negotiators face a situation of "positional" bargaining in which the stakeholders insist on the inviolability of the principles that undergird their positions and are therefore unwilling to look at a proposed solution unless it fits squarely within their conceptual framework, or is, at best, a slight variant of the core of their conceptual framework. This obduracy creates stasis.

But stasis does not last forever. Historical observation shows shifts on many issues under negotiation that result from changes in exogenous factors. The world economy may change, making economic factors more salient to decision makers than environmental factors and forcing a different trade-off. World decision makers may discover that more or less animal protein is available for human consumption, causing them to rethink the relationship of the consumptive versus the nonconsumptive uses of whales. A realignment of states in the world political system may cause a realignment in the balance of forces in the I W C as a result of a search for new allies. A positive vote in the I W C might be tied by a stakeholder state to a reciprocal vote on an entirely different issue in another negotiation.

More often than not, stasis collapses of its own weight. The "real" world problem under negotiation has changed, while the proposed solutions have remained fixed. At some point the parties recognize that what they are doing by sticking to their position is irrelevant. A new solution is needed. Sometimes ripeness occurs out of sheer exhaustion. All sides are tired and want to end the constant contentiousness. They become more "reasonable." In short, they see "a way out" of the present dilemma.[6]

In the meantime, despite doubts that more appropriate tactics alone could turn the situation around, I offer below some observations on the nature of the bargaining problem and some ideas concerning tactics useful to the circumstances. Perhaps these might be suitable for breaking up stasis, or helping to seize the moment of "propitious change" when and if it occurs, and moving toward ripeness.[7] Negotiators are not simply prisoners of their circumstances. Strategy, tactics, and leadership count. It might be possible to improve the outcome with smarter bargaining.[8] I will borrow freely from a large literature and show how the ideas I adapt can illuminate I W C discourse.

We will begin by looking at the IWC as a bargaining arena that creates both constraints on the bargainers and opportunities for creative resolutions.

THE IWC AS A BARGAINING ARENA

The International Whaling Commission, which meets annually and makes decisions binding upon most of its members, is an arena for large-scale, multilateral negotiations. It is arguably (and the counterargument will be found below) a real negotiating venue, since decisions can be made only with the consent of the affected parties. If they do not consent, a party feeling itself to be made worse off by a joint decision can leave the organization,[9] enter an objection to a decision to which it refuses to be bound,[10] or partially avoid being bound by using a right (e.g., to conduct scientific whaling and continue to use the whales caught for human consumption[11]) to avoid an obligation, such as a moratorium on commercial whaling. Negotiations occur at many levels within the IWC—on a person-to-person level; within and between NGOs, observers, and government parties; within committees and working groups; and between governments prior to a vote on major issues. These negotiations range from cooperative to very adversarial.

The IWC shares a number of characteristics with other large-scale multilateral negotiations, but its distinctive attributes are critical in affecting the outcomes of its proceedings. Most of the major common attributes of multilateral negotiation have been summarized by I. William Zartman: (1) multiple parties are involved, (2) multiple issues are raised, (3) multiple roles are played by participants, (4) the purpose for meeting is rule making, and (5) the decision-making process is dominated by coalition formation.[12] However, the IWC differs from other major multilateral negotiations in that its decision-making process is not driven by consensus (defined as coalescing around an outcome preferred by a dominant coalition, with dissenters—if any—acquiescing by being silent; thus unanimity is not required). In addition, even some of the general characteristics the IWC shares with other multilateral negotiations have been shaped quite differently in this context. Below, I will describe these attributes and point out the similarities to and differences from other international negotiating venues.

All of the attributes of multilateral negotiations add up to a whole that is noted most for its complexity: many negotiators; many stakeholders; many issues, often technically complex; the difficulty of integrating the preferences of multiple actors into a commonly accepted outcome; continuous negotiations over a prolonged period of time; and preparatory work performed by an international organization. These force the use of simplifying mechanisms such as coalition for gaining outcomes. They also can limit the scope of outcomes, since often it is the "least-common-denominator" outcomes that can gain support under a preferred consensus-based rule.

The IWC has all of these attributes. For example, at the forty-seventh meeting in June 1995, thirty-two of the forty contracting governments participated, as did observers from four nonmember governments, eight intergovernmental organizations, and ninety-one nongovernmental organizations.[13] State participants are quite asymmetric in terms of the usual attributes of political power,[14] including the remaining hegemon, the United States; Japan; the major European states such as Russia, the United Kingdom, France, Germany; a number of middle powers such as Norway, Denmark, Australia, New Zealand; and a number of small countries such as St. Vincent and the Grenadines, Dominica, Oman, and Monaco. As a result, important roles are often delimited by the power of the participating state.[15] Nevertheless, since all have a right to participate, to be heard, and to vote, and all have to be accounted for, decision making is cumbersome.

Even more telling in recent years, the nongovernmental stakeholders also have to be heard and, indeed, have a right to participate in virtually all phases of the IWC's business. They run the gamut from scientific organizations interested in technical issues, to lobbying groups representing those with direct economic interests in the outcome, to those whose interest is primarily normative.

The IWC also seemingly deals with many issues, or at least many manifestations of one issue. There are numerous working groups, workshops, and committees dealing with a revised management procedure and scheme, catch limits, moratoriums, sanctuaries, humane killing, management of small cetaceans, aboriginal subsistence whaling, whale watching, and data collection and analysis. Sessions of the IWC are busy times. But because all IWC issues are highly interrelated, it is difficult to promote trade-offs between issues, and in this regard the IWC is different from many other multilateral negotiations that try to resolve a wider variety of issues. Nevertheless, there have been rumors about issue linking, especially about linking issues on the agenda with issues off the agenda. For example, rumor has it that France introduced the notion of an Antarctic or Southern Ocean sanctuary as a device to mitigate the impact of negative publicity associated with its nuclear testing in the South Pacific.[16]

The ı w c considers issues that require specialized knowledge. Therefore persons with specialized knowledge have always played an important role in ı w c decision making. But lately tension has arisen between key groups. In an earlier period, it was whaling industry managers who controlled the data on whale catches and industry profitability that often moved decisions.[17] More recently, it has been scientists who have provided not only technical expertise concerning the nature of the problem but also recommendations concerning its solution. Scientists expect that their recommendations will be taken seriously and translated into authoritative decisions.[18] When they are not, some important scientists become frustrated, as they have in recent years when scientific recommendations have been ignored on normative grounds.[19]

Recent social science scholarship on international environmental decision making has pointed out that, in the cases studied, when scientists form a consensus on the cause and solution of a problem, they participate in an "epistemic community" and their unified view drives the outcome.[20] However, other cases, including the present one, limit the degree to which this observation can be generalized.[21] Based on the whaling case, Ronald Mitchell suggested that a scientific consensus demonstrating an activity's instrumental harm strengthens support for environmental measures, while a scientific consensus demonstrating the absence of harm will weaken support for environmental protection.[22] Organized environmental groups and governments committed to a "moral" solution pay attention to concerns about harm and ignore observations showing no harm, as they did on the question of establishing a Southern Ocean sanctuary. Currently, one of the major sources of dissensus in the ı w c is the distrust between some scientists and representatives of N G Os and governments that have ignored their advice. This tension will make negotiating a solution more difficult.

As in many other multilateral negotiations, the interactions of the ı w c's participants are on a periodic basis. The ı w c meets once a year, usually in late spring, with workshops, committees, and working groups meeting as required. Since it is not a one-time meeting of the parties, there is a "shadow of the future."[23] Because participants may meet again, although the future is discounted relative to the present, governmental negotiators do not want to be judged for present tactics that might impair their ability to function effectively in the future. Therefore deceptive tactics, bold threats, personal attacks, and so on are rarely used by governmental representatives. This observation does not apply to representatives of some N G Os, who do not seem to worry about the shadow of the future and will say or do virtually any-thing to advance their present cause, including launching physical as well as verbal attacks on their opponents.

There is a good deal of continuity of people and issues year-to-year. Thus the IWC has some internal stability, and substantial momentum is gained in lines of policy development. But the flip side is also present—there is little urgency to resolve a problem in the short run since everyone knows it can be considered next year. Decisions on important matters seem to move at a glacial pace in the IWC. While interpersonal relationships between official participants is usually good, the same cannot be said for the relationships of NGO representatives with representatives and governments with whom they disagree. The internal atmosphere of the IWC is a curious blend of the sta-ble and unstable, the orderly movement of business through the agenda, and a three-ring circus.

The meetings are well served by a small secretariat that is technically competent and adroit at getting along with the contending parties. How-ever, the overall organization is quite fragile from a financial point of view. It depends upon the assessments of a limited number of states, some of whom claim to be damaged by the decisions of the majority. If the minor-ity were to withdraw their financial support, the IWC would be badly hurt financially. There is no way to know whether the majority would make up a shortfall, but some observers doubt that they would in the light of major de-veloped states' demands to make international organizations leaner.

The decisions sought are rule-making decisions concerning access to, al-location of, and use of a scarce natural resource, or as others see it, a unique world treasure. But in practice, the core of the IWC's business involves amending a "Schedule" of rules concerning how whale stocks are classified, whether a stock can be exploited and to what extent (a quota), whether cer-tain whale species can be captured and by what means, and how the entire process can be supervised and controlled. In addition, an increasingly im-portant part of the IWC's annual business involves passing resolutions con-cerning all aspects of whales and whaling in order to provide guidance for (if one wishes to be polite) or to put pressure on (not polite) the parties to conform their behavior to the resolution's demands. These resolutions have no formal standing in international law, and in theory they are not binding upon the parties. But those who endorse such resolutions hope they will be-come part of "soft law," purportedly reflecting commonly accepted norms, and thus will be obeyed.[24]

The principal difference between the IWC and many other international organizations is the parliamentary nature of the IWC's predominant decision-making system; other organizations rely upon consensus rather than votes (or if there are votes, they are formal confirmations of decisions reached by consensus).[25] Decisions in the IWC are made by a minimum three-fourths majority positive vote. In the strictest sense, one might characterize the IWC as a legislative rather than a negotiating system.[26]

The underlying problem has to do with state sovereignty. Even if the world political system is considered less than anarchic,[27] or has evolved since the Treaty of Westphalia (1648), political entities called nation-states still believe they have an attribute called sovereignty,[28] so that on important issues they are bound only by those measures to which they consent. Consensus-based international organizations and conferences are a concession to that perception of sovereignty. The IWC's founders in 1946 also made their bow to sovereignty, but in a different way. They created Article V(3), which allows a member state to file an "objection" to a Schedule amendment so that it can opt out from a rule accepted by a majority. Article VIII gives member states an absolute right to conduct scientific whaling under special permits and to process the whales caught. Further, Article VIII states that "the killing, taking, and treating of whales in accordance with the provisions of this Article shall be exempt from the operation of this Convention." This has not stopped a present majority of states from passing resolutions at recent sessions condemning Japan for exercising this right.

Representatives of the majority in the IWC either have overlooked the underlying sovereignty problems or have a "hidden agenda"[29] and are using the IWC as a venue for attacking sovereignty. For some stakeholders, saving whales is the objective, for others reducing sovereignty is what they are after, and for still others these ideas reinforce each other and form what has been called an "overlapping cleavage."[30] As a result of this buttressing, no compromise has been possible. IWC politics have evolved toward the parliamentary, with states behaving like political parties in a domestic legislature, but without having the underlying social compact that assures that the minority will bow to majority policy demands. Behavior is crudely majoritarian—resolutions and amendments to the Schedule are passed without concern for whether the minority that voted against is seriously discommoded. By behaving in this manner, an anti-whaling majority has controlled the general policy direction of the whaling regime, but it has not been able to control the behavior of some important dissenters. It is clear that an anti-whaling majority sets the agenda, allows its opponents few victories, and indeed harasses them through resolutions to give up their sovereign rights. Since those in the minority have not acquiesced completely in the face of this assault, it can be said that the present whaling regime is not a negotiated regime. The minority is made worse off by majority decisions, a fact the minority is well aware of.

The IWC is a coercive regime dependent upon enforcement measures not agreed to by the minority. It certainly does not pass the test of a regime of mutual coercion mutually agreed upon.[31] I believe the regime is fundamentally coercive in nature and is an example of the "tyranny of the majority." Very likely, this pattern of behavior will come back to haunt some of its chief sponsors.[32] The chief public enforcer is the United States through the Pelly amendment[33] to the U.S. Magnuson Fisheries Act,[34] which authorizes the president to certify any state whose actions diminish the effectiveness of international environmental agreements as being in noncompliance with the act.[35] Miscreants can, in the first place, have their fishing rights in the U.S Exclusive Economic Zone (EEZ) cut or eliminated, and if that is not enough, they can be subject to unilateral trade sanctions. The former threat has been hollow for some years, since foreign fishing quotas in the U.S. two-hundred-mile EEZ have been eliminated for other policy reasons, but any real imposition of trade sanctions may well place U.S. sanctions squarely in the sights of the World Trade Organization dispute-resolution procedures. But thus far the Pelly amendment has served as a deterrent, even though actual punishments meted out have generally been slaps on the wrist. On the other hand, private coercion through consumer boycotts may have had more impact. For example, in 1988 Greenpeace mounted a campaign in the United States, the United Kingdom, and Germany to convince individuals and institutions to stop buying Icelandic fish.[36] They reported success in their U.S and German campaigns, and failure in the United Kingdom.

A critical feature of all multilateral negotiations is their dependence upon the formation of winning coalitions to mold outcomes. This is especially true of the IWC, where the process more nearly resembles legislative than bargaining behavior. But in any organization in which three or more parties are trying to devise an outcome acceptable to all, the first task is to persuade one other party that you should create a common position because it advances your common interests. Then you can go into the next phase. If the second phase is legislative, you outvote the third party. If it is a bargaining phase, you try to create a consensus based upon the formula you already worked out to persuade the third party to join the other two.[37] Regardless of whether we judge the IWC process as legislation or negotiation, the efforts of the parties has been to form a dominant coalition.

Coalition opens up a rich lode of tactical maneuvers whose purpose is to build and defend a winning coalition that unites around a particular substantive formula. To do so, since about the middle of the 1970s, key members of a group of states opposed to all forms of whaling have used a bandwagon effect, logrolling, and other such measures. They have formed a dominant coalition that has changed little in composition since it was formed. The coalition has persuaded wavering states, or those with low salience on whaling issues, to "come aboard" so that they can be on the winning side (bandwagon), and it is said to have provided tangible rewards (logrolling) to those who will go along with its preferred paradigmatic notion of terminating whaling forever. The coalition has held together in all the various issues relating to whales and whaling—unlike coalitions in many other multilateral negotiations. For example, the United Nations Law of the Sea negotiations involved a richer array of coalitions based on UN General Assembly caucusing groups, economic interests, security blocs, and the existence of certain physical features such as a continental "margin." Individual states moved from one coalition to another, depending on the issue.[38]

A coalition must have a "formula" idea to back: a principle of justice or a set of referents or underlying values that gives meaning to the items under discussion.[39] This idea is discovered in a diagnostic phase; the process moves through establishing the formula phase and culminates when all details are worked out in a final phase. Bringing an end to whaling is the formula idea that has appealed to the broadest number of members of the IWC. It clearly purports to be a principle of justice, and it was established as the formula idea of the majority in a second phase before the 1982 moratorium vote. It was subsequently refined in a third phase that culminated in the Southern Ocean sanctuary vote. Since it is a simple idea, it has not been as subject to refinement efforts as have formula ideas developed for other international environmental negotiations.

The coalition backing the no-whaling formula has shown no signs of breaking up, nor has it demonstrated willingness to accept anything but its maximum demands. If a negotiating party disagrees with the dominant coalition's formula, it has only two choices if it chooses to remain in the negotiating arena and not defect: (1) try to break up the dominant coalition;

that is, replace it with its own new, dominant coalition espousing a different formula idea, or (2) try to get the dominant coalition to mitigate its demands, that is, accept less than a maximum version of its policy idea. Both
are difficult and usually costly strategies, but the attempt to form a new
dominant coalition is the more costly. But it might be possible. Some tactics
for attempting or at least threatening to do so are explored in the next section. The other alternative is less costly, but it remains to be seen if it is any
more feasible than smashing the dominant coalition. Mitigation efforts
begin with naysayers recognizing that the dominant formula notion will
represent the main policy thrust in the issue area. But if the majority is willing to mitigate, minority participants willing to accept the formula might
ask the majority for vaguer language under which the minority can still
maneuver, exceptions to the rule, application of the formula notion only to
specific geographic regions, extensions of time for application, different
language that makes it at least appear that the minority has not surrendered,
or some form of compensation for acquiescing.[40]

These concessions are much easier to come by in a multilateral negotiation with a variety of issues. As noted, IWC negotiations are characterized
by a limited range of issues—all relating to whales and whaling. In many
other multilateral negotiations the major effort goes toward constructing a
grand package that includes trade-offs within it. There is no grand package
in the IWC. As far as I can tell, there are no major trade-offs. Each separate
but related whale issue is expected to be handled on its merits. There is an
expectation of "sincere" rather than strategic voting. Sincere voting involves
making a choice based on the merits of the issue regardless of how it might
affect other interests. Strategic voting involves making a choice based on
whether the measure's supporters or detractors will in turn support or oppose the issues salient to the voter.[41] I find it amazing that, for example, the
United States expects positive votes from Japan and Norway on U.S. requests for an indigenous whaling quota just because Japan and Norway believe the quota is justified, and that is in fact the case, while the U.S. votes
consistently against Japan and Norway's request for quotas for their smallholder coastal whaling efforts![42] Unexploited strategic possibilities exist in
the IWC. Despite the lack of variety in issues, issue linking may be possible.

Not unique but quite distinctive to the IWC is the degree of participation
of NGOs in the decision-making process. One side perceives that the issues
are largely moral, and the other side that the preferences of the majority
would unfairly destroy cultures and livelihoods. This polarization of opin-

ion accounts for much of the bitterness of participants on all sides. Whaling is an extremely divisive issue. It has resulted in a situation of close to zero trust. As a result, at this stage, it is virtually impossible to counsel that the parties try to move the discussions from the strictly distributive (in which the "pie" is divided so that one side's gains result in the other side's losses) to the integrative (in which both sides cooperate to increase the pie before it is divided).[43] Each side sees the other as immoral and untrustworthy.

Opponents of whaling are viewed by those who support it as racists and lunatics. The anti-whaling majority is seen as trying to impose as a universal standard a set of "ethical" requirements based on Western urban values. But proponents of whaling point out that many of them do not share one or more of the majority's social attributes—their state of affluence, urbanization, ethnicity, culture, or even race. Anti-whaling spokespersons are not seen as "responsible"; they are viewed as fringe members of Western urbanized societies. The rhetoric of the pro-whaling parties is filled with the hope that someday important developed countries will "wake up" and sweep aside the lunatic fringe members of their societies who have captured their country's position on whaling.[44]

On the opposite side, the anti-whaling forces combine several major viewpoints that allow all the subgroups of their coalition to come together to attempt to ban all whaling forever. Many nonnormative radical observers of whaling view the pro-whaling states as untrustworthy predators with no sense of limit. They cannot be trusted not to overreach because they have a "bad" history as industries and states.[45] For much of the early IWC history they acted in their short-run self-interest and killed whales in large numbers even after the destructiveness of their behavior was obvious to all. Whaling states should not be allowed to begin the cycle again by permitting the resumption of commercial whaling. Those against whaling may be amenable to nonadversarial discussions of the issues, but their suspicions make genuine negotiations that might lead to a bargained outcome very difficult. But some coalition partners are much more difficult to bring to the bargaining table. They are true believers who might be characterized as a religious left,[46] who insist that there is a new moral standard applicable to all, whether the dissenters are willing to recognize it or not. They argue that compromise is not the way to bring the dissenters around. Whalers must be shamed into conceding that whales are sentient beings with rights. They must be harassed and spit upon, and their vessels must be sunk. At the very least, some radical (and, it is claimed, not the most radical[47]) environmental groups such as

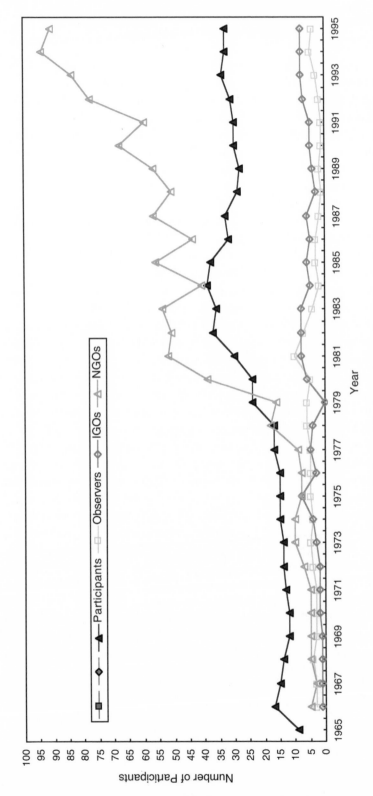

FIG. 7.1. Participation in IWC Meetings, 1965–1995. (SOURCE: IWC records)

Greenpeace are obligated to try to change "the ideational and effective level of human experience and activity."[48] But, as with all true believers, tolerance for the position of others is in short supply.

I do not mean to imply that proponents of whaling have been especially tolerant, or that whalers in the past did not get out of control, but the situation today is a product of history. That history changed fundamentally after the structure of the bargaining arena changed. In 1978, the United States proposed that all IWC sessions, and not just plenary sessions, be opened to participation by NGOs.[49] Figure 7.1 illustrates the consequences.

With a right to participate assured, it was worthwhile for numerous NGOs to join. More states also joined, and as Elizabeth DeSombre has demonstrated, some were provided inducements to join, or once members, were provided further inducements to assure their votes. In the arena that evolved, civility became scarce. It is difficult to negotiate in a bargaining arena with over ninety NGOs intervening in many parts of the process, with close to forty states that have to be brought to consensus if there is to be a negotiated outcome. Indeed, to even hope for a negotiated outcome may require that a new set of fundamental alterations to the IWC be part of the negotiating demands.

The above description should demonstrate what the IWC is not. It is not a Habermasian arena in which people discuss matters, understand one another, and try to persuade each other and modify views to meet counterarguments. In short, it is not an ideal legislature; it is not a perfect democratic meeting place. There is very little effort to persuade opponents, only to outvote them. There is very little effort to listen, only to repeat arguments. In sum, the IWC does not embody "deliberative politics" or "deliberative democracy."[50]

ADVICE TO THE PRINCES

What can be done to make bargaining smarter? Outside observers and analysts may be helpful to negotiators, considering that, on their own, the negotiators seem to have reached a hopeless stalemate. Can outsiders come up with a fresh approach? I think so, even if some of the points raised may dismay some present negotiators. There are few new ideas that will not raise the price of solution for some stakeholders—states and NGOs. Indeed, an "advisor" would have to be Machiavellian in approach, prescriptive rather than descriptive; the bold advisor would have to offer advice to the

"Princes."[51] Note the plural. I believe both sides would benefit by rethinking the problem if there is to be a negotiated solution, although obviously the losing side would gain more from movement away from the present situation than the winning side.

The advice below is Machiavellian to the extent that it emphasizes the underlying structural nature of the problem while recognizing that the logic of the advice is subject to moral objection.[52] But it is offered in the spirit of raising the right questions and not necessarily giving answers that a "vulgar" Machiavellian would approve.[53] As Roger Fisher noted, every statesman wants to win but also has an interest in a continuing future working relationship with the other players that will limit the severity of the necessarily Machiavellian elements that might otherwise be introduced.[54]

Eleven points will be made below that provide some ideas about how to break the stalemate. These categories are for the convenience of presentation. Many of the points overlap, so theoretical neatness is lacking, but they may provoke some of the stakeholders to rethink a situation described by one of our colleagues, David G. Victor (and Julian E. Salt), as "empty law." At stake is not only the IWC but also other international environmental treaties in which targets are pushed too far ahead of states' willingness to implement them.[55]

SALIENCE

There will be no movement until the salience of the whaling issue changes among the stakeholders. By salience I mean a preference for achieving a favorable outcome on one issue as compared to a favorable outcome on other issues valued by a stakeholder. Salience is always a comparative notion; it is a measure of relative importance.[56] Almost all individuals and collectivities have multiple objectives in the world they face. Except in the rare instances in which they value each of their objectives equally, they usually favor one or some over others. When they discover they cannot win favorable outcomes on all issues, they are forced to accept a trade-off—a less favorable outcome on an issue of lesser salience in return for a more favorable outcome on the issue of higher salience. Choosing a policy involves both expressing a substantive preference for a particular outcome and being willing to pay the cost of achieving that outcome. In attempting to press another party to change salience, there is some hope of persuading that party that your preferred outcome is substantively superior (in the manner of Jurgen

Habermas or other democratic theorists who emphasize deliberative principles).[57] There is a better chance of changing that party's perception of how
much it will cost to achieve a particular objective. Salience can be manipulated. But it may be even more important—albeit difficult—to reexamine
how much one is willing to pay to achieve one's own objectives. In effect,
one must demand that other internal stakeholders accept a trade-off of
some of what they value so that you have a better chance of gaining what
you value. International negotiation has been characterized (at least
metaphorically) as a two-level game.[58] In many cases, domestic discussions
and negotiations to change national salience must be concluded successfully
before one can negotiate externally to change others' salience, but such internal negotiations can be especially painful.

As it stands, the stalemate on whaling reflects both normative notions of
the correctness of allowing or preventing whaling and notions of how much
it would cost the antagonists to prevent or restore whaling. Until these perceptions are changed, the protagonists' policies and their bargaining stances
will not change. Iceland seemed to signal that, by withdrawing from the
I W C, the resumption of whaling was a highly salient issue. But when it came
to following through by resuming whaling, they hesitated, although perhaps
only temporarily.[59] The costs seemed too high. Norway and Japan have said
that whaling is an important issue but have not demonstrated a willingness
to pay a higher price to resume full-scale whaling. For them, too, the cost
is too high to achieve their maximum preferred position, partly because
they do receive benefits from limited whaling, however much they are
scorned by anti-whaling states and N G Os. Conversely, the United States, the
United Kingdom, France, Australia, New Zealand, and others claim that
ending whaling is a matter of great substance to themselves. They pay a low
price to achieve it. One may speculate on how much they would be willing
to pay if others raised the costs to them. They have no whaling industries,
few internal pressure groups that could influence them to change. They
have to consider only external costs, and these have proved to be low. They
can avoid being forced into a legal solution by denying jurisdiction.[60] The
only way to change their salience is externally (again, assuming that they are
not swayed by substantive arguments).

One anti-whaling observer noted that "Japan is more interested in exporting Toyotas to the United States than resuming whaling." This is probably correct; Japan does not want to exacerbate trade relations or, worse,
provoke a trade war with the United States, a possible outcome if the issue

gets out of hand. Of course, a matter of critical importance such as exporting automobiles for an export-dependent economy has to be more important than the resumption of whaling, which, at best, could provide only a minuscule portion of Japan's animal protein needs and certainly very little hard currency to substitute for American dollars. But the relative balance must change if Japan hopes to change the position held by the United States or other anti-whaling states. In short, Japan and other pro-whaling states must increase the risks inherent in tougher policies. They must create a sense of urgency for compliance with their demand. If done properly, there should be positive as well as negative inducements (to be discussed in several other categories).[61] You need not threaten to wage war or break off relations, you need only assert that the other party will have to pay a higher price to achieve its objective at your expense. There are multiple objectives even in the whaling negotiations. The anti-whaling coalition must be reminded that the price of obduracy may be dissolution of the IWC. The destruction of the IWC would be a very bad precedent for other international environmental negotiations, a demonstration of where "empty law" could lead. Unless the present winning coalition sees that some losses are possible, they have no reason to give up their gains.

LEADERSHIP

The IWC sorely lacks leadership. I do not mean that competent persons have not led their delegations or NGO participants in efforts to fulfill their mandates, or that competent international civil servants from the IWC or other participating intergovernmental organizations have not done their duty well. They have; they have pushed for, and some have achieved, the outcomes consistent with their instructions—but at the expense of finding a negotiated solution. An individual must emerge "who endeavors to solve or circumvent the collective-action problems encountered by parties seeking to reap joint gains through the formation of new regimes."[62] To put it differently, one or more persons must emerge who will attempt to bring the parties together to accept a solution that makes them all better off by agreement than by disagreement. To do this, they must go beyond the narrow mandate of the stakeholder organization that provides them official credentials.

Michael Canny tried, and I for one appreciated his efforts, but either he must try again or another entrepreneurial leader must emerge if the IWC is to end its division.[63] Unfortunately, if Canny cannot follow through, I cannot suggest how to bring forth a successor as entrepreneurial leader. Clearly, the IWC would benefit from the emergence of someone trusted by all sides, who can convince others that he or she has "found the way,"[64] who has good ideas, and who has both the substantive and political skills to bridge the gap between the protagonists.[65] Such persons have emerged in some important previous environmental negotiations, but not in the IWC. In some of these negotiations, entrepreneurial negotiators have even been delegation heads of some of the interested parties and were still trusted by negotiators with an opposing substantive position. Perhaps it will be necessary to tap someone outside the negotiations to help break the stalemate.

AN INTERVENOR OR MEDIATOR

What might help break the deadlock is a person or persons who could act as an informal mediator, a facilitator, a point of contact; someone who all sides could agree has good ideas, is fair-minded, and is an authority figure; someone whose behavior is "reflective" and therefore can act confidentially and who is nonjudgmental.[66] Unfortunately, if the IWC situation is not ripe for a negotiated agreement, it is also not ripe for formal, binding mediation or arbitration. Since the environment is so hostile, it does not even appear likely that a "rule manipulator" from outside would be trusted.[67] Establishing the worth of a mediatory figure would probably be slow and painful before it could garner the possibility of success.

Finding a mediatory leader or leaders outside of a negotiation is not mandatory, but it would be very difficult to find one within. A cornerstone of Canadian and Norwegian foreign policies has been determination to play the role of intermediary during the Cold War between the superpowers, and between the United States and the Third World in many international organizations. They successfully played such a role in the third UN Law of the Sea (UNCLOS III) negotiations, even though they were parties at issue.[68] Alas, while trusted to be fair-minded in these other situations, they would not be so regarded in the IWC. In everyone's eyes, they are too engaged.

But there are rich possibilities for transforming a dyad into a triad—that is, adding a third party to the two antagonists, a party (or parties) who can help restore communications, formulate new ideas for resolution, and even manipulate the situation.[69] Individuals of international prestige might be asked to act as facilitator, perhaps Tommy Koh, president of both UNCLOS III and the Rio Earth Summit, or Maurice Strong (though a Canadian). I'm sure there are others whose reputations for probity were earned outside the ocean or environment nexus who would also be suitable.

Finally, if the parties at issue do not want to trust an individual to provide a path out of the thicket, it might be possible to form an "Eminent Persons Group" to suggest ways out.[70] There are many possible configurations for such a group. One might include citizens of both "neutral" states and major quarreling states as long as they have taken no (or no strong) position on whaling issues. Another configuration would be to include only individuals from states that had no interest at stake. Consideration might also be given to finding "fair-minded" participants from some of the more moderate NGOs from both sides. What might be done is limited only by one's imagination and the goals one hopes to achieve.

CIVILITY AND TRUST

It might seem contradictory to suggest that proponents of whaling become "tougher" while acknowledging that a major deterrent to resolution is existing incivility and distrust. But, it is possible to be both firm and fair-minded, to have a reputation for "good faith."[71] It is important to have a reputation for a willingness to listen to others. I suggest that active efforts be made to listen to the concern of others.

Too often interactions are merely occasions to restate one's own position. Too often the parties to the quarrels in the IWC sponsor meetings or seminars that preach to the converted. It is now time to ask oneself—what in others' positions is a legitimate demand that I could accommodate? What do we share in common? Can we hold a discussion that starts from common ground and works outward until we reach that border area where discussions slip from agreement to disagreement so that we can narrow the range of disagreement? After all, no one in the current whaling debate begins from the position that unrestricted whaling should be restored. It should be possible to borrow from environmental mediation efforts or arms control confidence-rebuilding techniques to bring the parties closer together.[72]

To be blunt, the anti-whaling forces are in the saddle; therefore whaling proponents must do more than try to meet them halfway. Anti-whaling governments and NGOs have no reason to change their position other than fearing that if they press too hard the whole system might collapse. They know they are in a superior bargaining position. I suggest that they have some legitimate concerns that must be assuaged. The problem must be approached from the perspective of what can be done to foster the breakup of the anti-whaling coalition. Not all of those opposed to whaling are amenable to a reasoned interaction on the issues; those whose position is primarily normative are likely to remain obdurate. If it is a matter of secular faith that whales are special sentient beings who should never be killed, then no arguments concerning sustainability or any other utilitarian concept will be persuasive. Those whose concerns about whales and whaling are based on utilitarian considerations or the historical record might be willing to listen. It might be insulting for whalers to be regarded as unreformed predators whose "predation"—if reauthorized—is likely to get out of control. But pro-whaling forces must recognize that whaling in any form will never be reauthorized internationally until they provide ironclad assurances that it will never again get out of control. Such assurances cannot come from the whaling proponents without consultation with members of the anti-whaling coalition. What do those who oppose whaling for historical or utilitarian reasons want? What would provide the assurances they seek? Ask them.

Although I think the burden of proof in demonstrating the sustainability of whaling rests on the shoulders of whaling proponents, they have legitimate concerns to which the anti-whaling forces must also be responsive. Whaling opponents should try to understand the impact of a rigid anti-whaling position on smallholders, indigenous and nonindigenous alike, who are attempting to work out a painful compromise between sustaining their cultures and lifestyles and living in a world system of urbanized values. They should also recognize the arrogance of assuming that one set of "moral" requirements fits all, assess the costs of "empty law," and address utilitarian concerns regarding the future availability of animal protein. Civility is a two-way street.

COMPLETE THE RMP AND RMS

In recent I W C meetings, the major pro-whaling states have played "after-you-Alphonse-Gaston" on completing the Revised Management Plan (R M P) and the Revised Management Scheme (R M S)—each waiting for the other to introduce something new leading toward completion of the necessary formal revisions to be entered into the Schedule. While the R M P is largely complete, there may be refinements not considered part of the scheme that could be introduced. They should be evaluated and, if they are a measurable improvement, submitted. While some whaling proponents believe that the scheme is "overkill," that is, elaborate detail regulations tacked on to the plan that are unrealistic, unwieldy, expensive, and unnecessary for proper management of whaling, they must submit some of their own ideas to complete the scheme. Not doing so will lead to a catch-22. Anti-whaling forces will submit nothing and claim that pro-whaling forces also submitted nothing and are not cooperating, so it is appropriate to *not* approve whaling. Or the anti-whaling factions will submit proposals that are so unrealistic and expensive that whaling proponents will have no choice but to reject them. To avoid endless delay or a very onerous regime, those who favor whaling must submit plans of their own. They may have to accept a control system that is more expensive, more bureaucratic, and beyond what science would require, but that is the cost of assuaging the concerns of those who fear that a "Whaling Olympics" could begin again. I will have more to say in the concluding chapter on how it might be possible to complete the RMP and RMS.

RETHINKING NATIONAL POSITIONS

Creative new ideas are unlikely to emerge from the business-as-usual way of thinking that has dominated the current interaction. Too often the participants are defending positions rather than reviewing which among a variety of available positions could foster their interests. Often there are alternative ways to achieve an outcome that protects fundamental interests if the focus is trying to understand the interest rather than worrying about whether varying one's present position ever so slightly would be perceived as retreat. Although it is difficult at this stage to revert to an earlier, essentially diagnostic stage to discover new ways of expressing interest, doing so is essential. A negotiation need not be perfectly linear. To be most effective, those making suggestions must ask not only what their sponsors care about but what do others care about? Could they invent options that would provide for mutual gains?[73] They could do so informally by, as appropriate to each national situation, appointing agency or interagency study groups within governments with a mandate of providing a fresh look, commissioning outsider studies to seek new concepts and new language, and asking for contributions from intergovernmental organizations with expert knowledge but no direct interest at stake, such as the Food and Agricultural Organization or the Commission of the Convention for the Conservation of the Living Marine Resources of the Antarctic. It might help if some of these groups included not only "neutrals" but also individuals from committed groups with a personal reputation for fair-mindedness and a willingness to listen and contribute. Again, there are many possible permutations, as long as the stakeholders do not "round up the usual suspects."

Some readers may consider these remarks another attempt to write a new constitution for the IWC, a new treaty. This type of effort failed before. It was discussed in Canberra, Australia, in 1977, in Copenhagen, Denmark, in 1978, and Reykjavik, Iceland, in 1981.[74] The United States even offered to host a constitutive meeting, but nothing came of it. Would it fail again? Perhaps, but the situation is about twenty years further down the road, it is even more gridlocked, and unless someone blinks, it won't change. With each passing year, there is a greater chance of defection. There are ways that fundamental changes can be made other than a formal rewriting, resigning, and reratification of a convention, but that step too must at least be contemplated.

If by some stroke of good sense stakeholders are willing to look to the premises that underlie their positions and see that there are other ways of expressing their interests, it also might be possible to adopt new negotiating tactics that could contribute significantly to a positive-sum outcome. For many years, Roger Fisher and associates of the Harvard Negotiating Project proposed a method called "principled negotiation." Many of the ideas expressed above are drawn from their publications—separate people from the problem; focus on interests, not positions; and invent options for mutual gain.[75] They claim that their methods are not hard or soft but oriented toward finding efficient ways to reach consensus on mutual gains. Below, I will outline some "hard" positions. Perhaps they will not be needed if others respond to an expressed desire to start fresh. If Fisher is correct, his bargaining methods can still be useful even if it becomes necessary to assume "tougher" tactics. In any case, all parties would benefit if they got away from defending positions and if, in so doing, they found a jointly acceptable outcome.

DISTRIBUTIONAL VERSUS INTEGRATIVE NEGOTIATIONS

If the delegates and NGO representatives to the IWC were to develop a set of principled strategies, it is likely that the IWC negotiation would move from a zero-sum distributive to a positive-sum integrative outcome. There would be more to share. I bring the discussion back to the distributive-integrative distinction for several reasons. First, even if an outcome is largely integrative, there will always be distributive elements. The minority in the IWC should recognize this. In practice this could mean that, under the best possible alternative, they may have to give up more, and the majority less, to come to a positive-sum outcome. Second, the conversion does not happen by itself. Good tactics will be necessary. I. William Zartman notes that there are only three known ways to foster the conversion—expanding the common pie, establishing trade-offs between differently valued articles, or producing side payments.[76] These tactics can succeed only if parties concede that the "other" has legitimate interests and concerns. The subtext of the whaling negotiation is that the other sides' interests are not legitimate; this might be a barrier that cannot be overcome. Finally, I return to the distributive-integrative distinction because, if by a stroke of luck or genius it is possible to make the transition, all parties should recognize that a transition is taking place. The opportunity should not be wasted.

ACT STRATEGICALLY

Until it is clear that the whaling opponents are interested in an integrative outcome, pro-whaling states and coalitions should act strategically: use tactics most likely to lead to the goal, even if they require that whaling proponents violate some of their substantive values.[77] To this point, the anti-whaling forces have counted on the "sincere" behavior of the pro-whaling forces; that is, it is they expect that whaling proponents will make and execute their decisions based on the merit of the proposal. Strategic behavior calls for judging the issue on whether a positive vote will further your goals regardless of merit. So, if the United States proposes a quota for its Alaskan native whalers, strategically your vote would depend upon what promises or actions the United States makes in relation to issues highly salient to you. Your vote need not even be directly tied to a specific issue. To make their point about their interests being ignored, dissident parties can vote no on everything, substantive or procedural. They can tie up the meeting if they are clever in manipulating the rules of procedure. They can refuse to pay or delay paying their dues until a new formula is developed that puts the IWC support burden on anti-whaling states, since the only income under the conditions of no whaling is from whale watching. Or they may propose a international tax on whale watching, and so on. Again, they need not come to the point of dissolving the commission, but demonstrating that minority needs must be considered could have a salutary effect.

Strategic behavior also calls for treating the IWC meetings as bargaining rather than problem-solving sessions. Bargaining is a combination of offers and threats.[78] There have been precious few offers or threats made by the proponents of whaling. What would they give to get a relaxation of a total whaling ban? Would they pick up all the costs of a demonstration project in which, say, a very tight inspection scheme is tested on a very limited type of whaling? Would they offer a limited entry scheme for whaling in return for extending the jurisdiction of the IWC to other types of cetaceans?

What kind of threat could be made that would be meaningful to the anti-whaling coalition? Obviously, whaling proponents can always threaten to withdraw from the IWC. Unrestricted whaling is not a credible threat, but the creation of a rival organization might be. The North Atlantic Marine Mammal Commission (NAMMCO) was created partially as a threat, but it has been held on a very tight leash. It is possible to loosen the leash. Again, creative moves could raise the cost of realizing all of the anti-whaling coalition's goals.

The politics of the IWC is based on coalition. As noted, the anti-whaling coalition is very stable. If whaling proponents are to get anti-whaling states to listen to their arguments, key members must see the pro-whaling coalition as a bit shaky and subject to possible defections. It is difficult and costly to break up a successful coalition, but it has been done. In the case of the IWC, it might be possible to "repack" the membership, that is, bring in enough new states favorable to one's position to form a new majority. As figure 7.1 demonstrates, the anti-whaling forces were successful in doing this after the mid-1970s. For years, each side has accused the other of "bullying" and "bribing" (by paying their membership fees or arrears), coercing small states to become members or threatening to withdraw benefits if a state voted against them.[79] Such accusations may well be true, but thus far the anti-whaling forces are the packing winners. However, there are many more potential whaling "stakeholders." A 1977 IWC study showed that 102 countries had whale stocks off their coasts.[80] Since their interests are not being represented, it would be "democratic" as well as self-interested to solicit their participation. Many are middle developing (formerly called Third World) states that have a strong interest in protecting their right to exploit their natural resources.[81] While whale resources are "common property" in areas beyond national jurisdiction, middle developing states and their UN caucusing group—the Group of 77—might view the push toward imposing Western urbanized values regarding whaling as a bad precedent and potential threat to their interests.[82] This would be a very large genie if let out of the bottle.[83]

PROMOTING AND RELYING UPON EXTERNAL CHANGE

Lobbying by representatives of one government before the administrative agencies, legislature, or public of another government was considered by traditional diplomats to be a grave violation of their proper role as diplomats. The French diplomat François de Callières (1645–1717) would never have approved,[84] but it is much done in our own age. The object of such efforts is to put pressure on the government under assault to take measures favorable to the lobbying government, or if it has not, to change its position. Promoting change in another government's policy is common today, but convincing another government to reverse a set position is quite difficult, costly, and risky and should be approached with caution. Anti-whaling states looked with hope on the so-called conservative trend in U.S. voting that brought in a Republican sweep in the 104th Congress. But the extreme rhetoric of the new majority in the U.S. Congress on environmental issues frightened many voters, provided ammunition to environmental lobbying groups, and helped bring about a Democratic victory in the 1996 presidential election. The U.S. position has not changed appreciably.

This leaves the waiting game—relying upon the weight of time to induce a change in position because the problem has changed, the personnel have changed, and so on. Relying upon a process one does not control amounts to accepting the weakness of one's position. This seems to me to be the present situation. I doubt if internal change resulting from shifting values will make any difference in the preferences of the officials or publics of the anti-whaling states in the foreseeable future. But just hanging on and hoping to wait out one's opponents appears to be a low-cost option. Thus far, it has not worked, and I doubt if it will.

RELYING ON THE UNITED STATES

The United States should not be relied upon to salvage the situation, much less provide entrepreneurial leadership in future whaling negotiations. In discussions with pro-whaling leaders, I detect both lingering hope that it might and exasperation that the United States so often takes contradictory positions. The hopeful point to the United States as the remaining hegemon, with a stake in assuring a positive outcome for whaling, and also that whaling does not put further pressure on the fragile political and economic alliance structures of the Cold War that the United States would now like to transform to fit the needs of a post–Cold War world. In this environment, the ability of the United States to command the loyalty of others is shakier than in the Cold War past. Conversely, so is the ability of its allies to put pressure on the United States to look favorably on their interests. For example, during the Cold War, both Iceland and the United States understood that facilities on Icelandic territory provided a vital submarine listening post, creating for the Icelandic government a useful bargaining chip. I expect it is less so today.

The main reason not to rely on the United States is because it is seriously pressured from both sides of the issue. It has within its voting public large lobbying organizations with a commitment to end whaling and a pro-whaling community and state delegation (Alaska) that insists that the United States accommodate their needs. Since an exception to the no-whaling rule is available under the whaling convention, the United States uses it. But some question how the United States can in good conscience lead the effort to end whaling and at the same time preserve whaling for some of its citizens privileged by an exception already written into the Schedule, or expand that privilege to other of its citizens—the Makah tribe of Washington State—who have treaty-based whaling rights.

Unfortunately the United States has a long record of establishing a principle and then either seeking exceptions to the principle for itself and its citizens or ignoring the substance of the principle when it does not suit it. Ask anyone who has participated in GATT proceedings whether they were exasperated when the United States advocated free trade and also managed trade—but only for its favored industries.[85] Or the Group of 77 in the Law of the Sea negotiations, in which the United States refused to negotiate further on a convention because of a provision that would have allowed a majority of states to impose by vote new state obligations, even if a state voted against those provisions.[86] Since the issue was seabed minerals, the United States would not have been forced to impose the Pelly amendment on itself, but others see the parallel and have reason to be cynical.

But advising not to rely upon the United States to salvage the negotiations does not mean that the United States can be ignored. Perhaps because it must protect the rights of some of its citizens to continue to take whales, it can be brought to realize that others have needs as well.

PREPARING A BATNA

Most negotiation theorists emphasize the need for preparing a BATNA, or Best Alternative to a Negotiated Agreement. As Roger Fisher and William Ury note, "If you have not thought carefully about what you will do if you fail to reach an agreement, you are negotiating with your eyes closed."[87] A negotiator must know whether what is on the table will make their principals better or worse off. If worse off, they must look to other options. Moreover, it is important to convey to those on the other side of the table that you have options. Such recognition may help to soften an opponent's position. While the development of NAMMCO was motivated partly to demonstrate that its members have a BATNA, more must be done to show antiwhaling forces how they might lose control of the situation unless they are willing to negotiate seriously. Indeed, exercising some of the preliminary steps (although possibly reversible if appropriate) of a BATNA may be called for, such as having NAMMCO "authorize" controlled whaling. This action might be combined with publicly available plans showing how NAMMCO could be the operating arm of a regional organization whose activities can be coordinated with the IWC. In any case, more detailed planning on BATNA options is now appropriate both for its own sake and as a bargaining tool.

CONCLUSION

I wish I could promise that, if followed carefully, this advice to the Princes would guarantee a successful conclusion to the effort to manage whales and whaling into the twenty-first century. Alas, I can only hope it has been helpful.

NOTES

1. "International Whaling Commission: Inter-sessional Meeting in February," *HNWNews*, 8 December 1997.

2. This is also the opinion of a closer observer of IWC affairs, Patricia Birnie: "It could be argued that the deadlock has suited both sides." Quoted in Alison Motluck, "Blood in the Water," *New Scientist* (22 June 1996).

3. Articles V(3)(a) and VIII, *International Convention for the Regulation of Whaling*, 2 December 1946, T.I.A.S 1849, 1716–1729 (hereinafter ICRW).

4. Fred Ikle, How *Nations Negotiate* (New York: Harper & Row, 1964), 59–75.

5. I. William Zartman, "Negotiation: Theory and Reality," in *International Negotiation: Art and Science*, ed. Diane B. Bendahmane and John W. McDonald Jr. (Washington, D.C.: Foreign Service Institute, 1984), 3.

6. I. William Zartman, "Lesson for Analysis and Practice," in *International Environmental Negotiation*, ed. Gunnar Sjostedt (Newbury Park, Calif.: Sage, 1993), 264.

7. I. William Zartman and Maureen R. Berman, *The Practical Negotiator* (New Haven, Conn.: Yale University Press, 1982), 50.

8. We must concede that it also might result in a worse outcome if "smarter" bargaining increases hostility or causes the negotiations to collapse. Since many measures I introduce will increase short-run tensions—threats being potent tactics—those willing to listen to what I have to say must recognize that some of what I offer as possibilities introduce a greater element of risk to IWC bargaining. Moreover, as Thomas Schelling noted, threats must communicate a commitment. The threatener must convey both a capability and an intention of carrying out the threat. At a lower level of tension between the parties sometimes a warning will do, but that does not seem to apply to the problem of negotiating a solution to the whaling issue. Thomas C. Schelling, *The Strategy of Conflict* (New York: Oxford University Press, 1963), 38–40.

9. Article XI, ICRW.

10. Article V(3), ICRW.

11. Article VIII, ICRW.

12. I. William Zartman, "Two's Company and More's a Crowd: The Complexities of Multilateral Negotiation," in *International Multilateral Negotiation*, ed. I. William Zartman (San Francisco: Jossey-Bass, 1994), 4–7.

13. Chairman's Report, *Proceedings of the 47th Meeting of the IWC*, November 1995.

14. P. Terrence Hopmann, *The Negotiation Process and the Resolution of International Conflicts* (Columbia: University of South Carolina Press, 1996), 101–2.

15. This is similar to an observation made by Christophe Dupont concerning disparity of power between states negotiating Rhine River issues. "The Rhine: A Study of Inland Water Negotiations," in *International Environmental Negotiation*, ed. Gunnar Sjostedt (Newbury Park, Calif.: Sage, 1993), 143.

16. Obviously such rumors are hard to verify and must be treated as what they are—rumors. But the best rumors are those that seem plausible. Very likely some are deliberately planted to embarrass opponents and therefore—though plausible—must not be given too much credence.

17. M. J. Peterson, "Whalers, Cetologists, Environmentalists, and the International Management of Whaling," *International Organization* 46, no. 1 (winter 1992), 149–153.

18. They have good reason to believe their recommendations should count, since the ICRW specifies that amendments to the Schedule "*shall* be based on scientific findings" (my emphasis), Article V(2)(b), ICRW.

19. Phillip Hammond, letter of resignation to Ray Gambell, secretary of the IWC, 26 May 1993.

20. Peter Haas, "Do Regimes Matter? Epistemic Communities and Mediterranean Pollution Control," *International Organization* 43, no. 4 (summer 1989): 377–403.

21. Karen Litfin has noted that "Epistemic dissension may be as likely an outcome as epistemic cooperation." *Ozone Discourses: Science and Politics in Global Environmental Cooperation* (New York: Columbia University Press, 1994), 197.

22. Ronald Mitchell, "Forms of Discourse, Norms of Sovereignty: Interests, Science, and Morality in the Regulation of Whaling," in *The Greening of Sovereignty in World Politics,* ed. Karen T. Litfin (Cambridge: MIT Press, 1998), 141–71.

23. This proposition is based on game-theoretic modeling. See Robert Axelrod, *The Evolution of Cooperation* (New York: Basic Books, 1984), 12–13. But his notion reflects upon reality. Thomas Pickering, former U.S. representative to the United Nations noted that a national representative had to look toward not only immediate problems but how to improve the situation of his country "down the road" through good personal relationships with the representatives of other states. Barbara Crossette, "A Strange Diplomatic Ecosystem Awaits Richardson, the New U.S. Envoy to the U.N.," *New York Times,* 6 December 1996, A7.

24. Lawrence E. Susskind, *Environmental Diplomacy: Negotiating More Effective Global Agreements* (New York: Oxford University Press, 1994), 11.

25. Or as Thomas Pickering put it, "they operate the way legislatures do," quoted in *New York Times,* 6 December 1996, A7.

26. The Norwegian commissioner, Kare Bryn, pointed out at the forty-eighth session that "It is also serious for this organization that the majority is trying to introduce responsibilities on all members of a treaty character by passing resolutions by majority vote. It is a well-established practice in the UN that a majority cannot force a country to comply with a resolution which that country has voted against." Statement by the Norwegian commission on draft resolution on northeast Atlantic minke whales, Verbatim Record, 48th Meeting of the IWC, 28 June 1996 (IWC 48/41).

27. Hedley Bull, *The Anarchical Society* (New York: Columbia University Press, 1977).

28. There is a very large literature on sovereignty. A helpful recent paper that links sovereignty with bargaining is James Crawford, "Negotiating Global Security Threats in a World of Nation States: Issues and Problems of Sovereignty," *American Behavioral Scientist* 38, no. 6 (May 1995): 867–88.

29. John W. McDonald Jr., "Hidden Agendas," in *International Negotiation: Art and Science,* ed. Diane B. Bendahmane and John W. McDonald Jr. (Washington, D.C.: Foreign Service Institute, 1984), 32.

30. Robert Axelrod, *Conflict of Interest: A Theory of Divergent Goals with Applications to Politics* (Chicago: Markham, 1969), 158–64.

31. Garrett Hardin, "The Tragedy of the Commons," *Science* 162 (1968): 1243–48.

32. If Richard McLaughlin is correct, the threat of having to defend fisheries trade sanctions, not only before World Trade Organization panels and Law of the Sea Tribunals, might be a deterrent to the United States ratifying the Third United Nations Law of the Sea Treaty. See "U.S. Accession of the United Nations Convention on the Law of the Sea and the Loss of Unilateral Trade Sanctions to Protect Living Marine Resources," in *Ocean Yearbook 11,* ed. Elizabeth Mann Borgese, Norton Ginsburg, and Joseph R. Morgan (Chicago: University of Chicago Press, 1994), 46–89.

33. Public Law 92–219, 85 Stat. 786 (1971).

34. Public Law 96–61, 3(a), 93 Stat. 407 (1979).

35. Notice that it does not require that noncomplying states be in violation of their obligations under international law. U.S. unilateral measures using trade sanctions are likely to be found contrary to U.S. international legal obligations if put to an adjudicative body the United States does not control. This was noted in the proceedings of a workshop I chaired. I chose the workshop participants because they had impeccable international law credentials and, before the workshop, had not taken a public position on whaling matters. See Robert L. Friedheim, chair and editor, *Report: International Legal Workshop, Sixth Annual Whaling Symposium* (Tokyo: Institute of Cetacean Research, 1996).

36. Johann Vioas Ivarsson, *Science, Sanctions, and Cetaceans: Iceland and the Whaling Issue* (Reykjavìk: Center for International Studies, University of Iceland Press, 1994), 91–121.

37. It was put somewhat differently by William Riker: "But regardless of the number of persons conventionally believed to be decisive, the process of reaching a decision in a group is a process of forming a subgroup which, by the rules accepted by all members, can decide for the whole. This subgroup is a coalition." When it persuades those outside the subgroup to form a group of the whole, the outcome is a negotiated one; when it chooses to impose the preference of the subgroup on the

whole, and it is done under rules accepted by all members, the outcome is a legislative one. Riker was, of course, the "father" of modern studies of coalition. See *The Theory of Political Coalitions* (New Haven: Yale University Press, 1962), 12.

38. Robert L. Friedheim, *Negotiating the New Ocean Regime* (Columbia: University of South Carolina Press, 1993), 47.

39. Zartman and Berman, *Practical Negotiator,* 98.

40. I deal with these tactics in some detail in *Negotiating the New Ocean Regime.*

41. This can also be called strategic misrepresentation. Howard Raiffa, *The Art and Science of Negotiation* (Cambridge: Harvard University Press, 1982), 332.

42. I recognize that these situations are not perfectly parallel. The United States can claim legitimacy for its indigenous whaling because it is sanctioned by the Schedule (Paragraph 13, ICRW), while small-type whaling by nonindigenous people is not so recognized officially.

43. Richard E. Walton and Robert B. McKersie, *A Behavioral Theory of Labor Negotiations* (New York: McGraw-Hill, 1965), 11–45, 137–42.

44. Here I am stating the arguments that I have heard, not analyzing them. From an analytic point of view, it is clear that those who support whaling, especially the Japanese, often lump together the "Mainstream Splinter Activists" and the "New Grassroots Activists." Sheldon Kamieniecki, S. Dulaine Coleman, and Robert O. Vos, "The Effectiveness of Radical Environmentalists," in *Ecological Resistance Movements: The Global Emergence of Radical and Popular Environmentalism,* ed. Bron Raymond Taylor (Albany: State University of New York Press, 1995), 320.

45. Harry N. Scheiber, "Historical Memory, Cultural Claims, and Environmental Ethics in the Jurisprudence of Whaling Regulation," *Ocean and Coastal Management* 38, no. 1 (1998): 5–40.

46. Deepak Lal, "Eco-fundamentalism," *International Affairs* 71, no. 3 (1993): 22–49.

47. Daniel Deudney, "In Search of Gaian Politics: Earth Religion's Challenge to Modern Western Civilization," in *Ecological Resistance Movements: The Global Emergence of Radical and Popular Environmentalism,* ed. Bron Raymond Taylor (Albany: State University of New York Press, 1995), 282.

48. Paul Wapner, "In Defense of Banner Hangers," in *Ecological Resistance Movements: The Global Emergence of Radical and Popular Environmentalism,* ed. Bron Raymond Taylor (Albany: State University of New York Press, 1995), 314.

49. Notes to Provisional Addenda, IWC Verbatim Record, 30th Meeting of the IWC, June 1978 (IWC/30/2).

50. Jurgen Habermas, *Between Facts and Norms* (Cambridge: MIT Press, 1996).

51. Niccolo Machiavelli, *The Prince and the Discourses* (New York: Random

House, 1940). For a helpful commentary on Machiavelli's thought, see William T. Bluhm, *Theories of the Political System* (Englewood Cliffs, N.J.: Printice-Hall, 1965), 224–59.

52. Steven Forde, "Internationalism Realism and the Science of Politics: Thucydides, Machiavelli, and Neorealism," *International Studies Quarterly* 39 (1995): 152–53. Perhaps one reason we are uncomfortable with Machiavelli was revealed by Max Lerner: "May I venture a guess as to the reason why we still shudder slightly at Machiavelli's name? It is our recognition that the realities he described *are* realities; that men, whether in politics, in business, or in private life, do *not* act according to their professions of virtue; that leaders in every field seek power ruthlessly and hold on to it tenaciously; that the masses who are coerced in a dictatorship have to be wooed and duped in a democracy; that deceit and ruthlessness invariably crop up in every state; and that the art of being ruled has always been a relatively easy one, the art of ruling ourselves is monstrously difficult." Max Lerner, "Machiavelli the Realist," in *Machiavelli: Cynic, Patriot, or Political Scientist?* ed. De Lamar Jensen (Boston: Heath, 1960), 11.

53. Gordon A. Craig and Alexander L. George note that "Those who oversimplify Machiavelli's position . . . may be referred to as . . . *vulgar realists.* . . . Machiavelli says that *not all ends*, but only some ends justify morally dubious means; they must be constructive, beneficent ends." *Force and Statecraft*, 3d ed. (New York: Oxford University Press, 1995), 280.

54. Quoted in Diane B. Bendahmane and John W. McDonald, eds., *International Negotiation: Art and Science* (Washington, D.C.: Foreign Service Institute, 1984), 60.

55. David G. Victor and Julian E. Salt, "Keeping the Climate Treaty Relevant: An Elaboration," *International Institute for Applied Systems Analysis* (April 1995): 22–23.

56. Ward Edwards and J. Robert Newman, *Multiattribute Evaluation* (Beverly Hills: Sage, 1982), 12. Also see Detlof von Winterfeldt and Ward Edwards, *Decision Analysis and Behavioral Research* (Cambridge: Cambridge University Press, 1986).

57. Amy Gutmann and Dennis Thompson, *Democracy and Disagreement* (Cambridge: Harvard University Press, Belknap Press, 1996).

58. Robert D. Putnam, "Diplomacy and Domestic Politics: The Logic of Two-Level Games," *International Organization* 42 (summer 1988): 427–60; Peter B. Evans, Harold K. Jacobson, and Robert D. Putnam, *Double-Edged Diplomacy: International Bargaining, and Domestic Politics* (Berkeley and Los Angeles: University of California Press, 1993).

59. But Iceland may be reconsidering. The national legislature, the Althingi, voted 37 in favor, 7 against, and 12 abstentions regarding the resumption of whaling, and it requested the government to prepare a plan to accomplish that end. *HNWNews*, 11 March 1999.

60. My colleagues at the International Legal Workshop (see note 35 above) were skeptical about the possibility of finding an acceptable legal solution because of the difficulty of finding a venue in which the anti-whaling states would be compelled to participate over their objections.

61. Alexander L. George, *Bridging the Gap: Theory and Practice in Foreign Policy* (Washington, D.C.: U.S. Institute of Peace Press, 1993), 119.

62. Gail Osherenko and Oran R. Young, "The Formation of International Regimes: Hypotheses and Cases," in *Polar Politics: Creating International Environmental Regimes*, ed. Oran R. Young and Gail Oshrenko (Ithaca: Cornell University Press, 1993), 18.

63. Ibid. Also see Oran R. Young, "Political Leadership and Regime Formation: On the Development of Institutions in International Society," *International Organization* 45, no. 3 (summer 1991): 281–308; Oran R. Young, *International Governance: Protecting the Environment in a Stateless Society* (Ithaca: Cornell University Press, 1994), 45–46.

64. Arild Underdal, "Leadership Theory: Rediscovering the Arts of Management," in *International Multilateral Negotiation*, ed. I. William Zartman (San Francisco: Jossey-Bass, 1994), 178–97. Also see Hopmann, *Negotiation Process*, 265–68.

65. Or, as Jeffrey Rubin and Walter Swap put it, "Effective leadership, however, can occur only if the group tolerates it; expressed another way, an effective leader is empowered by the group's members, even as those members are empowered by an effective leader." Jeffrey Rubin and Walter Swap, "Small Group Theory," in *International Multilateral Negotiation*, ed. I. William Zartman (San Francisco: Jossey-Bass, 1994), 134.

66. Jacob Bercovitch, *Social Conflicts and Third Parties: Strategies of Conflict Resolution* (Boulder: Westview, 1984), 43.

67. Raiffa, *Negotiation*, 23.

68. Friedheim, *Negotiating*, 344.

69. Saadia Touval and I. William Zartman, eds., *International Mediation in Theory and Practice* (Boulder: Westview, 1985), 10–14.

70. Scott Snyder, "The South China Sea Dispute: Prospects for Diplomacy," *Special Report* (United States: Institute of Peace, 1996). We will revisit the idea of an Eminent Persons Group in the concluding chapter.

71. Alain Plantey, "Paradigms in International Negotiation: The Example of 'Good Faith,'" in *Processes of International Negotiations*, ed. Frances Mautner-Markhof (Boulder: Westview, 1989), 287–92.

72. The literature on confidence and arms control is quite large. For an interesting application of the notion of "confidence building" to ocean issues, see Jozef

Goldblat, ed. *Maritime Security: The Building of Confidence* (New York: United Nations, 1992).

73. Roger Fisher and Scott Brown, *Getting Together: Building Relationships as We Negotiate* (New York: Penguin, 1988), 70.

74. Chairman's Report, *Proceedings of the 29th Meeting of the IWC,* Canberra, June 1977, 14–15; Chairman's Report, *Proceedings of the 30th Meeting of the IWC,* Copenhagen, July 1978, 32; Chairman's Report, *Proceedings of the 32d Meeting of the IWC,* Reykjavĺk, May 1981, 34.

75. Roger Fisher and William Ury, *Getting to Yes: Negotiating Agreement without Giving In* (New York: Penguin, 1981), 3–14. For a "mutual gains" approach see Lawrence Susskind, Paul Levy, and Jennifer Thomas-Larmer, *Negotiating Environmental Agreements* (Washington, D.C.: Island Press, 1999).

76. I. William Zartman, "Development of the Concepts of Negotiation and Mediation and Their Mechanisms in the Contemporary International Community," *Kyoto Conference on Japanese Studies* (1994), 1: 1–224.

77. At times, strategic behavior of the sort illustrated can be viewed as threatening by negotiators from other cultural systems. See Robert L. Friedheim, "Moderation in the Pursuit of Justice: Explaining Japan's Failure in the International Whaling Negotiations," *Ocean Development and International Law* 27 (1996): 349–78.

78. Barry O'Neill, "Conflictual Moves in Bargaining: Warnings, Threats, Escalations, and Ultimatums," in *Negotiation Analysis,* ed. Peyton Young (Ann Arbor: University of Michigan Press, 1991), 87–107.

79. For example, Japan was accused of threatening to withdraw its sugar purchases if Panama did not drop its sponsorship of a moratorium resolution. Japan denied the accusation. IWC Verbatim Record, 30th Session, London, 1978, 60–61, 69–70. On the other hand, it was rumored in a more recent session that a small Caribbean anti-whaling state paid its dues in arrears with a suitcase full of cash provided by an environmental NGO. See chapter 6 of this volume for Elizabeth DeSombre's analysis of these tactics.

80. This figure should be used with care. Since 1977 some former nonmembers have become members; others have ceased to exist as states, for example, the German Democratic Republic. IWC Verbatim Record, 29th Session, Canberra, June 1977, Appendix A (IWCC/29/14/Appendix A).

81. Principle 2 of the Rio Declaration on Environment and Development is very explicit: "States have the sovereign right to exploit their own resources," found in *Agenda 21, Rio Declaration, Forest Principles, the Final Text of Agreements Negotiated at the United Nations Conference on Environment and Development* (New York: United Nations, 1992), 7.

82. Compare: "Some nations are redefining the environment as a territory-free, non-geographical issue in which supranational institutions may intervene. They seek to mediate when necessary between other nations, and to force them to follow particular policies. Apparently their aim is to impose the economic and political norms and lifestyles of the North on the rest of the world, instead of allowing other nations to develop their own norms. The outcome will be a still greater tilt in favor of those that already hold economic and political power." Somaya Saad, "For Whose Benefit? Redefining Security," *Eco-Decisions* (September 1991): 59–60.

83. Much of present international environmental politics is already dominated by a North-South split. Marian A. L. Miller, *The Third World in Global Environmental Politics* (Boulder: Rienner, 1995).

84. François de Callières, *On the Manner of Negotiating with Princes* (South Bend: University of Notre Dame Press, 1963).

85. Daniel C. Estey, *Greening the GATT* (Washington, D.C.: Institute for International Economics, 1994); Richard McLaughlin, "UNCLOS and the Demise of the United States' Use of Trade Sanctions to Protect Dolphins, Sea Turtles, Whales, and Other International Marine Living Resources," *Ecology Law Quarterly* 21, no. 1 (1994): 1–78.

86. Friedheim, *Negotiating,* 291.

87. Fisher and Ury, *Getting to Yes,* 104.

8 / The Whaling Regime

"Good" Institutions but "Bad" Politics?

STEINAR ANDRESEN

I have a very demanding task in this book, to address the following two questions:

1. What kind of a regime should the international whaling regime be operating under?
2. How can the world community institute such a regime?

While the chapter by DeSombre essentially deals with what is wrong with the present regime and how it imperils other environmental regimes, I shall go a step further and ask what would be the *right* regime, from a regime-analyst perspective, one that would meet the standard of *equity* and *efficiency?*

My short answer to the first question is that there is essentially nothing wrong with the present institutional setup of the regime. On the contrary, it is quite good. My answer to the second question is that as long as the members of the IWC cannot agree on the very purpose of the regime the world community will not be able to get there. There is no reason to expect that this will change during the next few years.

I would like to thank the following for having provided very helpful comments to an earlier version of this paper: Milton Freeman, Robert Friedheim, Alf Hakon Hoel (University of Tromso), Jon B. Skjaerseth (Fridtjhof Nansen Institute), Arild Underdal (University of Oslo), David Victor, Jørgen Wettestad (Fridtjhof Nansen Nansen Institute), and Oran Young (Dartmouth College). I also benefited from discussion with Michael Tillman and Erica Keen Thomas of the U.S. delegation to the IWC.

In principle, it *may* be possible to design an "ideal" regime on paper, but the real challenge is to make it work in practice.[1]

Economists who deal with regime design often provide illustrations of how international environmental and resource regimes should be designed to meet the criteria of equity and efficiency. As economists have more powerful models than most (softer) political scientists, they have more tools to deal with questions of efficiency.[2] However, when confronted with "political realities," the "ideal approach" tends to remain on paper, with limited practical significance. Although the parties may have common interests in principle, there tends to be a gap between collective and individual rationality.[3] An additional problem in the recent period of the whaling regime is the conflict over *values*, usually not captured by models or approaches when rationality plays a key part. The question of equity is certainly relevant to the International Whaling Commission in this sense. However, it is more difficult, if even possible, to negotiate over values (whaling versus nonwhaling) than it is to do so over numbers.

In addressing the first main question, I will discuss some of the main institutional features of the IWC. I will concentrate on what is politically feasible—not what is technically possible. In addressing the second question, I will focus primarily on the positions, interests, and preferences of the main actors—governmental and nongovernmental—in order to discuss a basis for breaking the IWC's present deadlock. First however, it is necessary to establish whether in fact there *is* anything wrong with the IWC, and if so, what?

EVALUATING THE MORE RECENT PERFORMANCE OF THE IWC

Although I am not supposed to deal with what is wrong with the regime, it is difficult to prescribe a potential *cure* before you *diagnose* the problem (or problems). The majority of IWC members would claim that overall the IWC is functioning well and essentially doing the job it should do. But this book's point of departure—that the regime needs to be "fixed"—is in itself controversial. Moreover, there is no consensus among the contributors to this book on this question.[4]

In another work on the IWC (a chapter in the book *Unlocking Effectiveness*), I argue that the effectiveness of the IWC in its last phase (around 1980 to present) is *low*.[5] According to one of the reviewers of that book, "this is a highly controversial characteristic."[6] Why do we assign the IWC a low effectiveness score? The short answer is that the more recent policies adopted by the IWC run counter to the purpose of the regime.[7] The purpose of the IWC is to "conserve whales in order to manage the whaling industry in an orderly manner." However, other indicators may give different scores. Considering the strong change in perceptions of whaling, the IWC could be seen as a highly effective regime, as commercial whaling is at present extremely limited.[8] That is, the effectiveness of the IWC in recent years could be considered high because it has seemingly induced behavioral change among its members and gotten them to quit whaling.[9] Arguments can certainly be made against using the official purpose of the regime as a measuring rod, but it is also problematic to use the majority view at any given time, implying that the "majority is always right"—to paraphrase a famous Norwegian author.

In recent years, there has been a consensus in the Scientific Committee that certain species may be harvested along the lines of the very conservative Revised Management Procedure, but the IWC majority has opposed even this limited harvest.[10] This runs contrary to the convention, which states that the IWC should follow "the best scientific advice." Thus, if we introduce the elusive criterion of *sustainability*, the present policy does not measure up. But the situation was much worse in the initial phase of the IWC. From a sustainability perspective, it is worse to *overutilize* resources, as was done in the 1950s, than to *underutilize* them, as in the 1990s. In this sense, there has certainly been progress in the management of whales from the 1950s to the 1990s. However, the IWC was on the right track before the 1982 moratorium. *Selective moratoriums* and a number of other conservation measures were in operation before the influx of new members.[11] These measures already had taken care of the endangered species, and the *blanket* moratorium was essentially a political step in the wrong direction, making agreements further down the road much more difficult. "If conservation means a sensible balance between the current use of a resource, and conserving it for possible use in the future, the moratorium was hardly a major victory. Some, myself included, consider it a setback."[12]

Any evaluation of the recent performance of the IWC is bound to be controversial, and different indicators give different results. From my perspective, catch within the limits accepted by the Scientific Committee should be resumed if the IWC is to become a better-functioning regime. This diagnosis may be controversial from a broader political point of view. The fact that the author is Norwegian may also raise suspicions in some circles that my analysis is biased in favor of the Norwegian position on whaling.[13] Given this caveat, what are the chances that a limited catch will resume within the present IWC?

AN INSTITUTIONALLY "ADVANCED" REGIME

I will not go into how different international relations schools consider whether regimes have an *independent* effect on the behavior of the members, or whether they are essentially mere reflections of the underlying power structures.[14] The point of departure for this chapter is that institutional design *does* matter. However, the case of the IWC offers an illuminating illustration of the insight into the study of international regimes conveyed by both the institutionalists and the realists. On the one hand, there are few regimes that have rules and regulations with such a strong impact on the policies adopted, and on the subsequent behavior of the target group. On the other hand, in large parts of its history, the IWC has essentially been a *captured* regime, a mere reflection of the power and interests of its most dominant players. Thus, both perspectives are highly relevant for understanding the dynamics of the IWC and its potential for change. In fact, the extent to which IWC policies can be changed is decided by the ability of the IWC as an institution to make the main contestants transcend their original preferences and find a new common ground.

The following features are generally considered important in deciding how regimes are able to deal with the problems they are set up to handle.[15]

Decision-making Procedures: Majority Voting

It is a common assumption, other things being equal, that majority-voting rules tend to be more effective than decision rules of unanimity or consensus. Policy makers have also called for majority voting to deal more effectively with international environmental questions, most notably the Hague Declaration.[16]

The rules governing the IWC are fairly traditional on this point. Each member has one vote, decisions require a simple majority, and a three-fourths majority is needed to amend the Schedule. Members also have the right to object to any amendment, and the party lodging an objection is not bound by it. These rules were no doubt designed to serve the interests of the Antarctic whaling nations; if the nonpelagic nations wanted to adopt a more conservation-oriented policy, the Antarctic whaling nations could either block it or, if that posed a problem, they could withdraw from the IWC. The early history of the IWC confirms this interpretation.[17] However, from a comparative perspective, the decision-making rules of the IWC are by no means particularly biased toward protecting the interests of the main harvesting nations. One of the more recent international fisheries regimes, the 1980 Convention on the Conservation of Antarctic Marine Living Resources (CCAMLR), praised by many for its novel ecosystem approach, in fact has a "double veto" built into it.[18]

De facto veto for reluctant parties is more the rule than the exception within international regimes. While seeking *consensus* is the normal approach within most international regimes, open *voting* has long been a practice within the IWC, reflecting the strong conflict since the mid-1970s. The fact that voting became *institutionalized* made it much clearer than is usually the case who was the "progressive majority" and who was the "blocking minority." Thus, negotiations in the IWC have been more similar to debates in national assemblies, in contrast with the more diffuse consensus-driven international negotiations. More recently, voting over Schedule amendments has no longer been very relevant, but a large number of majority resolutions are adopted at the yearly meeting. In principle, the use of majority voting is in line with the call for more and better "international governance," but the way it has been used in practice in the IWC may indicate that it is not necessarily a panacea to international problem solving.

Scientific Advice—from Simple and Uncertain to Advanced and Certain

Rational policy making requires that decisions be based upon the best available knowledge about the problem itself, and on possible cures. Knowledge is necessary, but not sufficient for rational international resource management.

Initially, knowledge was poor, scientists were few, and the ties to their respective governments were often strong.[19] Lack of contact with international scientific organizations was identified as a major weakness of the IWC in the 1950s.[20] However, more recently representatives of all relevant scientific organizations have been present at IWC meetings. The IWC was also one of the first international organizations to bring in more independent expertise, and hence scientific methods improved considerably in the 1960s.

In the 1970s, the United States was the leader in a successful attempt to secure the more independent element on a permanent basis by adopting procedures for inviting in individual scientists and scientists from various international organizations. Overall, the scientific process became much more open, peer review was introduced, publication procedures were improved, and overall participation increased sharply.[21] However, the positive effect of this development was upset to a large extent by the strong infusion of politics and conflicts in the Scientific Committee. More recently, however, because of the use of innovative organizational procedures, as well as a massive scientific effort (especially by some of the nations formerly engaged in whaling), scientific consensus on the status of key stocks has finally been reached.[22] In fact, the Revised Management Procedure is among the most technically advanced *and* conservation-oriented procedures that has ever been produced for the management of marine living resources.[23] It has been stated that, "viewed as a whole, its status is near the best, and best determined, of any of the world's marine resources."[24] Thus, the overall scientific approach of the IWC Scientific Committee could probably serve as a model for most international fisheries commissions, but it is uncertain whether it will ever be put to use by the IWC.

Access and Participation:
Increased Openness and Transparency

As the question of access for various whaling actors has mostly historical interest, I will be very brief on this point. For quite some time now, limited entry has been seen as the best way to optimize the harvest of marine resources as a sufficient but not necessary condition for effective management.[25] When the IWC was conceived, this was not common knowledge, and the Whaling Convention forbade entry restrictions in line with the liberal ideas of the time, paving the way for the well-known "whaling Olympics." Had the drafters of the convention known what we know today, the history of the IWC *might* have been different. More limited entry might have reduced the wasteful competition and reduced the problem of massive overexploitation.[26] *If* this had been the case, the rationale for "capturing" the IWC by the "green forces" might never have arisen. Today, this represents an example of a *missed opportunity*,[27] reminding us of the potentially powerful impact of institutional design when crafters design new conventions.

The present trend is toward increased participation by states, as well as nonstate actors, in international regimes. Generally, this trend is welcomed by analysts and policy makers as a step toward a more dense "world community." The IWC was designed as an open organization from the outset. It has been argued that the negotiators who wrote the convention "probably assumed that only states possessing indigenous whaling industry would join."[28] The following statement, made by the acting U.S. Secretary of State at the 1946 Washington conference, shows that some parties saw it otherwise: "Whale conservation must be an international endeavor. Each nation, whatever its direct or indirect interests in whaling, will ultimately participate actively in the great task of fostering and developing this common resource."[29] Some thirty-five years later, his idea proved to be at least partly true. Had the IWC been designed along more traditional lines, with some kind of material interest or user-criteria needed to become members, like the Antarctic Treaty System and most fisheries agreements, the moratorium would never have been adopted and it is doubtful whether commercial whaling would have stopped.[30]

In an analysis of the effectiveness of the whaling regime,[31] I conclude that the best performance of the IWC took place when there was a rather balanced representation of whaling and nonwhaling nations.[32] However, a "balanced" regime can hardly be designed. The regime is either open or closed. When an issue is politicized and the members pour in, you cannot close the regime.

The crafters of the convention did not know much about environmental organizations, but the increasing presence of such organizations has had a significant impact on the whaling regime.[33] Generally, the role of the nongovernmental organizations (NGOs) is to push reluctant (state) actors to adopt more far-reaching measures than they would otherwise. This situation is different in the IWC. NGOs are well represented in the delegations of key nations, and a rare case in which NGOs, usually conceived of as the "underdogs," are allied with the most powerful players. For institutionalists hailing the significance of NGO access, participation, and influence as a means to achieve better transparency, again the IWC is a model for other international regimes. It may be argued that in this particular case, though, many NGOs, in alliance with powerful states, have been less constructive by making the issue into a simple battle of whaling versus nonwhaling. The open structure of the IWC has also made feasible both "bullying" and "bribing," which are not welcomed in international politics.[34] Nevertheless, it would be neither possible nor wise to restrict NGO access. If the whaling nations are dissatisfied with the NGO majority, *mobilization* rather than restriction is the way to go. It may even be argued that polarization would increase if the NGOs were kicked out.[35] However, as witnessed during a recent IWC meeting in Oman, it may be a problem that some players are both NGOs and delegates.[36] (As will be discussed later, more shielded forums are sometimes necessary to reach agreement.)

The Secretariat: Lacking Institutional "Geist"?

The secretariats of international organizations may have an important role to play within an international regime.[37] Is there room for improvement by the IWC secretariat to facilitate increased understanding among the parties? Secretarial facilities were initially extremely limited. In fact, the IWC did not get its own secretariat until 1976, some twenty-five years after it was established.[38] Fees were also largely symbolic, "and the IWC budget was operated out of a cigar box"; there was only enough to cover the costs of meetings and a few publications.[39] This gradually changed when the secretariat was established. However, in the early 1990s, it was claimed that, "unlike many other specialized intergovernmental institutions, the IWC has always been a thing to be captured rather than an actor in its own right. [It] has not been able to persuade the commissioners to adopt an "organizational ideology" that would dilute national influences."[40]

It is certainly true that the IWC has been captured through large parts of its history, and that the secretariat has not been able create an "organizational ideology" over the last two decades. However, in judging the role of the secretariat, we have to consider its room for maneuvering and the question of timing. The secretariat's role may be significant in initial phases when institution building and learning dominates the agenda, or if conflicts are not too strong between the parties. However, it is hard to envisage how to create an "organizational ideology" or a common "geist" among the members on an issue that is so politicized as that of the whaling issue. It is my impression that the secretariat has done its utmost to be *neutral*, which has probably been the only feasible strategy to secure necessary legitimacy. One could argue that the secretariat should have played the role of an *activist*, like the United Nations Environmental Program (UNEP) and Food and Agriculture Organization (FAO) secretariats did. If this strategy had been used, it is my guess that the whaling nations would have left the organization. However, more recently, the secretary of the IWC seems to have become more active in trying to build bridges among the contending parties.[41] The new dialogue represents a potential "window of opportunity" for the secretariat.[42] As to the size of the secretariat, fees to the IWC are no longer symbolic. Japan pays the highest membership fee, presently some 80,000 British pounds a year.[43] The annual budget for 1999 was close to 1.3 million British pounds, and there are three professionals and twelve support staff. These may not be impressively high figures, but international secretariats tend to be rather small, and in comparative terms, the IWC secretariat is average, considering its functions and scope of membership.[44]

It is a paradox that when thousands of blue whales were killed, the IWC was administered out of a cigar box. Now, when there is virtually no management of whales, it has grown into a fairly large organization with a professional secretariat and hundreds of participants in various categories at the yearly meetings, illustrating the growth of international organizations, sometimes irrespective of the functions ascribed to them.

All things considered, I cannot see that there is much to be gained by increasing the size of the secretariat or by giving it roles other than those it already performs. However, it is important that the secretariat play a role if the process of informal negotiations is to continue.

Reporting and Verification: Better than Average

Another mode of increasing the effectiveness of an international regime is to improve its procedures for reporting on, and verification of, relevant data received from the members. Overall, the record of the international community is not impressive.[45] National reporting is the main rule, and independent verification is the exception, but overall progress has been noted recently.[46] As inaccurate reporting of catch by some actors was believed to be *one* major reason for the rapid depletion of Antarctic whales, a neutral inspection system was suggested by Norway as early as 1955.[47] However, because of procedural stalling and political opposition, mainly from the USSR, it took *twenty years* to establish an inspection system, and then it was only in a modified form. It has been argued, and rightly so, that the IWC still lacks "enforcement authority and [has a] weak central monitoring capacity."[48]

The question, however, is whether the comparison should be made with some kind of management *ideal* or with other international fisheries and environmental regimes. The above observation is certainly valid from the first perspective, but the IWC was among the first regimes to establish any kind of inspection system. Initially the system did not function very effec-tively, but over time it improved considerably through the establishment of various subcommittees, which investigated infractions as well as methods of catch. In fact, the fewer whales taken, the more thorough was the scrutiny. This was not only because it was easier to track the more limited catch but also because more political energy was invested in the system—not least by the anti-whaling states.

Presently, a main issue of contention between whaling and nonwhaling nations is the so-called Revised Management Scheme. While nonwhaling na-tions argue for the need for stronger inspection procedures, pro-whaling na-tions tend to see this as a deliberate tactic to stall implementation of proce-dures to resume commercial whaling. Just as the management procedure adopted is very conservative, there is every reason to believe that an inspec-tion system, if ever adopted, would also be very strict.

Recently the IWC has stood forth as an institutionally rather advanced regime, much in line with the call for transparency, accountability, a solid knowledge base, advanced enforcement procedures, and even extensive use of majority voting. This institutional development, achieved essentially within the framework of the original 1946 convention, which has been amended only once,[49] has been made possible by actively utilizing the flex-ibility inherent in the convention. Thus, instead of adopting protocols re-flecting new commitments, in line with practice within more modern envi-ronmental regimes, the same has been accomplished through changing the Schedule.

In part, this process of modernization reflects the general trend within international regimes, as well as the activity and creativity of the anti-whaling forces. One piece seems to be missing, however: it would seem log-ical that the majority also changed the very purpose of the convention to ex-plicitly forbid commercial whaling. As this would demand the consent of all parties, it is clearly not attainable. The result would probably be that the na-tions previously engaged in whaling would leave and try to set up an *alter-native* to the IWC.[50] Thus, the majority prefers to continue to live with the present seemingly schizophrenic situation: to prevent commercial whaling within a legal framework that demands the parties "secure an orderly de-velopment the whaling industry."

Although it may be argued that some of the institutional features have been "misused" by the majority, there is nothing wrong with the IWC institutions as such. Thus, we are not able to "design" a way out of the current stalemate, because the problems encountered by the IWC are *political*, not *institutional*.[51]

SLIM CHANCES FOR A POLITICAL COMPROMISE

If a political compromise is reached, the present organizational structure of the IWC is probably better equipped than most others to serve as a body for the management of a global resource.[52] In chapter 7 of this volume, "Negotiating in the IWC Environment," Robert Friedheim suggests some points "on how to break the stalemate" that I will use as my point of departure. They can, somewhat impressionistically, be lumped into three broad categories:

1. Higher bargaining profile for whaling states (strategic actions, preparation of a BATNA[53])
2. Traditional means to break a stalemate (leadership, an intervenor or mediator, civility and trust, quiet diplomacy, rethinking national positions, promoting and relying on external change)
3. Issue-specific factors (completion of the Revised Management Plan and the Revised Management Scheme, reliance on the United States)

Will the Whaling Nations Become "Tougher"?

The main pro-whaling countries (Japan, Iceland, and Norway) have adopted different strategies toward the more recent developments in the IWC.[55] While Iceland did not lodge a formal objection to the 1982 moratorium resolution, Japan and Norway did, but Japan withdrew its objection because of pressure from the United States. Iceland conducted whaling for scientific purposes until 1989 but has since then conducted no whaling. Iceland withdrew from the IWC in 1992. Norway quit commercial whaling in 1987 but continued whaling for scientific purposes and resumed commercial whaling in 1993. Japan has been conducting scientific whaling only since the 1987–88 season. Japan has been most *loyal* in the sense that it withdrew the objection to the moratorium, conducted only scientific whaling, and remained in the IWC. Iceland chose the *exit* strategy, due to frustration with developments in the IWC. Norway has loyally remained in the IWC but made its voice heard by resuming commercial whaling.[56] What does this snapshot of strategies indicate regarding the call for a more decisive and strategic position?

Iceland seemingly comes closest to a tough position because of its exit strategy. Moreover, Iceland has been the main architect in establishing NAMMCO. Thus, Iceland has demonstrated the high saliency of the issue and has indicated that it may be willing to pay the price and take some risks in order to resume whaling. This would have been a valid observation until the early 1990s. However, taking a closer look at the *domestic* game in relation to the whaling issue more recently, one gets an impression of *indecisiveness* and a lack of *unified strategy*. So far the domestic actors, fearing negative economic consequences of resumed commercial whaling, have outvoted the forces in favor of resumed whaling. This tendency has been strengthened by lack of support from Norway in making NAMMCO into a real alternative to the IWC. However, in contrast to Iceland, Norway was able to elaborate a unified high-level political strategy, orchestrated by the Norwegian Prime Minister Gro Harlem Brundtland. In this process, Norway demonstrated the high salience of the issue and its willingness to take risks, in line with the call for tougher and more strategic action by the pro-whaling nations.

Somewhat paradoxically, however, so far this successful unilateral national strategy has not strengthened the position of the pro-whaling forces as such. Norway demonstrated that it *was* possible to conduct commercial whaling and still be an IWC member, although the majority does not approve of it. As Norway achieved its main goal *within* the present IWC, the credibility of the threat to exit has been reduced. Norway still strongly rejects the policies of the IWC, but just as the protests of the IWC against Norwegian whaling have become routine, so have the Norwegian protests against the IWC become "the same procedure as last year." This development has also reduced Norway's need for NAMMCO, and thereby reduced the credibility of this card. Less-than-friendly relations between Iceland and Norway over fishing rights also have prevented a closer collaboration between the two on the whaling issue.

Japan seems to be in much the same position as Iceland, not being able to hammer out a consistent high-level strategy to deal with the issue.[57] The forces that fear potential negative actions from a tougher position are much stronger than those wanting a hard line in the IWC. Not least because of the culture of its political system, Japan seeks consensus and continues with large research programs. Probably partly because of its inability to come up with a unified political strategy, Japan is very preoccupied with international *legal* matters. It may well be that the adoption of the Southern Ocean sanctuary, as well as other IWC decisions, runs contrary to the convention as well as to international law. However, within the present international system, violation of international law rarely means that political realities can be changed.[58] Looking forward, one observer concludes that "nothing that Japan could offer [could] make the opponents of whaling more amenable to Japan's wishes."[59] Since the whaling issue simply is not regarded as important enough, compared to wider Japanese interests, the occasional Japanese threat of leaving the IWC does not seem credible. Thus, the only *major* actor on the pro-whaling side presently does not seem to have much to offer in the sense of a more decisive or "tougher" policy toward the IWC majority.

Compared to the much larger and still very unified anti-whaling coalition, the pro-whaling coalition has been loose and fragmented.[60] Those hoping that a strong and unified action on the part of the pro-whaling states will change the cost-benefit calculations of the anti-whaling nations and thereby their positions will probably be disappointed. Although on the same side, their specific situations seem sufficiently different to prevent joint and forceful action. So far, the anti-whaling forces have achieved their main goal in the sense that both Iceland and Japan have been split domestically, and the fear of negative economic consequences is greater than the wish to resume whaling. The issue is not salient enough.

Traditional Means to Break a Stalemate: Will They Work?

There are few regimes in which there is more *need* for the traditional means to break a stalemate, but the chances that they will work in the IWC are slim because of the complete lack of trust and understanding among the parties. In the following section I will discuss briefly two of the points suggested by Friedheim in chapter 7 of this volume: quiet diplomacy and a somewhat revised version of "rethinking national interests."

QUIET DIPLOMACY

Recently, IWC negotiations have not been conducted in an atmosphere conducive to reaching compromises. Multilateral negotiations with close media scrutiny and a strong presence of NGOs tend to highlight differences rather than possible room for compromise. In international negotiations, there will always be a need to play for the gallery, and this can put pressure on reluctant actors. But there is also need for more quiet forums to try to resolve knotty issues. Thus, within most international regimes some sessions are open and others are closed. Often the parties also see the need to meet between sessions within a more limited group of participants.[61] This has not been a prominent feature within the "new" IWC—there was such a process some years ago, but that did not amount to anything. As the anti-whaling side, to a large extent, has been driven by the "green" NGO agenda, which is not known to appreciate quiet diplomacy, there has been no basis for it.

However, over the last couple of years, many of the IWC commissioners have retreated to peaceful places in the Caribbean, where more exclusive intersessional negotiations have been conducted. The basis for these negotiations has been the so-called Irish Proposal, suggesting a ban on all high-seas (pelagic) whaling and on all trade in whale products, but allowing limited coastal whaling for local consumption. At the first meeting in Grenada in January 1997, the secretary to the IWC was optimistic: " My hope is that a compromise will be reached by the fiftieth anniversary of the IWC in 1998."[62] However, the British commissioner was quick to denounce this optimism, and no progress was noted on the issue at the IWC meeting in 1997.

At the commissioners' meeting in Antigua in February 1998 it appeared that things were back to normal. The Norwegian commissioner stated that "there is almost no hope whatsoever of a compromise."[63] The chairman of the Norwegian Whalers Union used stronger language in a comment to the meeting: "The IWC has been dead for a long time; it is now high time to bury it once and for all."[64] The IWC was not buried prior to its fiftieth anniversary in Oman, but no progress was achieved. At present, it is uncertain whether this process remains alive. The Irish proposal may have been an effort to "fix the regime" by the fiftieth anniversary. Because it did not succeed, the willingness to seek a negotiated outcome may well die.

The premise for the Irish proposal is that the anti-whaling states will have to give the most, in principle, as they have to accept that commercial whaling can be conducted, implying a fundamental change of policy. For their part, the whaling states will have to abandon any plans for a real revitalization of the whaling industry; what would be allowed would be a *minimum version* of whaling, which might be very difficult for the whaling nations to accept. It may be even more of a challenge to convince the anti-whaling states that they should accept commercial whaling. As nongovernmental actors play a main role on this issue, I will include them in the ensuing discussion.

RETHINKING ROLES AND POSITIONS

While the *scientists* set the ball rolling in the mid-1960s by calling the world's attention to the overexploitation of whales, recently most of them have been mute in public debate. Those I have referred to as "footnote scientists" have tended to dominate the scene, but with the emerging scientific consensus, even most of them have become more silent. There is a real danger that many of the most able scientists, having elaborated the Revised Management Procedure, may avoid future involvement in the IWC. Some scientists left in the 1980s, and the chairman of the Scientific Committee resigned in 1993. For any good architect of new important structures, it would be demoralizing never to see your work put to use. Scientists have essentially finished their most important work through completion of the comprehensive assessment and the Revised Management Procedure. Thus, if the present policy is upheld, the role and position of the committee seems uncertain. Apart from scientific questions related to aboriginal whaling, more recently they have been kept busy by examining issues like the consequences of global warming on marine mammals. This and other general and long-term issues may be important, but more basic scientific work on whales can surely be conducted within the framework of the International Committee for the Exploration of the Seas and other international scientific bodies. As there are other regimes and other research questions that might welcome their skills, why waste them on an organization that does *not* want to utilize their most important work?

In contrast to the scientists with close connections to the "green" movement, the large majority of scientists have preferred to remain in their ivory tower. This may be in line with traditional scientific standards, and it may be heretical to suggest any change, but perhaps some of them could show more active involvement in policy issues. Experience from other regimes has shown that "activist" scientists may still keep their scientific integrity, even if they take a stand on an issue, provided their arguments are scientifically sound.[65] In the words of a South African member of the IWC Scientific Committee: "The time seems overdue for scientists to speak out against the near-farcical pronouncements of some international organizations regarding endangered species."[66]

As noted, the vast majority of *nongovernmental actors* have lobbied against whaling, but they are not a homogenous group. One group consists of the traditional large environmental groups like Greenpeace, Friends of

the Earth, and World Wildlife Fund, active in virtually all environmental issues. Among these, Greenpeace has been particularly active on the whaling
issue.[67] It is worth noting, however, that more recently Greenpeace has been
less active; that is, Norwegian commercial whaling has taken place almost
unnoticed and without much disturbance, which would have been unthinkable a few years ago. This may in part be a result of Greenpeace's severe financial problems,[68] but it may also be because whaling is not as hot
an issue as it was in the 1980s. A few years ago it was also reported that some
of the major NGOs were ready to accept a resumption of commercial whaling, but nothing came of it.[69]

Considering the magnitude of real environmental threats like global
warming, loss of biodiversity, the thinning of the ozone layer, acid rain, and
radioactive waste, and given that these groups have limited resources, the
relative high priority given to the whaling issue may seem strange. Sidney
Holt's statement more than a decade ago that "saving the whales is for millions of people of the world a crucial test of the world's ability to halt environmental destruction"[70] made at least some sense at the time. Now, however, it may be time for environmentalists to realize that *they have won the
battle to save the whales.* The whaling issue was an excellent first target for
the young international environmental movement in the early 1970s; the
whaling industry had diminished greatly and was therefore very weak, the
problem was easy to visualize, and massive failure to conserve the largest
animals on earth could be easily demonstrated.

In the 1990s, when *sustainability* is the key word among environmentalists, in *rational* terms it is difficult to understand the massive opposition toward carefully managed harvesting of species that are clearly *not* threatened.
It is of course perfectly legitimate to be against commercial whaling for
moral or other reasons, as are the various animal protection groups. Some
major environmental groups, however, still cling to the "whales are threatened" argument. Ideally, they should *rethink* their roles and positions on the
issue. Their position on the whaling issue may also, in the long run, destroy
their credibility on other issues, where they generally use scientific findings
as the main reason for action.[71]

As long as the whaling issue is a big moneymaker and the large majority
of people in anti-whaling countries prefer to *watch* whales rather than *eat*
them, no sudden shift should be expected. Since organizations like Greenpeace are much more influential on the issue than are most IWC members,
the likelihood of any compromise in the IWC is significantly reduced. The

commissioners may be "allowed" to negotiate in relative peace in the Caribbean, but the NGOs will be there when the commissioners return.[72]

The current IWC membership of some forty state parties is fairly small, considering that the IWC is a global organization. Usually representatives of some thirty states show up for meetings, less than the number of parties to the regional acid rain regime. The large majority of the world's states do not find the issue important enough to join the organization.[73] Moreover, key members are few and the largest group is made up of "bystanders." They may be important when it comes to voting, but they have only marginal interest and involvement in the issue.[74]

Underlining the marginal nature of the issue; public opinion on whaling may be strong in some quarters, but knowledge is extremely limited.[75] This might make a "rational" outsider with an eye for cost-effectiveness conclude that the whole IWC should be dissolved and the issue left to the few with some kind of interest in the issue. However, as demonstrated in this volume as well as in a number of other analyses, those who do have an opinion have a very strong opinion.

Some of the more fundamentalist states are the United Kingdom, Australia, and New Zealand. The two newcomers, Austria (1994) and Italy (1997), also appear to belong to this category. France also seems to be a rather uncompromising actor, playing a key role in getting the Southern Ocean sanctuary adopted. As these states are essentially against whaling, irrespective of what scientists say, it is hard to see that they will have any interest in compromise.[76]

Most of the anti whaling states cannot brag about particularly favorable international environmental profiles, making the whaling issue a free ticket to improve their environmental images.[77] Moreover, except for New Zealand, where the Maori have shown support of aboriginal whaling, none of these countries has domestic constituencies to consider, and the environmental movement is very influential on the issue. Unless the environmental movement changes its position, the "fundamentalists" are not likely to change their policies.

Some of the countries on the anti-whaling side, like Sweden, Germany, Switzerland, and Ireland, seem less uncompromising. Ireland has demonstrated that initiatives can be taken by smaller players in an attempt to break the impasse. So far, however, a distinct middle ground has been absent in the IWC, illustrating the polarization of the issue. As the issue is so marginal in most countries, some changes may come about as a result of new persons entering the scene, but such shifts are often unstable. Clearly, if a more distinct middle ground emerged, the possibilities for compromise would increase. The potential reward for the brokers would be success in breaking the decade-long stalemate in the IWC. As the chances of success are so slim, and the public and the NGOs continue to oppose whaling in these countries as well, potential domestic losses outweigh potential international gains.

The next category of countries within the IWC is the aboriginal whaling countries, or the *acceptable whaling countries:* the United States, Denmark, Russia, and St. Vincent and the Grenadines.[78] It has been stated that "the disintegration of the international whaling regime stems from the persistence of a method of distinguishing between permissible and impermissible whaling that is now regarded by many as arbitrary."[79] Be that as it may, there could be reason to expect that these countries, as "acceptable" whaling nations, could represent a bridge between the "unacceptable" whaling nations and the anti-whaling countries. To some extent this role has been played by Denmark, a significant player in the IWC at recent meetings. However, as Denmark's main goal is to protect its aboriginal catch (in Greenland), it is not likely to be a driving force in changing the status quo. Denmark also has a "green" image to foster, and it is also influenced by the fact that the European Union opposes commercial whaling.

Russia played a passive role for many years after it quit commercial whaling in 1987, usually sending only one delegate to IWC meetings. More recently it has become more active, as witnessed by the deal with the United States on aboriginal catch at the 1997 IWC meeting. Although Russia has been rather passive so far, it is a potentially interesting actor because of its whaling tradition, its observer status in NAMMCO, and its reservation against the 1982 moratorium, which has not been withdrawn. However, in the present situation, Russia is not a very predictable player. Given its crowded agenda, the issue of whaling is hardly among its most pressing issues.

In short, neither "bystanders," "fundamentalists," "potential middle-grounders," nor "acceptable whaling nations" at present seem to have sufficient motivation to work actively for a compromise on the issue.

Issue-Specific Factors: Relying on the United States?

Friedheim argues that "The United States should not be relied upon to salvage the situation."[80] Based on the role played by the United States in the IWC so far, this makes sense. The United States has been by far the most important actor in changing the IWC to bring about an end to commercial whaling. In order to protect its own interests, the United States was also the main architect behind the "double standard" system: distinguishing between permissible and impermissible whaling. Although the United States has been the *group leader* for the anti-whaling forces, it hardly qualifies as an *entrepreneurial* leader, as all this has taken place at the expense of an opposing minority.[81]

Although this leads one to think that not much should be expected from the United States in breaking the current stalemate, I will argue that *if* a compromise is to be achieved, the United States is the only actor that can bring it about. Or, if the United States does not play a key role in the process, nothing will come out of it unless the outcome is acceptable to the United States. When I was studying the whaling policies of Norway and Iceland since the 1982 moratorium, it became crystal clear that it was *not* the IWC as a regime that was considered the main opponent by these countries.[82] Rather, it was the combined weight of the political, legal, and economic threats presented by the United States.

There is every reason to believe that the same goes for Japan. The main reason for Norway's successful resumption of whaling was *not* increased understanding for the Norwegian position by the environmental movement or the IWC majority, but rather, it was a *tacit but de facto acceptance by the United States*. In return, Norway probably had to promise that it would continue to operate within the IWC.

In chapter 7, Friedheim mentions the pressure the United States is under from constituents on both sides of the issue as a main reason not to rely on the United States. Nevertheless, the fact that such pressure exists indicates that there are some forces more favorable to whaling, a community that is essentially absent in most other anti-whaling countries. The United States also has a much larger scientific community involved in the issue than other anti-whaling states. In principle, the United States also accepts that whaling can be conducted when done on a scientifically sound basis but is not yet ready to accept it in practice. As the leading scientific nation of the world, in the long run this contradiction may pose difficult dilemmas, and precedents from other issue areas should be considered.

In comparative terms, the salience of the issue is quite high in the United States, as reflected in the very large delegations the United States has sent to the IWC meetings since the moratorium.[83] There is also some debate over the whaling question in the United States, demonstrated by this book and the fact that three previous U.S. commissioners have publicly spoken in favor of carefully resumed whaling.[84] Also, as a whaling nation, the United States cannot use the same moral arguments against whaling as most other anti-whaling nations can. Finally, as the remaining superpower, the United States, at least in principle, should be concerned with the orderliness of global governance as well as its relations to allies like Japan, Iceland, and Norway.

Thus, the role and interests of the United States is quite different from most other anti-whaling states. While interests are essentially monolithic in most anti-whaling countries, in the United States they are broader and more diversified. Moreover, the United States is the only actor capable of changing the present rules of the game. Does this mean that the United States will *rethink* its national position on the whaling issue? The fact that Norwegian commercial whaling is de facto accepted indicates that there has been some change. The issue does not seem to be quite as salient as it used to be. This opens up a kind of "catholic situation" in which commercial catch (sin) is forbidden but an "acceptable level of sinning" is tacitly accepted.[85] Moderate "sinning" may be tolerated from Norway, since this is "legal sinning"; it is less likely that commercial whaling by Japan and Iceland will be tolerated. Moreover, the U.S. position is decided by national legislation, and the likelihood of a "real" change at present is small, although the legality of its unilateral measures has been questioned. As the U.S. whaling policy, along with most other U.S. foreign policy, is basically domestically driven, there has to be a major change within and among domestic coalitions for a new policy to come about on the whaling issue. Although the whaling issue is not so "hot" today, observers from the U.S. whaling delegation have pointed out that this may well change.

CONCLUSION

New students of the International Whaling Commission are often shocked and surprised over what they find when they start studying the IWC. With the amount of tension present and the mutual lack of understanding among its members, as well as harassment of the minority, how can this regime continue to stay alive? Consequently new observers tend to predict that, unless the regime is mended in some way soon, the "fall," or at least breakup, of the regime is just around the corner, usually at the next IWC session.

I have followed the development in the IWC fairly closely since the mid-1980s. This is not an especially long time, but it is long enough to understand that the IWC may well continue its present course for many years, without mending or breaking, as it has essentially done over the least decade or so. As one member of the U.S. delegation to the IWC pointed out, the present situation in the IWC hardly qualifies as a crisis.

Clearly, the present regime is not satisfactory for the pro-whaling nations, but after all, Norway is "allowed" to whale commercially and Japan is "allowed" to whale for scientific purposes and thereby to keep its whaling fleet intact. The main reason the whaling nations have not broken out or created an alternative is probably that the issue is not *salient* enough for them. In economic terms, whaling has no significance today. Nevertheless, it touches upon important questions regarding the role of science in international resource management and the self-determination of states, and it is a bone of contention between otherwise close allies. Thus, ideally, some sort of compromise should be worked out.

In first discussing what kind of regime the IWC should work under, I concluded that institutionally the IWC is a rather good regime that does not need to be fixed. Other design criteria, like giving more influence to the major stakeholders, may be quite effective and dynamic within other regimes, but it would not be feasible for the IWC. Although some of these institutions may have been misused by the majority to achieve their main goal, the fundamental problem of the IWC is not bad institutions but political conflict, as the parties do not agree on the goal of the regime.

In discussing whether there is a chance for a compromise within the framework of the present institutional structure of the IWC, I conclude that at least from a short-term perspective, the chances are slim. As the pro-whaling forces are weak and fragmented, they are not likely to be able to force any viable compromise on the much larger, better unified, and more powerful anti-whaling coalition. Recent attempts to break the stalemate through more quiet diplomacy have been unsuccessful. As the compromise proposal suggests that limited commercial whaling should be permitted, the anti-whaling forces will have to abandon their principal view.

In discussing the positions and interests of these various groups, I conclude that it is not likely that they will change their position. The only actor on the anti-whaling side with a possible incentive and ability to change the IWC is the United States, because of its more varied interest and unique power base. Although there has been some softening of the U.S. position, it does not seem likely that it will go much further. Thus, I do not foresee any quick solutions to the problems of the IWC, but I may well be wrong. As witnessed by the fall of the "iron curtain," international politics can be full of surprises—not only for casual observers but also for the social scientists giving the issue close scrutiny.

In a longer time frame, the picture is more uncertain. Regarding some environmental issues one can always hope that "technological fixes," or "external shocks," will open unexpected "windows of opportunity" and pave the way for solution or progress on knotty issues. "External shocks" were instrumental in visualizing the slaughter of whales and thereby changed the rules of the game within the IWC. It is more difficult to imagine that pictures of abundant and healthy whale species will increase the likelihood of a public call for commercial resumption of whaling in key Western countries. These animals have been symbols for so many, and for so long, that a "demystification" will take time. Ideally, a new and more flexible perception is needed; some whale species should keep their role as objects of whale watching only, while others should be allowed to be utilized for other purposes.

Another possible future scenario is the one presented by David Victor (in chapter 10 of this volume); sooner or later the demand for whale products will diminish and the problem will thereby be "solved." An alternative view is that killing of whales for human consumption will continue, but the international legal framework for continued whaling is more uncertain.[86] At the end of the 1980s it seemed that whaling was about to end, at least within the framework of the IWC. Since then, however, Norway resumed commercial whaling, Japan has increased its scientific catch, and even the aboriginal IWC quotas are increasing. There are no signs of decline in catch of cetaceans outside the realm of the IWC, but rather the contrary. Considering that this was all "allowed" to happen, it may well be that pro-whaling forces will gain in strength, while anti-whaling forces will decline.[87] Consider also that a country like Japan is usually much more patient and has a longer time perspective than Western societies. This may be a weakness in the present situation but an asset in the long run.

Finally, although there are no viable alternatives to the IWC today, in a longer time frame this may be different. Although the prospects for alternatives still are meager, the longer it it takes for the IWC to be "fixed," the higher the likelihood that some kind of alternative will see the light of day.

NOTES

1. A regime meeting the "tests of sustainability and equity" has been suggested by Oran R. Young et al., "Subsistence, Sustainability, and Sea Mammals: Reconstructing the International Whaling Regime," *Ocean and Coastal Management* 23 (1994): 117–27. Their assertion, along the lines of "small is beautiful," is that small-type whaling, be it commercial or not, should be allowed, provided certain conditions are met. Their suggestion is not very different from the so-called Irish proposal that I will comment upon later on.

2. Within the field of environmental regimes, economists have long argued for the need to increase cost-effectiveness and have suggested taxation on emissions and a system of tradable permits toward that end. John Dales, *Pollution, Property, and Prices* (Toronto: University of Toronto Press, 1968). Failures of international fisheries regimes are much more plentiful than successes, since fisheries are a classic case of market failure, as has been vividly demonstrated in relation to whaling. Again the economists have the remedy: avoid overcapitalization and privatize resources. C. Heck, "Collective Arrangements for Managing Ocean Fisheries," *International Organizations* 29, no. 3 (summer 1975).

3. Mancur Olson, *The Logic of Collective Action: Public Goods and the Theory of Groups* (Cambridge: Harvard University Press, 1965).

4. In chapter 10 of this volume David Victor argues essentially that although the regime is far from any "ideal," overall it is "good enough" for most of the parties. In chapter 6 of this volume, Elizabeth DeSombre identifies a number of the IWC's weaknesses, although she correctly points out that many of them are not unique to the IWC. However, she also argues that more recently the IWC has been a rather "effective" regime, but *effectiveness* is not defined.

5. Edward L. Miles et al., *Explaining Regime Effectiveness: Confronting Theory with Evidence* (Cambridge: MIT Press, forthcoming).

6. Because of the controversial nature of this regime, we discussed this ranking extensively at our last project meeting in Seattle, January 1998, and decided to keep the score. But as the final manuscript has not been delivered, it may still be adjusted. However, the fact that the IWC manages only a tiny fraction of the cetaceans currently caught indicates a rather low score also from a problem-solving perspective.

7. For an elaboration of the definition of *effectiveness*, see Miles et al., *Regime Effectiveness*, chap. 1. For an earlier contribution along essentially the same lines, see Arild Underdal, "The Concept of Regime Effectiveness," *Cooperation and Conflict 27*, no. 3 (1992): 227–40. For other definitions of *effectiveness* see Young et al., "Subsistence"; Marc Levy et al., "The Study of International Regimes," IIASA Working Paper, November 1994, WP 94–113; T. Bernauer, "The Effect of International Environmental Institutions, How We Might Learn More" *International Organization 49* (spring 1995): 351–77.

8. Norway is the only IWC member conducting commercial whaling, and the quota for 1998 is set at 671 minke whales.

9. The IWC cannot take credit alone for this development. Reduced demand for whale products has clearly made a difference. Moreover, the IWC as such would probably not have been able to make its members quit whaling without the bilateral means used by the United States. Steinar Andresen, "The Making and Implementation of Whaling Policies: Does Participation Make a Difference?" in *The Implementation and Effectiveness of International Environmental Regimes: Confronting Theory with Practice*, ed. David G. Victor, Kal Raustiala, and Eugene B. Skolnikoff. (Cambridge: MIT Press, 1998), 431–75.

10. Whether there is in fact scientific consensus will be discussed later.

11. Alf Hakon Hoel, *The International Whaling Commission, 1972–1984: New Members, New Concerns* (Lysaker, Norway: Fridtjhof Nansen Institute, 1985).

12. John A. Gulland, "The End of Whaling?" *New Scientist* (29 October 1988): 42. Gulland was among the group of so-called wise men brought into the IWC in the early 1960s as independent scientific experts.

13. In fact, this was an argument used by another of the reviewers of Miles et al., *Regime Effectiveness*, and this "problem" is also hinted at by Robert L. Friedheim in a comment regarding our earlier work on the effectiveness of the IWC. Robert L. Friedheim, "Moderation in the Pursuit of Justice: Explaining Japan's Failure in the International Whaling Negotiations," *Ocean Development and International Law 27* (1996): 349–78. I guess we are all to some extent affected by core values, if there are any, in the societies in which we live. Moreover, it should be noted that I have taken part in working group sessions (not commission meetings) on the Norwegian delegation twice (1989–90) to increase my insight into the issue. In part because the meetings tended to be "the same procedure as last year," and in part to avoid close connections with the government on the issue, I have not participated since. Finally, I may well be biased, but if so, I believe I am in good company with a number of other analysts of the whaling issue, be they from the United Kingdom or the United States.

14. Robert O. Keohane, *After Hegemony: Cooperation and Discord in the World*

Political Economy (Princeton, N.J.: Princeton University Press, 1984); Peter Haas, *Saving the Mediterranean: The Politics of International Environmental Cooperation* (New York: Columbia University Press, 1990).

15. These (and other) features have been used in relation to the climate regime by myself and Jørgen Wettestad, "International Resource Cooperation and the Greenhouse Problem," *Global Environmental Change* (December 1992): 277–91. They have been elaborated since then by Wettestad in *Designing Effective International Environmental Institutions: The Key Conditions* (Cheltenham, England: Edward Elgar, 1999). On the question of access and participation, see David Victor et al., eds., *The Implementation and Effectiveness of International Environmental Commitments: Theory and Practice* (Cambridge: MIT Press, 1998).

16. Dan Bodansky, "The United Nations Framework Convention on Climate Change: A Commentary," *Yale Journal of International Law* 18, no. 2 (1993): 451–558. One of those pushing the hardest for majority-voting procedures was the then Norwegian Prime Minister, Gro Harlem Brundtland, leader of the UN Commission on Environment and Development. However, she has also been the main architect behind the Norwegian resumption of commercial whaling (Andersen, "Whaling Policies," 1). In this setting, she has *not* been very happy with the position of the majority. This raises the problem of *consistency:* "international democracy" is fine when one belongs to the majority but is more frustrating when one belongs to the minority.

17. Johan Nicolay Tønnessen and Arne Odd Johnsen, *The History of Modern Whaling* (London: C. Hurst, 1982).

18. Olav Stokke, "The Effectiveness of CCAMLR," in *Governing the Antarctic: The Effectiveness and Legitimacy of the Antarctic Treaty System,* ed. Olav Stokke and Davor Vidas (Cambridge: Cambridge University Press, 1996).

19. For a much more extensive coverage of the role of science in the IWC, see chapter 3, "Science and the IWC," by William Aron, in this volume. See also Steinar Andresen, "The Whaling Regime," in *Science and International Environmental Regimes: Integrity and Involvement,* Arild Underdal, Steinar Andresen, Tora Skodvin, and Jørgen Wettestad (Manchester, England: Manchester University Press, 2000). Tore Schweder and Patricia Birnie give somewhat different accounts of the role of science in the 1950s. Tore Schweder, *Intransigence, Incompetence, and Political Expediency? Dutch Scientists in the International Whaling Commission in the 1950s: Injection of Uncertainty* (Oslo: University of Oslo, 1994); Patricia Birnie, *International Regulation of Whaling: From Conservation of Whaling to Conservation of Whales and Regulation of Whale-Watching,* 2 vols. (Dobbs Ferry, N.Y.: Oceana, 1985).

20. Schweder, *Intransigence.*

21. For discussion of this transformation of the IWC Scientific Committee, see

James E. Scarff, "The International Management of Whales, Dolphins, and Porpoises: An Interdisciplinary Assessment," *Ecology Law Quarterly* 6, no. 2–3 (1977): 323–438; Birnie, *International Regulation;* M. J. Peterson, "Whalers, Cetologists, Environmentalists, and the Management of Whaling," *International Organization* 46, no. 1 (winter 1992): 147–86; Schweder, *Intransigence;* Ronald Mitchell, "Forms of Discourse, Norms of Sovereignty: Interests, Science, and Morality in the Regulation of Whaling," *Global Governance* 4(1998): 275–93; Andresen, "Whaling Regime," as well as chapter 3, "Science and the IWC," by William Aron, in this volume.

22. For an interesting account of the five competing procedures developed in connection with the development of the Revised Management Procedure, see Schweder, *Intransigence.* As to the question of scientific *consensus,* traditionally there has been a small group of "footnote scientists" (a small minority of anti-whaling scientists tending to oppose the majority view) with close ties to the green movement. It was long believed that full consensus would be achieved on the latest estimate of the Northeast Atlantic minke whale stock, but in the end, one member expressed reservation. Andresen, "Whaling Regime."

23. Phillip Hammond, letter of resignation to Ray Gambell, secretary of the IWC, 26 May 1993; Ray Gambell, "International Management of Whales and Whaling: An Historical Review of the Regulations of Commercial and Aboriginal Subsistence Whaling," *Arctic* 46, no. 2 (1993): 97–108.

24. D. S. Butterworth, "Science and Sentimentality," *Nature* 357 (18 June 1992), 532–34.

25. Arild Underdal, *The Politics of International Fisheries Management: The Case of the Northeast Atlantic* (Oslo: Universitetsforlaget, 1980).

26. When national quotas were finally introduced in the early 1960s, they did little or nothing to improve the situation as they were introduced too late. Tønnessen and Johnsen, *Modern Whaling.*

27. Although there were those who argued for the establishment of national quotas, it does not make sense to blame the creators of the IWC for adopting a total quota. This approach was quite novel at the time; they thought it would be an effective limit on catch, and it does not make sense to judge the past with the eyes of the present. At least initially, this was more a case of knowledge failure rather than policy failure.

28. Peterson, "Whalers," 147.

29. Birnie, *International Regulation.*

30. Although the United States has seen to the implementation of the moratorium, it would have been much more difficult for the United States to stop whaling without the political and legal basis provided by an international organization.

31. Andresen, "Whaling Regime."

32. As this was not the only factor explaining its better performance during this period, it cannot be claimed that there is a causal link between the pattern of participation and the outcome produced.

33. For an account of the increase in the number of scientists, affiliated with NGOs as well as states, in the IWC, see Andresen, "Whaling Policies." See also Friedheim in the introduction to this volume.

34. For an elaboration of these points, see chapter 6, "Distorting Global Governance: Membership, Voting, and the IWC," by Elizabeth DeSombre, in this volume.

35. A small minority of the NGO community, that is, the Sea Shepards, have used what some would label terrorist means. This tendency could increase if the NGOs were not included in the ordinary political game.

36. High North Alliance, Reine, Norway, 11 June 1998.

37. Wettestad, *International Institutions*.

38. Until then, the Ministry of Agriculture and Fisheries in the United Kingdom had provided part-time administrative assistance.

39. Communication with William Aron, U.S. delegate and commissioner to the U.S. delegation in the 1970s, 5 January 1998.

40. Peterson, "Whalers," 156.

41. Ray Gambell, "The International Whaling Commission Today," in *Whaling in the North Atlantic: Economic and Political Perspectives*, ed. Gudrun Petursdottir (Reykjavík: Fisheries Research Institute, University of Iceland Press, 1997), 47–67.

42. Oran R. Young, "Political Leadership and Regime Formation: On the Development of Institutions in International Society," *International Organization* 45, no. 3 (summer 1991): 281–308.

43. The membership fee is differentiated and calculated in relation to the size of the delegation and type of involvement in whaling. Presently no member pays less than 14,000 British pounds.

44. The secretariat for the Long Range Air Trans-boundary Pollution regime (LRTAP), with about the same number of members as the IWC, consists of five professionals and two support staff. The secretariat consisted of five professionals and six assistants. Finally, the secretariat for Montreal Convention in Nairobi consists of five professionals and eight support staff. *Yearbook of International Co-operation on Environment and Development (1999/2000)* (Earthscan Publications, Ltd.).

45. Jesse Ausbel and David Victor, "Verification of International Environmental Agreements," *Annual Review of Energy and the Environment* 17 (1992): 1–43.

46. Victor et al., *International Environmental Commitments*.

47. Later on it was confirmed that the Soviet Union was cheating on a massive

scale. Peter Stoett, "The International Whaling Commission: From Traditional Concern to an Expanded Agenda," *Environmental Politics* 14, no. 1 (1995): 130–35. Also see Tønnessen and Johnsen, *Modern Whaling.*

48. Peterson, "Whalers," 158.

49. The convention was amended in 1958 to allow for the elaboration of an inspection system. Tønnessen and Johnsen, *Modern Whaling.*

50. Iceland left the IWC in 1992 because it was dissatisfied with the organization. However, its attempt to create an alternative to the IWC, the North Atlantic Marine Mammal Commission (NAMMCO), has so far not been successful. That is, the organization fulfills important functions, but no whaling quotas are allocated. Steinar Andresen, "NAMMCO, IWC, and the Nordic Countries," in *Whaling in the North Atlantic: Economic and Political Perspectives*, ed. Gudrun Petursdottir (Reykjavík: Fisheries Research Institute, University of Iceland Press, 1997), 75–89.

51. Young et al. "Subsistence," 125, concludes that "problems with the regime, while important, are not the primary threats to the continuation of the regime."

52. Robert L. Friedheim, "Fostering a Negotiated Outcome in the IWC," in *Whaling in the North Atlantic: Economic and Political Perspectives*, ed. Gudrun Petursdottir (Reykjavík: Fisheries Research Institute, University of Iceland Press, 1997), 145.

53. BATNA is the Best Alternative to a Negotiated Agreement. See chapter 7, "Negotiating in the IWC Environment," by Robert L. Friedheim, in this volume for a further discussion of BATNAs.

54. Robert L. Friedheim, "Negotiating in the IWC Environment," chapter 7 of this volume.

55. The arguments and discussion on this point in relation to Norway and Iceland are based on research I have conducted, reflected in Andresen, "NAMMCO, IWC, and the Nordic Countries," and Andresen, "Whaling Policies." As to Japan, I rely on secondary sources, primarily Friedheim "Moderation," and interviews and conversations with various members of the Japanese delegation to the IWC. In a comment regarding a previous draft of this chapter, I also received very useful information on the whaling politics of Japan, as well as on other issues, from Milton Freeman.

56. Albert O. Hirschman, *Exit, Voice, and Loyalty* (Cambridge: Harvard University Press, 1972).

57. Japan and Iceland are of course in a much more difficult position than Norway as they have no reservation toward the 1982 moratorium resolution.

58. Moreover, there will always be *some* international lawyers there, ready to defend the interpretations of the other side.

59. Friedheim, "Moderation," 361.

60. The lack of cooperation between Iceland and Norway has already been noted. The fact that Norway did *not* vote against the Southern Ocean sanctuary but preferred to abstain did little to improve Japanese-Norwegian relations on the issue.

61. Robert L. Friedheim, *Negotiating the New Ocean Regime* (Columbia: University of South Carolina Press, 1993); Miles et al., *Regime Effectiveness.*

62. *Aftenposten* (Oslo, Norway), 5 March 1997.

63. High North Alliance, 5 February 1998.

64. Ibid.

65. Scientists with ties to key international organizations played an important role in getting both the climate issue and the ozone issue high on the political agenda.

66. Butterworth, "Science and Sentimentality," 533.

67. Johann Vidar Ivarsson, *Science, Sanctions, and Cetaceans: Iceland and the Whaling Issue* (Reykjavík: Center for International Studies, University of Iceland, 1994); Andresen, "Whaling Policies."

68. *New York Times*, 16 September 1997. It may also be that internal conflicts between environmental and animal rights groups in the United States have weakened the opposition to whaling.

69. *Aftenposten* (Oslo, Norway), 25 April 1994.

70. Sidney Holt, "Whale Mining, Whale Saving," *Marine Policy* 3 (1985): 192–214.

71. The credibility of organizations like Greenpeace and to some extent also w w f has been seriously undermined in Norway because of their position on the whaling issue. All national Norwegian green groups of any significance support sustainable whaling. Just for the record, the same groups are highly critical of other aspects of Norwegian environmental policy, for example, the Norwegian climate-change policy.

72. Some countries also expressed concern that the intersessional diplomacy would not be transparent enough.

73. Based on the fact that there are whales outside of more than a hundred coastal states, it has been suggested that active recruitment of new members is one strategy to change the whaling/nonwhaling ratio. Friedheim, "Moderation." The idea is interesting, but it hardly seems feasible in the short run—considering Japan's modest success in recruiting new members. At least so far the n g o community has been better equipped to deal with recruitment.

74. Andresen, "Whaling Policies."

75. Milton M. R. Freeman and S. R. Kellert, *Public Attitude toward Whales: Re-*

sults of a Six-Country Study (University of Alberta and Yale University, 1992); "New Poll on Public Attitudes to Whaling," *INWR Digest (International Network for Whaling Research)* 14 (December 1997): 3.

76. Within other environmental regimes these countries, not the least the United Kingdom, are very concerned about the scientific message, not the least when it is uncertain and can be used as a basis for lack of action.

77. As Australia stands out as one of the chief "laggards" on the climate issue, receiving the scorn of the environmental community, no softening on the whaling issue can be expected. The United Kingdom has long had the image of being the "dirty man" of Europe, using the whaling issue extremely actively to make some friends among the environmentalists when there is a chance. Similarly, France is eager to propose sanctuaries both in the south and in the north (Arctic) when media attention and political pressure increase on issues like nuclear testing.

78. The yearly catch by St. Vincent and the Grenadines is only one or two humpbacks. The United States will be dealt with separately.

79. Young et al., "Subsistence," 119.

80. See chapter 7 in this volume.

81. Arild Underdal, "Solving Collective Problems—Note on Three Models of Leadership," in *Challenges of a Changing World,* Festschrift to Willy Ostreng (Lysaker, Norway: Fridtjof Nansen Institute, 1991), 139–153. For other discussions of leadership, see Young, "Political Leadership," and R. Malnes, "'Leader' and 'Entrepreneur' in International Negotiations: A Conceptual Analysis," *European Journal of International Relations* 1, no. 1 (1995): 87–112. For a discussion of leadership in relation to the whaling issue, see chapter 7 of this volume.

82. Andresen, "Whaling Policies."

83. Only Japan tends to send larger delegations than does the United States. Andresen, "Whaling Policies."

84. John Knauss wrote that "continuing the moratorium would be a mistake." Some of his ideas have probably been important for the Irish proposal. John A. Knauss, "The International Whaling Commission: Its Past and Possible Future," *Ocean Development and International Law* 28, no. 1 (1997): 79–99; William Aron, in "The Commons Revisited: Thoughts on Marine Mammal Management," *Coastal Management* 16 (1988): 99–110, wrote that the United States should shift its whaling policy in order to "help right some of the wrongs of international whaling regulations." The third U.S. commissioner is Richard Frank; see Richard Frank, "The Paradox of the American View on Utilization of Marine Mammals," *ISANA* 6 (May 1992): 11–13.

85. I owe this point to Arild Underdal in a comment regarding a previous draft of this chapter.

86. Young et al., "Subsistence," 119.

87. One illustration that the killing of marine mammals is no longer able to catch the attention of the public as it used to is the fact that hundreds of thousands of seals are taken every year, mostly by Canada, but you do not see many pictures of the hunt anymore.

PART III

TESTING OUR ARGUMENTS
AND FINDING A SOLUTION

9 / Summing Up

Whaling and Its Critics

CHRISTOPHER D. STONE

R obert Friedheim launches our volume on a somber note. The cur rent regime, in his view, is stalled. The proponents of a sustainable harvest of whales and the protectionists are at loggerheads. The former have some good grounds to cry "foul," but the latter have the votes. In Friedheim's view, the atmosphere is so fractious that there appears to be "no chance of reaching a consensus that would allow the commission to go forward." By the time of his conclusion (chapter 11), Friedheim nonetheless offers some sensible and modest agenda options. But what does one find, in the intermediate chapters, to inspire any hope for, or perhaps even diplomatic reason for, "going forward"? What bars progress is not each side's exaggerated and sometimes silly claims. (Do the Japanese really "fear that their entire ocean-based food system and taste preference is under attack" by "Anglo-Saxon meat eaters"?) Were everyone suddenly to mind their manners and leave ersatz vials of blood and paranoia at home, it would not make much difference. The preservationists have a majority. If "going forward" means opening the way to commercial harvest, *going forward* is exactly what the preservationists are not going to consent to. They have the power to stand pat. Why budge?

"Justice" and "fairness" might be bases of appeal. Granted, moral coinage has uncertain buying power in international relations. But the moral perspective should not be abandoned, whatever its limitations in motivating diplomats. Part of morality invites examination of animal rights. And part asks whether what the International Whaling Commission (IWC) is doing is a just (or otherwise defensible) use of power. The latter is a question of interest not to whalers alone, nor is it solely casuistic. How would Americans respond if the parties to the Framework Convention on Climate Change (FCCC) were to come under the domination of fish producers and order a phasing out of cattle on the grounds that cattle produce methane, and therefore ranchers had to close down and beef be dropped from the American diet?

Obviously, the contributions are too rich and wide-ranging to respond in full to each. Let me comment on a few of the major focal issues.

IS THE WHALING REGIME HONORING THE ORIGINAL INTENT?

Whether the commission is doing its job need not be referred to some abstract international morality. Most of the contributors would grade the IWC's performance in light of its original mandate. If so, it can fairly be said that none of the diplomats who met in Washington in 1946 likely intended (or even contemplated) a zero quota on abundant stocks, much less a full moratorium in the Southern Ocean. But has the commission therefore exceeded it mandate (acted ultra vires in the legal vernacular)? Not necessarily. Surely none of the framers of the U.S. Constitution, in giving Congress power over interstate commerce in 1789, had in mind a Civil Aeronautics Board or a Securities and Exchange Commission. Some plasticity is expected of organic documents such as international conventions, no less than of constitutions.

On the other hand, the charge against the IWC is not that the International Whaling Convention has been flexed to accommodate changed circumstances, such as a new generation of harvest technology. William Burke and Jon Jacobson (among others) maintain that the focus has swung from securing the profits of whalers to protecting the lives of individual whales. How else to construe a zero quota on abundant stocks? The original (or "authentic" or "primary") aim, in this view, has not been fleshed out or tinkered with as much as betrayed in its basics.

At what point does a deflection become betrayal? In conventional legal terms, the issue might be put this way: Is the current zero quota (and the Southern Ocean sanctuary) so ultra vires—so far beyond the commission's powers as authorized in the 1946 agreement—that it cannot be accomplished legitimately without express authorization through amendments to the International Convention for the Regulation of Whaling (ICRW) by the convention's parties?[1] Similar questions of agency power come up frequently in the law, and typically involve, as they do here, complex matters of degree.

It seems to me that the critics slightly overstate their point by invoking "the unambiguously stated purpose of the ICRW and IWC to conserve whales for the benefit of the whaling industry and its customers."[2] No one doubts that the predominant historical motivation was, as Steinar Andresen puts it, "to secure a stable and long-term income for the whaling industry."[3] But in fact, the first stated purpose that appears in the preamble is a little more ambiguous and expansive: "Recognizing the interest of the nations of the world in safeguarding for future generations the great natural resources represented by the whale stocks. . . ." And what the treaty says about the industry and its consumers is, more exactly, that the commission "*shall take into consideration* the interests of the consumers . . . and the whaling industry."[4] To take a factor into consideration is not to make it conclusive.

Hence, a preservationist-inclining IWC commissioner might defend against allegations that he is acting ultra vires on the grounds that the mandate is to consider industry health *along with other things*. The commissioners might be banking on a moratorium not only (negatively) to avoid the risks of collapse but also (positively) to permit stocks to rebound. Even if there are enough of a certain stock to avoid collapse now, a yet higher level might, in the long run, be consistent with taking consumers and the industry "into consideration" along with future generations and—why not?—the burgeoning whale watching industry. One study estimates that 5.4 million people worldwide are now going whale watching, spending an estimated U.S.$504 million for the tour plus food, travel, accommodations, and souvenirs.[5]

Indeed, it is not unthinkable that some day the commission may have to take into consideration the effect of whales on *food security*. The ugly, little-discussed fact is that marine mammals, probably whales in particular, compete with humans for fish on an unclear but clearly huge scale. In its opposition to the Southern Ocean sanctuary, Japan estimated that whales in the Antarctic region alone currently consume 240 million tons of feed annually, an amount calculated to rise (given anticipated population growth under a moratorium) to 670 million tons by 2050.[6] Elsewhere, Japan has estimated that for all whales, all oceans, the whale diet already exceeds 500 million metric tons annually.[7] "Feed" includes, as well as marketable fish, some biomass not directly consumed by humans. Nonetheless, to put the rough magnitude in perspective, humanity's marine harvest has reached approximately 80 million tons a year.[8] Will an expanding humanity in the mid-twenty-first century by tempted to regard whales as ranchers do wolves—with public sentiment shifting accordingly? And might we not someday be looking to the IWC to—flexibly—account for food needs by culling out excessive competition?

The point is that it is hard to assess what is intra vires and what is ultra vires for an organization, particularly if, as here, the commissioners do not own up to mutinous motives. In other words, if the commission took an authoritative position as blunt as that of some nongovernmental organizations (NGOs) ("We're against whaling because whales have rights, period"), there would be a clear—and, in the traditional legal view, a clearly wrongful—straying, absent formal amendment. But there is little frank acknowledgment of the preservationist shift.[9] The publicly professed explanation is facially consistent with the convention: that harvesting has to be *suspended* to allow the collection of more data; the preferred rationale for the additional data is the precautionary principle, buttressed by the claim that to permit commercialization of nonendangered stocks would complicate efforts to police protection of stocks that are truly endangered.[10]

Those who argue that the IWC has strayed from its original purpose understandably point to the proven abundance of minke stocks and the commission's apparent disregard of the Scientific Committee.[11] On the other hand, the fact that the commission did not *adopt the committee's recommendations* is not proof positive that the Schedule amendments are not (in the language of the ICRW) *"based on scientific findings."*[12] What, after all, does "based on" (not "adopt") mean? Questions of *acceptable* risk are not purely scientific; whether a policy that yields, say, a 1 percent chance of collapsing a whale stock is acceptable is ineluctably a question of policy, or, if you will, politics. Moreover, the commissioners might *take into consideration,* even *accept,* the scientists' assessment of the probabilities but approach the (admittedly slim) prospect of population crash in a more risk-averse spirit. The commissioners (who, unlike American appellate judges, are not obligated to issue an opinion) might be assigning a higher value to the existence or option value of whales than do the scientists, or they might believe they ought to give greater weight to the projected interests of future generations.

I am not saying that I find these constructed defenses convincing. In fact, my own view is that the IWC *has* substantially deviated from its original, hunt-oriented charter. But is this departure so ultra vires that a formal rewriting of the ICRW is called for? This strikes me as an open question more than it does Jacobson, who thinks "a good case can . . . be made for the proposition that the Southern Ocean sanctuary simply does not exist. It can and perhaps should be ignored."[13] This is the conclusion I am, with respect, stopping short of. (But, as we will see in a moment, it is a question that might well be submitted to dispute resolution).

Moreover, whether or not the commission is legally ultra vires, one must be troubled by the unmistakable *appearance* that the IWC has been kidnapped, and by what this presages for international environmental laws and institutions. Although we can rationalize the conduct of the commission— soften our judgment with generous presumptions of good faith—the fact remains that what has happened within the IWC may chill the development of international environmental accords in various other areas. At its worst the "moral" of the IWC's straying is this: any nation that signs a global environmental or resource convention may find itself in a (frankly) "stuffed" regime that tosses aside its original premises and pays little heed to its own scientific advisors. That is not a lesson that we should want broadcast to a global community already wary—certainly the United States is—of extending international commitments.

WHALING AND UNCLOS

Thus far, I have reviewed the contributors' criticisms of the IWC in the context of its authorizing treaty, the ICRW. But in chapter 1 of this volume Burke reminds us that the IWC and ICRW have to be read in conjunction with the UN Convention on the Law of the Sea (UNCLOS), which, both as the later convention and the more universal, is the dominant source of obligation. Burke draws several implications. For one, he makes a powerful case that neither UNCLOS nor customary law compels all whaling to take place under the ICRW. It is not that international law lacks a general duty to cooperate and to conserve on whaling. But Burke argues that the modality of discharging those duties is open. A whaling nation may, for example, drop out of ICRW and cooperate with other nations directly or through regional whaling organizations, quite consistently with the UNCLOS.

Indeed, Burke suggests that some of the same arguments supporting Regional Fisheries Management Organizations (RFMOs), which are marshaling to play a larger role after negotiation of the Straddling Stocks Agreement, could be put behind regional whaling agreements. If RFMOs work for highly migratory fish stocks, such as tuna, why not for whales? Burke notes that regional (perhaps stock) management may be superior to a plenary global regime both in data gathering and in sensitivity to specific ecosystems. Perhaps. But on the other side a proliferation of whaling organizations (organized on regional or other lines) would invite increased pressure on stocks. Is there a way to prevent the most permissive organization from attracting—and being dominated by—the most aggressive fleets with the shortest temporal horizons? Indeed, the RFMO record on tuna (not exactly the same situation) is not consistently encouraging. Nonetheless, Burke is right that multiple organizations, no less than a renegotiated ICRW, remain a legal option.

A major part of Burke's analysis addresses another undeveloped connection between the IWC and UNCLOS. He takes us on a tour of the labyrinthine dispute settlement procedures of the UNCLOS, tracing the various circumstances under which a conflict over whaling might wind up in various forums (the International Tribunal for the Law of the Sea, the International Court of Justice, an Article V Arbitral Tribunal, or a special Article VII Arbitral Tribunal). Were the ICRW to be renegotiated today, it might capitalize on the many opportunities, in ways Burke indicates.

But Burke's most provocative idea is how to use UNCLOS to challenge the IWC's exceeding its powers under the ICRW. His thinking is this: All states have the right to whale consistent with customary law and UNCLOS. UNCLOS (Articles 65, 87, 116) authorizes states and international organizations (through the ICRW or other conceivable conventions) to set more stringent standards than would otherwise prevail under UNCLOS, of its own force. But the continuing authority of the ICRW today, at least as between parties to both ICRW and UNCLOS, has to be regarded as a delegation from UNCLOS, as the subsequent (and therefore, in cases of conflict, dominant) treaty. Hence, derogations from powers parties would otherwise enjoy "can be imposed only in accordance with [the UNCLOS-authorized] agreement." UNCLOS retains a guardian role. On this view, "Members of [the ICRW] who impose restrictions in violation of their basic charter not only infringe [the ICRW] but are in violation of . . . UNCLOS."[14] The very allegation, moreover, triggers compulsory dispute procedures.

If Burke's view is sustained, then a member of the ICRW who takes exception to the Southern Ocean sanctuary, or to the zero moratorium as applied to minke stocks, as ultra vires the whaling treaty, has a complaint against, if not the IWC as an organization, members of the ICRW who are also members of UNCLOS.

It is a pregnant proposal. As I indicated above, everyone interested in protecting global resources and the environment has a stake in assuring the good-faith performance of the relevant accords. Ways to challenge excesses ought to be fostered.

While I therefore support Burke's proposal, I am doubtful that a UNCLOS forum would be inclined to second-guess the IWC on the merits. There is the fact, first, of what U.S. lawyers call administrative expertise: a third party is likely to defer to organs such as the IWC that can claim special expertise in the need for, say, a Southern Ocean sanctuary. Second is the fact that members can usually plausibly claim ICRW-consistent motives for their actions, for example, that more data is needed. Whatever its private doubts, a tribunal is not likely *publicly* to impugn the motives of the IWC and its individual members. Third, any member dissenting from an IWC regulation has the right to opt out of the regulations or, for that matter, quit the ICRW. These escape valves are apt to reduce a third party's incentive to step in.[15]

It is true that the International Tribunal for the Law of the Sea recently (August 1999) stepped into a dispute among the parties to the Commission for the Conservation of Southern Bluefin Tuna (CCSBT), at least to the extent of handing down a provisional measure restricting Japan's "scientific" harvest of southern bluefin tuna in excess of the last agreed-to allocation. The proceedings are certainly relevant. But it is worth cautioning that while the tribunal's acceptance of jurisdiction is consistent with Burke's analysis, it came out deferring to the regulatory regime, consistently with mine: when push comes to shove, an arbitral tribunal will be wary of undermining the IWC.

Even so, Burke's stratagem deserves to be tried, even if nothing more is available by way of remedy than a declaratory judgment.[16]

IS THE WHALING REGIME "EFFICIENT"?

The goodness or badness of the commission need not be judged by its adherence to its boundaries as originally conceived. David Victor, in his excellent contribution (chapter 10), points out that whatever inconsistencies, even flaws, one can find in the IWC's current position, there is no obvious material change that would not make some parties to the arrangement worse off. Implicit is the view that the regime should be assessed not by reference to its fidelity to original intent ("originalism" in the U.S. Constitutional literature) but, more flexibly, by reference to the efficiency of out-come. This is quite in accord with Jon Jacobson (chapter 2), who sympa-thizes with the plight of whales but acknowledges that any further tough-ening of the IWC (the elimination of "scientific" whaling, for example) would be just enough to tilt an exodus of whaling states; operating outside the IWC, and absent any new regime, whalers would be constrained only by "general principles" of international law and UNCLOS—which would un-doubtedly be less restrictive.[17]

So: who stands to gain from change? As Victor says, the environmental-ists are being accommodated (to put it mildly) by a near universal freeze. Notwithstanding the freeze, the Japanese are getting some minke under a permissive application of the "scientific" exception. The aborigines—even the Makah, who until 1999 had not hunted in seventy years—are being ac-commodated under the aboriginal subsistence provisions and an apparent side deal with the Russians. The nations that have withdrawn from the IWC are hunting, but, to avoid U.S. sanctions for undermining the IWC, doing so gingerly. True, the environmental (or, at least, animal rights) groups would like the harvest to be smaller; the Japanese and other whalers would like it to be larger. I am not certain (as Victor suggests) that demand for whale meat is being met. But by and large, given the conflicts, he seems right that the regime is likely to be as good as it can get and still retain its stakehold-ers. At the least, before we "fix" the IWC, the critics owe us an answer: what is the better outcome that the present institutional arrangements obstruct?

IS THE WHALING REGIME WELL STRUCTURED?

Efficiency analysis takes as its starting point exogenously fixed welfare func-tions: the Japanese are pictured as going into sessions wanting x; the Amer-icans, y, and so on. The critical question is only: are there any remaining outcomes such that at least one party is better off and none is worse off? An institution is good or bad depending upon whether it facilitates or impedes moves toward the Pareto frontier.

But we need not accept the parties' conflicting interests as *given*—rather we may consider them as variables subject to the regime's influence. Hence, a competing critical platform might judge the IWC by its ability to change conflicting views, not merely to accommodate them. The notion sounds idealistic; after all, the policies international diplomats put forward are pre-sumably devised in their capitals. But the total policy process, including feedback loops, is sufficiently rich to permit policy revision, particularly if the institutional framework is artfully constructed. For example, whatever the failures of the climate change regime thus far to modify domestic policies, one could argue that the meetings of the parties to the FCCC and, in particular, the efforts of its Intergovernmental Panel on Climate Change have altered some parties' perceptions of their interests.

There is a well-known literature suggesting that international regimes can foster genuine ideation and consensus by, for example, empowering epistemic communities to formulate policy somewhat buffered from constricted state interests.[18] But one thing our authors agree on is that whatever features a regime requires for such ideal harmony to flourish, the IWC lacks.[19] As Milton Freeman says, the conflicts—and distrust, insensitivity, and impassioned exaggeration—are far more blatant than any cooperation.[20] Where one might hope to find persuasion as a source of agreement, Elizabeth DeSombre finds "bullying and bribing."[21] Meetings have been marred by verbal and even physical attacks.[22]

Perhaps the truculence within the IWC is beyond the balm of any institutional therapy. There are reasons to be pessimistic. For one thing, the more the issues are conceived as ethical and symbolic, rather than empirical or pragmatic, the less pliant the disputants. To a large extent, national positions have been hardened—or, more accurately, captured—by inflexible domestic constituencies. The root cause of the abdication of policy by central governments is that for the general public in most countries—including the United States and Japan alike—whaling is not a high-salience issue.[23] Those who care, care a lot—and amplify their influence accordingly. For U.S. administrations, charged with foot-dragging on climate change and biodiversity, concession to whaling opponents is a cost-free sop. Ditto for the French, who presumably still owe Greenpeace something for having dynamited the Rainbow Warrior.[24] Moreover, as Friedheim astutely observes, the very fact that the IWC has so few issues under its wing cramps opportunities for the sorts of linkages and trade-offs that enable compromises in more expansive arenas.[25]

That is why one should not be too optimistic about changing anything via bureaucratic reform. Friedheim, in fact, ultimately points us in another direction, claiming that the parties have been surprisingly unadroit in exploiting the opportunities of the structure as it exists. He maintains that whaling proponents have failed to make standard bargaining offers and threats[26] and elsewhere has cited the Japanese as having failed—for whatever reason—to orchestrate normal coalitions with their natural allies, including Iceland and Norway.[27] I do not know enough to evaluate these claims; one hears that the Japanese have been known to trade economic aid for whaling support, a claim that, if true, doesn't quite square with the charge of naivete.[28] In all events, to the extent Friedheim is right, the "stalemate" (if that is what it is) owes less to defects in structure, which reform could address, than to those in individual nations' strategic behavior, which is harder to reach.

Supposing, however, that the structure is to open for reform, there are a number of suggestions in the text, explicit and implicit, as to what measures might improve the effectiveness and atmosphere. Andresen addresses a number of options, and his chapter deserves close attention. But on his own account, most of the possible reforms turn out to be of uncertain value. One can imagine a different secretariat, but "it is hard to envision how ... an 'organizational ideology' or a 'common geist'" might have been created on an issue as politicized as whaling.[29] It sounds plausible to improve the inspection system, but Andresen also suggests that, given the low level of hunting, it would not make much difference. Indeed, the demands for an improved system come in part from whaling opponents looking for excuses to extend the moratorium.[30] On the one hand, NGO participation may be rigidifying positions and eroding civility; on the other hand, who is to say that the conflict would not "be even more polarized if the NGOs were kicked out"?[31]

William Aron, as would be expected, has an insightful insider's critique to offer regarding the history of the Scientific Committee. But when, in the last analysis, he puts to himself the question of whether "the [Scientific Committee] and its operation can be improved?" his answer is a frank "Of course, but ... the changes would be small."[32]

WHY THE INTRANSIGENCE?

One fascinating "finding" of this volume is that most of the contributors imply or even state that there is really nothing to "fix" in the whaling regime. Victor, as indicated, considers it "good enough" in that there is no attainable (and material) change that would not make at least some of the major constituents worse off. In fact, there is the distinct possibility that the ungluing of the IWC would set in motion a chain reaction that could make almost everyone worse off.

For all the weaknesses DeSombre identifies, I do not think she finds the

IWC in the last analysis much different from comparable international organizations. As Freeman puts it, "the current nonresolution of the whaling problem is the desired outcome";[33] or, as Andresen puts it, "this book's point of departure—that the regime needs to be 'fixed'," turns out to be at least "controversial."[34]

Why, then, do so many speak, frustrated, of a "stalemate" and why is there such intemperance? The answer, of course, has less to do with the whaling commission than with (popular views of) whaling and whales.

Therein lies a story that no single volume—perhaps nothing in the sum of world literature—can clarify with satisfaction. I like the stab Arne Kalland offers elsewhere: whales have undergone a totemization, no less by their would-be saviors than by their aboriginal hunters.[35] Why else, other than that both groups invest whales with mythical potencies, does the world's media rivet on the rescue of an errant whale grasped in the ice?[36] After all, the killing of most other animals is generally lawful (if done "humanely"). An estimated 6.5 million of the European Union's 370 million people regularly kill wild mammals for *sport*[37]—this on top of all the mammals, from cattle to rabbits, that are raised and killed for food and medical research. At the international level, protections for nonhuman life, if any, are generally triggered only when there is evidence that a *species* is endangered.[38] Some of the earliest international conventions protected migratory birds "useful to agriculture."[39] But whale hunting is banned even for species that are not credibly threatened.

The reason most commonly given for the special treatment of whales (and other cetaceans) is their reputed intelligence, which, in the face of science,[40] has been drummed up to fabulous proportions among the credulous public.[41] Oddly, the killing of nonhuman primates—200,000 monkeys were sacrificed to test the Salk vaccine alone[42]—draws far less widespread public outcry.[43]

But that is only the beginning of the story. The whaling debate has become a forum for defining and acting out a wide range of important social and ethical issues that have little to do with whales or whaling as such. For one example, during the controversy over whether the Makah should resume whaling after a seventy-year hiatus (a controversy that divided even the tribal members), a reporter concluded

> To understand why a hunt for five whales is so important to the tribe today is to know something about the treaty of . . . 1855 [which] secures the tribe's

right to whale. It is more than a contract, a land settlement, or a legal agreement. It is what sets the Makah apart from other tribes, from the rest of America. It holds the Makah tribe's sovereignty, their identity.[44]

And of course what is true of the Makah is true of the pro-whaling forces generally: the passions and sensitivities of sovereignty—much more than a taste for whales, or the economy of whaling—are stimulating the fray. Whaling is not just about whaling; it is about the right to choose, or at least not to be dominated.[45]

Two of our authors advance other linked controversies that may have tended to inflate the whaling issue beyond manageable dimensions.

Milton Freeman points out that the whaling debate—as if it did not have enough baggage to carry—has been clouded with fuzzy indictments of money and capitalism, wrapped up in capricious notions about noble savages. The trouble stems from the special rules for aboriginal subsistence whaling, under which (irrespective of the general schedule) an allotment is made to certain "aboriginal, indigenous or native peoples who share strong community, familial, social and cultural ties related to a continuing traditional dependence on whaling."[46]

Application of this exception has been uneven and controversial. The Environmental Investigation Agency revealed that the indigenous populations in Chukotka (the Inuit's Siberian cousins, the Chukchi) had been using allotted gray whale meat to feed foxes on fox farms—perhaps 175 whales a year, far more than they killed, or wanted to kill, traditionally. Apparently, fox farming aside, gray whales are not worth the trouble. One hunter from the nearby Yupik said,

> We never asked for the gray whales. Not even the Chukchi people did. Gray whales were taken in the old days by the local people, maybe fifteen a year. But only occasionally . . . [they] are dangerous. They attack to protect their young. Sometimes the leading bull will attack whaling boats as soon as he sees them. And the meat is no good.[47]

No good, indeed; the Yupik name for gray whale translates into English as something like "the one that makes you shit fast."[48]

In the United States, none of the present Makah members, whose allotment is from the same stock of grays, has any idea how to hunt or butcher their quota; they have dispatched a few members to be trained by the Inuit.

At the same time, other communities of people with a long and uninterrupted whaling tradition are denied quotas because they do not qualify as "aboriginal, indigenous or native, etc." These include groups with long traditions of whaling among the Faeroe Islands, Iceland, Japan, northern Norway, and Greenland.[49] The appearance of arbitrariness does little to shore up the commission's credibility.

To make matters more questionable, the IWC has sorted out which groups are and which are not qualified for a quota, and there is a further catch: The quota can be used only for *local aboriginal consumption*—which is defined to mean "traditional uses of whale products by local aboriginal, indigenous or native communities in meeting their nutritional, subsistence and cultural requirements." Somehow this allows fox farms to be operated in the name of the Chukchi (although killing whales for fur coats is environmentalism's counterpart to serial murder!), but neither the Chukchi nor any other group can sell the meat.

What sense does this make? Put it this way: Assuming a tribal group has had an approved quota—say, five whales—why not let them do what they want with it—eat it or sell the meat for something they value more highly? If the meat in Tokyo will fetch the price of a fleet of snowmobiles, and the tribal elders prefer a snowmobile to the meat . . . well, that's what makes the rest of the world go round.

But they cannot. It is this that agitates Freeman. He regards adopting being-outside-a-money-economy as the criterion for "subsistence" as born in ignorance and unevenhandedly applied. He finds the roots in a tradition that regards

> the exchange of commodities outside of household-based exchange as being unnatural and dangerous . . . [and that views] "money [as] the root of all evil." . . . According to Marx, the "noble savage," living in balance with nature, engaged only in altruistic, nonmonetized barter relations with neighbors, until becoming corrupted and greedy after contact with agents of capitalism.[50]

In effect, by a prudent and paternalistic manipulation of doctrine, we save the aboriginal soul from any Faustian bargains that might tempt it—and at the same time smugly conserve the world's portfolio of colorful cultures.

I am not sure this is entirely so—at least, the original motivation may be more complex. The practice of making special but limited allowances for aborigines goes back to the now succeeded Convention for the Regulation of Whaling (CRW 1931).[51] Article 3 of this original whaling regime declared:

> The present convention does not apply to aborigines dwelling on the coasts of the territories of the High Contracting Parties provided that:
> (1) They only use canoes, pirogues or other exclusively native craft propelled by oars or sails;
> (2) They do not carry firearms;
> (3) They are not in the employment of persons other than aborigines;
> (4) They are not under contract to deliver the products of their whaling to any third person.

The way I read these restrictions (perhaps as a former antitrust lawyer rather than as an anthropologist), I would guess that the predominant original motivation was that of a cartel anxious to close off the threat of commercial competition. In other words, as long as the aborigines used primitive methods, and did not become agents of real trading powers, they posed no threat of what cartelists refer to as "over-supply" and "price destabilization."

At any rate, whatever motivated the original framers of the predecessor section (no doubt different diplomats came to the bargaining table with different aims and prejudices), I agree with Freeman that the persistence of the system, and its wobbly application, is tainted with "phony analysis and demeaning stereotypes."[52]

I have only one thing to add—quite in line with Freeman's analysis. The United States, in support of restoring whaling to the Makah, submitted an anthropologist's paper aimed at demonstrating the dietary and cultural needs of the Makahs for whales. The paper claimed, for example, that the Makah "evolved as a biological population" over two thousand years within a nutritional context that included whale meat and oil.[53] The implication is just plain silly—that a group's gastrointestinal tract could develop a genetic dependence on whale meat in the evolutionary wink of two thousand years. But why it is worse than silly—why it is tragic—is that the Makah and other Native American tribes apparently *do* suffer from dietary-related diseases and these conditions *are* correlated to the introduction of Western eating habits, including refined flour, sugars, and beef.[54] But it is not a problem that is likely to be addressed by adding whale meat to the table (since the same pattern appears in other Native American populations including the Navajo). Bad diet is a serious problem among Amerinds—and to drag it into the whaling issue as an antidote—even to invoke a "nutritional need" for whales—can only deflect remedial efforts from more productive public health responses.[55]

There is the same risk of deflection from real problems in the U.S. submission regarding the Makah's cultural need for whaling. The report laments that

> the introduction of American values worked against the traditional subsistence pursuit. For example, the American philosophy of social equality made it difficult for Makahs to continue to staff and organize whaling canoes, and therefore households, according to the ancestral patterns. Whale hunting was no longer the sole avenue to a position of ceremonial and political importance as the headman of a large household.[56]

Now, I can understand why an anthropologist would regret the passing of a colorful primitive lifestyle. But my own reaction is more ambivalent. Is it really good for the Makah to step backward and revive whaling as "the symbolic heart of the culture"?[57] To prepare for the twenty-first century by reinstating whaling skills as the "sole avenue" to positions of leadership? What the Makah do now is their choice, but I'm hopeful some of them will teach their children to make heroes of computer geeks.

Finally, Russel Lawrence Barsh maintains that the whaling struggles are entangled not just with filthy lucre but with global power and domination.

> Privileged societies have acquired the power to determine what the world eats and to impose their own symbolic and aesthetic food taboos on others. Placed in . . . historical context, . . . efforts to abolish whaling and sealing are exposed as [the result of] Western European domination of world food supplies. . . . Peripheral societies have had to change what they produce and what they eat.[58]

Now, this is an odder and probably more complex—and probably less defensible—thesis than first appears. But Barsh has provided one of the most consistently provocative contributions to the present volume, full of more ideas than can be addressed justly in a brief comment.

To begin with, the many connections between political power and food supply are fascinating. But what exactly is being claimed about the relationship between them here? There may be historical instances of powerful nations forcing the weak to eat their preferred foods, much as conquered peoples have been forced to worship the conqueror's gods. But any such menu forcing has little relation to anything going on at the IWC. The ban *affects* diet. But taking whale off the plate is a far cry from affirmatively forcing anyone to eat anything, for example, Muslims to eat pork. Moreover the anti-whaling faction is not motivated by questions of diet, as such, anyway. The opponents of whaling would be equally (perhaps even more) opposed if whales were killed for nondietary purposes, such as lamp oil or clothing. The defenders of dolphins object not to the eating of dolphins but to their being killed incidentally in the catch of tuna.[59]

Most of the history Barsh cites involves strong powers influencing what the weak powers *produce*. Some of these shifts in production have involved the exercise of direct political power, such as colonial nations establishing coffee and tea plantations to supply home demand. And the sheer purchasing power of powerful nations undoubtedly shapes land use decisions in poor countries, so that, for example, South American forests are converted to grazing land in order to expand beef exports to the United States. What the weak produce, in turn, must gradually affect what they themselves eat and drink, locally.

But I am not sure what conclusions to draw from this: that the strong *want* to impose their tastes on the weak? In most of these instances the strong have been (I suspect) rather indifferent to their effect on the consumption patterns of the weak. That is, why should the British be any more eager for the colonials to adopt the British taste for tea than the British taste for theater? There may be a partial answer, depending upon whether the powerful state's interests are principally as *producer* or as *consumer*. As consumer, there is an advantage in seeing that others do not share your tastes, because their competing demand for the product lowers your own supply and raises the price. On the other hand, wherever the strong are producers and sellers they have a conflicting interest in expanding the taste for their product as a means of increasing the market for sales. To illustrate, suppose that the whale-eating states were all-powerful politically and ask yourself whether they would prefer everyone to share their taste for whales. As consumers of whales, wouldn't they prefer the rest of the world to eat cake and leave the whales for themselves? By contrast, whaling nations whose seller interests (especially to "outsiders") dominated their domestic consumer interests would presumably emerge as Barsh's taste imperialists.

In like vein, I am unsure what to make of Barsh's point about Rome implanting its food production system in Iberia, Britain, and the Lower Rhine. He sees this as a case of "Roman tastes defin[ing] what Europeans chose to plant and eat."[60] But what, if anything, motivated the "defining"? Drawing inferences from historical evidence is tricky. Efficient foods and methods of food production are among the most critical factors in the emergence of a powerful civilization: Other things being equal, the nations that can extract the most calories per acre can support the densest populations and largest armies. This, at least, is quite persuasively maintained by Jared Diamond in *Guns, Germs, and Steel*.[61] If so, to the extent that Roman colonies adopted Roman crops, it may have been because the conquered discovered (too late!) that the Romans were onto more efficient agriculture (which is why the Romans were the conquerors).

Indeed, while Barsh's claims about the role of food security in French and German colonial power strikes me as credible,[62] the notion that conquerors in general impose their food on the conquered flies in the face of what we know as the great "Columbian exchange."

> When Columbus returned from his voyage of exploration he brought not only news of his discovery, but maize seeds. The next year he was back in the New World carrying planting material for wheat, olives, chick peas, onions, radishes, sugar cane and citrus fruits . . . maize, the common bean potatoes, squash, sweet potatoes, cassava, and peanuts went east, while wheat, rye, oats, and Old World vegetables went west.[63]

I think it would be hard to say, based on who emerged eating whose food, which were the real conquistadors, the Spaniards or the Mesoamericans.

At any rate, there are many reasons to doubt that the IWC represents the triumph of Western food hegemony. If Western food hegemonists were truly dominant and oppressive, wanting us all to share a taste for white bread, sugar, and hamburgers, why would they recognize an aboriginal exception to the zero quota, rather than forbid whaling outright? Indeed, if the IWC were in the hands of Western food hegemonists, you would expect that if the aborigines were allowed a small harvest, they would not be prevented from selling their catch for cash, with which they could accelerate their inevitable craving for McDonald's. And, as I've indicated, to the extent that international politics is food-driven, it is not clear why nations that consume beef, at least qua beef consumers, would not want the IWC to *encourage* others to consume whales, to get their calories that way, so there would be more beef for the beef eaters. It seems to me that the ban on whales is more persuasively explained by differences in morals than by (literal) tastes.

Then, again, the moral issues about whaling, the activity, have dominated the discussion. It is to Barsh's credit that he alerts us to the more obscure food dynamic that may underlie the moral discourse and, over time, give it shape. Moreover, even if I have doubts about parts of Barsh's lively thesis—which I would gladly turn over to the historians to arbitrate—it is easier to swallow in a slightly diluted form: whether or not the world's eating patterns reflect imperialists imposing their food tastes on others, those of us in power have to be continuously on guard against insensitivity.[64] Sunlight striking land grows grasses; striking the ocean it produces plankton. Is there any principled justification, or is it mere hubris, that we should with clear conscience eat large land mammals that graze on the one, while excoriating foreigners for eating large sea mammals that graze on the other?

NOTES

1. I address this issue in Christopher D. Stone, "Legal and Moral Issues in the Taking of Minke Whales," *Report: International Legal Workshop, Sixth Annual Whaling Symposium,* ed. Robert L. Friedheim (Tokyo: Institute of Cetacean Research, 1996): xviii–xxiii. See William Burke's discussion of the Vienna Convention on the Law of Treaties, Article 31, in chapter 1 of this volume. The article reads as follows: "A treaty shall be interpreted in good faith, in accordance with the ordinary meaning to be given to the terms of the treaty in their context and in light of its object and purpose." Official Records, United Nations Conference on the Law of Treaties, Documents of the Conference 293, 1969, UN Doc. A/CONF. 39/27.

2. Chapter 2, "Whales, the IWC, and the Rule of Law," by Jon L. Jacobson, in this volume.

3. Chapter 8, "The Whaling Regime: 'Good' Institutions but 'Bad' Politics?" by Steinar Andresen, in this volume.

4. Article V(2), *International Convention for the Regulation of Whaling,* 2 December 1946, 62 Stat. 1716, 161 U.N.T.S. 74 (hereinafter ICRW), emphasis added.

5. Erich Hoyt, *The Worldwide Value and Extent of Whale Watching: 1995,* (Bath, England: Whale and Dolphin Conservation Society, 1995), 3. See, generally, International Fund for Animal Welfare (IFAW), "Report of the Workshop on the Socioeco-

nomic Aspects of Whale Watching," 8–12 December 1997 (Kaikoura, New Zealand: IFAW, 1998).

6. Letter of Kazuo Shima, IWC commissioner for Japan, to Ray Gambell, secretary of the IWC, 19 January 1994, 5 and appendix 1. The Marine Research Institute noted that unchecked whale populations around Iceland could cause a decrease of up to 10 percent in the annual sustainable yield from cod stocks, costing Iceland thirty-five thousand tons of cod annually. Translated into economic terms that equals 4 billion kronur in export revenue. See Gudrun Petursdottir, ed., *Whaling in the North Atlantic: Economic and Political Perspectives* (Reykjavík: University of Iceland Press, 1997), 41.

7. The Institute of Cetacean Research, *What We Can Do for the Coming Food Crisis in the Twenty-first Century* (Tokyo: Institute of Cetacean Research, 1998).

8. United Nations Food and Agricultural Organization (FAO), *Global Fishery Production in 1994* (Rome: FAO, 1995). See also FAO, *A Global Assessment of Fisheries Bycatch and Discards*, FAO Fisheries Technical Paper No. 339. (Rome: FAO, 1994). *http://www.fao.org/waicent/faoinfo/fishery/catch/catch94a.htm.*

9. However, in chapter 1 of this volume, "A New Whaling Agreement and International Law," William Burke indicates that the United Kingdom and other ICRW members have announced that they will not support resumed commercial harvest even if the scientific data show that harvest can be taken without endangering stocks.

10. It is surely not irrational, in the interests of enforcement efficiency, to impose a blanket ban on trade of a class of products, both legitimate and illegitimate. In its effort to protect eagles, the United States prohibits commercial transactions in feathers taken from all eagles, even those legally killed before the relevant statutes came into effect. The U.S. Supreme Court upheld the regulations in the face of a constitutional ("takings") challenge by a trader wanting to sell his inventory of legitimate feathers, in part because a blanket ban avoids the acknowledged difficulty in ascertaining their age. *Andrus v. Allard*, 444 U.S. 51. (1979). The problems of operating a system that permits trade in minke whale meat while banning trade in products of other whales can presumably be overcome but are not trivial.

11. See chapter 3, "Science and the IWC," by William Aron, in this volume.

12. Article V(2), ICRW, emphasis added.

13. See Jacobson, chapter 2, in this volume.

14. See Burke, chapter 1, in this volume.

15. The "exit" and, indeed, objection procedures both probably win the commissioners *more* latitude than they would have in a framework convention that allowed, say, majority control on only some issues (such as the FCCC). They should not have

carte blanche, because the options are not pain-free. As Friedheim reminds us, the objector has to face disabilities under U.S. domestic law for "reducing the effectiveness" of the ICRW. This objection can perhaps be bracketed on the grounds that the legitimacy of various modes of U.S. "bullying" (otherwise known as its indispensable leadership on environmental issues) would seem to be a matter for the World Trade Organization to sort out. In my mind, the more serious "pain" is that, understandably, the whaling state members do not want to be labeled as the defectors, when they genuinely believe it is the majority that is defecting.

16. It is not clear how useful a form a declaratory judgment might take. It might be merely a ruling that the IWC must conform to the ICRW (a sort of reminder). Or it might indicate that some specific actions were ultra vires, for example, that the order for a Southern Ocean sanctuary requires a specific endorsement by the Scientific Committee (does it?). Or, a judgment might clarify, for example, whether it is appropriate for a member to favor preservation in all circumstances, that is, in the face of any data as to stock abundance. See note 9 above. This strikes me as an open and intriguing question.

17. In chapter 4 of this volume, "Is Money the Root of the Problem? Cultural Conflict in the IWC," Milton Freeman remarks in like vein, "It would be naive not to recognize that for the majority of participants . . . the current nonresolution . . . is the desired outcome."

18. See Peter M. Haas, ed., "Knowledge, Power and International Policy Coordination," *International Organization* 46 (special issue, winter 1992): 1. See also Oran R. Young, *Drawing Insights from the Environmental Experience (Global Environmental Accords)* (Cambridge: MIT Press, 1998).

19. In addition, see M. J. Peterson, "Whalers, Cetologists, Environmentalists, and the International Management of Whaling," *International Organization* 46, no. 1 (winter 1992): 147.

20. See Freeman, chapter 4, in this volume.

21. DeSombreSee chapter 6, "Distorting Global Governance: Membership, Voting, and the IWC," by Elizabeth DeSombre, in this volume.

22. See the introduction, by Robert Friedheim, in this volume.

23. As Friedheim reports, Japan is more interested in exporting Toyotas than resuming whaling. See Friedheim, chapter 7, in this volume. In fact, at a meeting in Tokyo, Friedheim and I were told that Japanese car sellers in New Zealand wanted the whaling, and the attendant anti-Japanese sentiment, to come to an end.

24. See Stanley Meisler, "France Admits Greenpeace Ship Bombing," *Los Angeles Times*, 23 September 1985, A1. The Rainbow Warrior was destroyed when demolitions set by French commandos exploded, critically damaging the vessel and killing

a Greenpeace photographer. Ironically, a French warship rammed and damaged the Rainbow Warrior II on 10 July 1995, the tenth anniversary of its predecessor's demise, this time near a nuclear testing rig at Mururoa Atoll. See Scott Kraft, "Greenpeace Ship Seized by French Navy," *Los Angeles Times*, 10 July 1995, A1.

25. See Friedheim, chapter ï, in this volume. On the other hand, Friedheim acknowledges that bilateral interests in nonwhaling matters can be leveraged into shifts of position at the whaling meetings (Friedheim, chapter ï, in this volume), which is borne out by DeSombre's narratives of "bullying and bribing" (chapter 6 in this volume).

26. Friedheim, chapter ï, in this volume.

27. Robert L. Friedheim, "Moderation in the Pursuit of Justice: Explaining Japan's Failure in the International Whaling Negotiations," *Ocean Development and International Law* 27 (1996): 349–78.

28. See DeSombre, chapter 6, in this volume. See also Mark Fineman, "Dominica's Support of Whaling Is No Fluke," *Los Angeles Times*, 9 December 1997, A1.

29. Andresen, chapter 8, in this volume.

30. Ibid.

31. Ibid.

32. Aron, chapter 3, in this volume.

33. Freeman, chapter 4, in this volume.

34. Andresen, chapter 8, in this volume..

35. See Arne Kalland, "Management by Totemization: Whale Symbolism and the Anti-Whaling Campaign," *Arctic* 46, no. 2 (1993): 124–33.

36. See Barbara Stanton, "'Freeing' Tells of a Whale of a Non-Event," *Orange County Register*, 24 December 1989, G22 ($6 million rescue effort for three trapped whales broadcast around the world by satellite).

37. See "Bang-Bang-and Tally-Ho: Field Sports in Europe," *The Economist*, 8 November 1997, 60.

38. And to some extent, cruelty, as in trade bans on fur obtained by methods that appear objectionable (clubbing, leg traps). See Finn Lynge, *Arctic Wars, Animal Rights, Endangered Peoples,* trans. Marianne Stenbaek (Hanover, N.H.: University Press of New England, 1992).

39. See Patricia W. Birnie, *International Law and the Environment* (Oxford: Clarendon Press, 1992): xviii.

40. See Margaret Klinowska, "Brains, Behaviour, and Intelligence in Cetaceans," in *Eleven Essays on Whales and Man*, 2d ed., ed. Georg Blichfeldt (Norway: High North Alliance, 1994), 21–26.

41. See Anthony D'Amato and Sudhir K. Chopra, "Whales: Their Emerging

Right to Life," *American Journal of International Law* 85, no. 1 (January 1991): 21; whales are "a species of animal life on earth that scientists speculate has higher than human intelligence," and speak "abstruse metaphysical poetry."

42. Estimates on the total number of monkeys killed to produce the vaccine range from 1 million to 5 million. See Bill Deitrich, "A High-Stakes Battle over Animal Testing," *Seattle Times*, 4 February 1996, A1. See also Andrew Rowan, *Of Mice, Models, and Men: A Critical Evaluation of Animal Research* (Albany: SUNY Press, 1984).

43. Of course, medical research carries a warrant more impressive than banqueting. But still, if *intelligence* is the characteristic that endows animals with rights—that is, withdraws actions that injure them from calculations of general utility—one would elevate chimpanzees to that status before one conferred rights on minke whales.

44. Paula Bock, "The Accidental Whale," *Pacific Magazine (Seattle Times)*, 26 November 1995, 14.

45. That does not distinguish whaling from other international controversies, of course, such as nuclear proliferation.

46. G. P. Donovan, "The International Whaling Commission and Aboriginal/Subsistence Whaling: April 1979 to July 1981," in *Aboriginal/Subsistence Whaling (with Special Reference to the Alaska and Greenland Fisheries)*, Reports of the International Whaling Commission 30,special issue 4 (Cambridge, England: IWC, 1982), 79, 83.

47. Erik Sander, "Whales for Foxes from Arctic Circle," (High North Alliance). *http://www.highnorth.no/wh-fo-fo.htm.*

48. Ibid.

49. See Friedheim, introduction, in this volume. Oddly enough, for the seasons 1993–95 and 1995–96 there was even provision for the Bequians of St. Vincent and the Grenadines to take two humpback whales. *Proceedings of the 46th Meeting of the IWC*, 1994, 13(b)(4). How such an exception was carved out may say something about the politics surrounding the IWC. According to the Encyclopedia Britannica (http://www.eb.com, as of 24 March 2000) "nearly three-fourths of the inhabitants are black, and another fifth is mulatto. A small minority is of European descent, and there is a small group of East Indian extraction; only a few are of Carib-Amerindian stock, but there is a small group of black-Amerindian mixture." Whaling was reportedly introduced by Scottish descendants in 1875. See note 30 above.

50. Freeman, chapter 4, in this volume.

51. See Patricia A. Birnie, "International Legal Issues in the Management and Protection of the Whale: A Review of Four Decades of Experience," *Natural Resources Journal* 29 (fall 1989): 907.

52. Freeman, chapter 4, in this volume.

53. Ann Renker, *Whale Hunting and the Makah Tribe: A Needs Statement,* February 1996, updated October 1997, (IWC/49/AS5), 29.

54. See Neil J. Murphy et al., "Dietary Changes and Obesity Associated with Glucose Intolerance in Alaska Natives," *Journal of the American Dietary Association* 6 (1995): 672–76.

55. This is consistent with the *Report of the Nutritional Panel,* that while the nutritional problems of the Inuit were significant, they did not uniquely require a bowhead solution. *Aboriginal/Subsistence Whaling (with Special Reference to the Alaska and Greenland Fisheries), Reports of the International Whaling Commission* 30, special issue 4 (Cambridge, England: IWC, 1982): 30.

56. Renker, *Whale Hunting,* 18.

57. Ibid.

58. See chapter 5, "Food Security, Food Hegemony, and Charismatic Animals," by Russell Barsh, in this volume. In a footnote Barsh defines *Western European* as the largest global industrial powers and thereby includes Japan.

59. Surely Barsh is right that the pressure to end sealing is driven by dominant society notions that seals are "too cute." Moreover, profit motivation may lie behind Western favoring of synthetic coats to replace natural furs. But the sealing controversy is, I take it, more about clothing than about food.

60. Barsh, chapter 5, in this volume.

61. Jared Diamond, *Guns, Germs, and Steel* (New York: Norton, 1997).

62. Barsh, chapter 5, in this volume.

63. Jack R. Klopenburg Jr., ed., *Seeds and Sovereignty: The Use and Control of Plant Genetic Resources* (Durham, N.C.: Duke University Press, 1988), iv.

64. Or consider Arne Kalland's related thesis, that anti-whaling represents a cultural imperialism. Kalland observes that totem prohibitions are not ordinarily expected of outsiders. The Hindus, for example, "in no way try to impose the prohibitions of killing cows on the rest of mankind [but] whale protectionists try to make the prohibition universal. In their zeal they continue a form of western cultural imperialism initiated by Christian missionaries." Kalland, "Totemization," 129.

10 / Whale Sausage

Why the Whaling Regime Does Not Need to Be Fixed

DAVID G. VICTOR

The international regime for regulating whaling is a mess. The International Convention for the Regulation of Whaling (ICRW), the regime's main legal instrument, calls for sustainable management of whaling to "make possible the orderly development of the whaling industry." Yet most members favor extinction of the industry. Rather than formally banning whaling, the ICRW's decision-making body—the International Whaling Commission (IWC)—imposed a moratorium against commercial whaling in 1982 to allow extensive assessment of whale stocks and development of a sustainable whaling management scheme. Today, after more than a decade of extensive scientific review, even sober plans for sustainable whaling developed by the IWC's own Scientific Committee have been rejected by the IWC's political process. The moratorium continues, and the IWC has even established a whale sanctuary in the Antarctic, although every reasonable assessment shows that limited minke whaling there could be easily sustainable.

International whaling law, it seems, has become a farce. Japan, the only Antarctic whaler, continues its operations under an "objection" to the inclusion of abundant minke whales in the sanctuary and supplies its market with whales caught for "scientific" purposes. Iceland has exited the IWC altogether but failed in its ambition to establish an alternative management process—the North Atlantic Marine Mammal Commission (NAMMCO). Norway has resumed commercial whaling in the North Atlantic under an "objection." Whaling for subsistence aboriginal purposes continues, but the

IWC allows only some communities that meet the definition to whale while banning others. Some aboriginal whaling is focused on endangered species and thus violates the ICRW's goal of sustainable management of whale stocks. IWC meetings have regressed to a circus; nearly as many observers and delegates ply the halls as whales are caught every year. The principal reason the regime doesn't unravel is that the United States threatens sanctions against spoilers.

The premise of this book is that this situation needs to be fixed. The present chapter argues the opposite. The problem of whaling is not scientific; rather, the whaling regime is in crisis because the core objectives of whale "preservationists" are incompatible with the "conservationists," who want to resume commercial whaling. Efforts to fix the regime, such as by negotiating a new legal instrument or massively reforming the IWC and ICRW, will make stark and inescapable the deadlock conflict over objectives.

Not only does the regime not need massive reform, it may be *Pareto optimal*—the interests of most of its participants could not be better satisfied by any feasible alternative to the status quo. The dwindling consumer demand for whale products is met with whales caught under the objections and scientific whaling. Preservationist groups probably benefit because the continuance of this stalemate keeps the whaling issue salient and allows the public to believe that commercial whaling threatens whales, which is good for membership. Some stakeholders fare poorly under the present regime. They include some whalers—especially in Iceland, in Japan, and in communities wrongly barred from practicing aboriginal whaling—and the approximately one thousand whales killed each year. But they are unlikely to do better under an "improved" whaling regime, and they could do much worse. The reform process would be dominated by the interests of preservationists, who constitute the IWC majority. Surely they would adopt reforms that would make it more difficult for whaling nations to disregard IWC decisions by such means as issuing legal objections and engaging in scientific whaling, practices that preservationists regard as dangerous loopholes that allow whalers to skirt IWC decisions. Whaling nations would exit (or not join) a reformed regime, which would lead to even more whaling than at present. Only substantial coercion would keep them inside. Sanctions and threats, such as those imposed by the United States, have already harmed some whalers; such tactics, which many view as unjust bullying, would grow more common if a reformed regime built on incompatible objectives were to be maintained.

Both conservationists and preservationists believe that reform would better serve their interests; some have focused resources on the development of new management schemes. In this chapter I argue that rather than speeding reform, the guiding principle should be to slow it down to a pace that just barely sustains the IWC's legitimacy as the only needed whaling management regime. Not only are reform efforts likely to founder, but they also waste time and resources on a problem that is being solved through changing norms. Preservationist values have spread and the world is losing its taste for whale products; the annual catch has plummeted from eleven thousand animals on the eve of the whaling moratorium adopted in 1982 to only one thousand animals today. Essentially, all of today's whaling, except in a few aboriginal communities, is sustainable by wide safety margins. Perhaps the only significant reforms that are possible and needed are to ensure that aboriginal subsistence whaling is directed to nonendangered species. But for the regime as a whole, the messy deviations between written law and actual practice allow the whaling agreement to survive and parties to protect their interests, despite deadlock. Like sausage, the production process is reviling upon close inspection, but the final product is useful—perhaps even most palatable.

This argument in favor of the status quo will be made by reviewing the main themes identified in Robert Friedheim's preparatory paper for this volume[1]—sustainable management, justice and human rights, and international governance. On all three themes the IWC appears to be a failure, but in practice, I argue, could not do much better. The final section of this essay will suggest that although the whaling debate is highly unusual, many of its attributes are generic to international governance. Unlike the other authors of this volume, I am no expert on whaling—this chapter merely suggests why the regime has its current form and cautions against attempting to improve what is both peripheral to world politics and practically as good as it can be. For reviews of the history of the whaling issue, readers are referred to other chapters in this book and other published literature.[2]

SUSTAINABLE MANAGEMENT?

One reason to reform the IWC might be to promote "sustainable development." Like many popular slogans, sustainable development means many (sometimes conflicting) things, and thus "sustainability" as a management principle has been appropriately attacked. Nonetheless, most visions of sustainable development include the need for rational assessment of stocks, harvest plans, and protection of species against extinction. Can the IWC be reformed to conform with these principles of sustainability?

Sustainable use of common pool resources is hardly a new goal. Long before the spread of environmental values, many sustainable management regimes were created for simple commercial reasons—to achieve maximum sustainable yield (MSY) and profits. While there have been some successes,[3] international fisheries regimes are marked principally by failure.[4] Scientific assessments have been ignored and conflicts have been papered over by raising quotas. In those cases in which rational scientific assessment has been employed, uncertainties were exploited by harvesters who sought short-term benefits. The result was quotas that were systematically excessive and set without regard for adequate safety margins. Ecosystem effects and other interdependencies, if understood at all, were ignored. Once quotas were set, cheating was typically rampant. The result has been the decline of many fish stocks below MSY levels, sometimes to commercial extinction. Gear restrictions rather than tradable quotas led to industry overcapitalization; a spiral of subsidies has been demanded to compensate fishermen for sitting at the banks.

The history of whaling is marked by similar failures. Prior to the ICRW, whaling regimes practically ignored scientific assessment, failed to set rational quotas, and didn't enforce rules. Whalers were agents harvesting a common pool resource—individually, they maximized their return by killing all they could; collectively, they failed to manage the resource for the long term. Since the end of the nineteenth century, the range of whaling fleets has grown to cover the globe, including the whale-rich Antarctic Ocean. Innovative and effective exploding harpoons were trained on the most convenient targets, such as blue and right whales. Factory ships allowed efficient division of labor between killing and processing. Disaster didn't strike earlier because rising demand for whales was interrupted when petroleum replaced whales as the chief source of lamp oil in the nineteenth century, and the supply of whales was interrupted when twentieth century wars sidelined or sank whaling fleets. Moreover, the oceans were abundantly stocked with whales.

The ICRW, adopted in 1946, made only tiny improvements. Quotas for the season in Antarctica, where most whales were caught, were set in "blue whale units" (BWUs) and thus did not distinguish endangered from abundant whales. The practice of regulating the season length rather than allocating quotas among nations and restricting the number of whaling ships led to the famous "whaling Olympics"—whalers worked around the clock to kill all they could find during the short season, and whaling fleets were overcapitalized. Conflicts within the IWC were resolved by raising the total quota; when conflicts were particularly severe, no quota was adopted at all. Thus formal compliance with IWC quotas has been high, even as whalers emptied the oceans.[5] Whales were big losers because of technological innovation and the peace of the Cold War. Blue and right whales neared extinction, and the maximum sustainable yield of whaling declined. When outsiders gazed inside the IWC in the late 1960s, they were shocked by what they saw: foxes had been guarding the hen house.

In the ashes of this disaster, should the whaling regime be reformed today to allow for more rational management? It has been. In the late 1960s, environmental and animal rights groups "discovered" whales, and soon thereafter the populace in many industrialized countries was buying records with whale songs and sending checks to Greenpeace to finance whale protection by motorized rubber raft. Spurred by this pressure, and by its own obvious failure, the IWC began in the 1970s to implement more rational schemes. It stopped managing "whales" as a single resource with BWUs as the currency and adopted species-specific quotas; zero quotas were set for the most endangered species. The New Management Procedure (NMP), implemented in 1975, increased the use of rational assessment in managing whale stocks and in principle allowed better enforcement of rules. In practice, the NMP proved difficult to apply, but it marked a crucial milestone: compared with the regulatory regime of a decade earlier, serious efforts at managing for sustainability were under way in the IWC. A golden age of whaling management dawned; whalers were pressed to do the right thing because anti-whaling forces were clamoring at the gates of the IWC. But that age did not last for long. The forces of preservation amassed inside the IWC, and when they had the three-fourths majority needed to change the Schedule of whaling regulations, they adopted the whaling moratorium to allow a pause for reassessment of whale stocks and to put whaling on a sustainable footing. The moratorium, although controversial, had a sound basis. Although the IWC was already moving toward rational management, the status of whale stocks was not well known and the IWC's management track record was poor.

During the moratorium the IWC and the main whaling states made new assessments of whale stocks and debated plans to improve the NMP. The result was a Revised Management Procedure (RMP) that was extensively reviewed and approved by the IWC's Scientific Committee. The RMP accounts for uncertainties in whale stock measurements and modeling, recognizes ecosystem effects, includes a monitoring scheme, and is built on vastly improved whale stock data. The Scientific Committee also reviewed and blessed national surveys and management plans. The objective of the RMP is sustainable whaling. But the RMP hasn't been implemented because the IWC has opposed any scheme that would allow the resumption of commercial whaling. Presented with the sober RMP, the IWC adopted the procedure but moved the goalposts again. The IWC demanded that a Revised Management Scheme, which would require still further data and assessments and rigorous control and inspection, be developed before commercial whaling would be allowed to resume. The cat and mouse game continues because arguing over science and technical matters is easier than confronting deadlock. But from an objective technical perspective there is little doubt that a sustainable harvest of at least minke whales, with adequate margins for uncertainties and extensive enforcement, could be resumed. Iceland, Japan, and Norway have all presented rational plans, based on extensive research, to do just that.

Objectively, better commercial whaling management schemes are not needed. The commercial whaling that does take place is well within sustainable margins. The purpose of the whaling regime is no longer to promote sustainable levels of whaling. One reason is that demand for the consumption of whales has declined. The catch of whales steadily declined from the peak in the 1960s until 1988, when all commercial whaling stopped. Some of that decline is the consequence of dwindling whale stocks and tighter management. But the decline in demand is probably the most important factor. Public interest groups benefit from the membership dues that follow the salient and exciting images of confronting whaling ships, but they have had their largest impact on whale stocks by spoiling the human taste for whale products.

Whale oil, blubber, and meat are still used in some communities, but in those few countries where whale meat is still consumed, the last generation of whale consumers is aging. In Japan, which has caught the most whales in recent decades, whaling resumed in 1946 to feed a population starved after war; those who were hungry teenagers then would be in their late sixties today. The potential demand for whale meat in Japan probably remains large—whale meat on the market today is priced as a delicacy—but even when whales were abundantly available they were replaced by other protein sources, including Big Macs. Whales have not been the largest source of meat in the Japanese diet since 1963.[6] Some Japanese whale products were imported, including from Iceland. Under threat of U.S. sanctions, Iceland reduced its exports during the 1980s, scaled back its lethal whaling research program, and stopped commercial whaling altogether after 1989. Even if Iceland resumed commercial whaling, perhaps a hundred jobs would be created. Norway kills about five hundred minke whales per year for commercial purposes, and whaling is a marginal economic activity. Demand for whales is low and whaling is a peripheral economic activity, and thus whale stocks are not under pressure.[7] Demand for whale meat might rebound in the aftermath of global war or catastrophe, or if tastes swing back to whale products; but those contingencies can be negotiated if they arise.

The other, more important reason that better management is not needed is that most IWC members abhor whaling. For these whale "preservationists," a scheme for managing whale stocks is akin to planning the sustainable harvest of dogs, cats, or people. Ever since the anti-whaling states constituted a decision-making majority in 1982, science has been ignored in favor of whale preservation. At that time it was possible to contend that uncertainties required better assessment and management procedures, but today that is no longer true. Moreover, the uncertainties were little more than a tactical ploy—a way to use the fig leaf of science to protect a moral argument. The point was proved when the commission moved the goalposts on the RMP. It was proved again when it adoped in 1994 the Antarctic whale sanctuary, which has no scientific basis.

Sustainable management must be built on scientific assessment. But IWC decisions are increasingly made in spite of scientific assessment. Nonetheless the science that is done today is useful because it confirms that whales are not threatened and it largely restricts the tiny amount of commercial whaling that remains to species that are extremely abundant. Even more important, it gives the IWC something to do so that it appears to be the only forum needed to address the whale problem. Efforts to reform the IWC should continue only at the minimum level necessary to sustain this veneer. Beyond that, resources devoted to the rational management of whales are probably wasted because politically the IWC can't adopt a more rational system. Indeed, excessive attention to sustainable whaling probably prevents biologists and governments from focusing more on fisheries, where better management is badly needed.

In short, the logical conclusion for policy is to do exactly the *opposite* of what the whaling nations, who have the most at stake, are pursuing today: Rather than accelerate the advancement of scientific assessment and reform, the regime and most interests would be better served if the pace of efforts to build an improved management scheme were more lethargic.

IMPERFECT JUSTICE

Although the present regime serves the interests of most participants, no party achieves all that it wants. Some are even seriously injured as fundamental principles of justice and human rights are violated. Some whales are killed, which violates the whales' possible right to life and freedom from interference. Failure to harvest a resource to its fullest potential violates a possible right of people to exploit a vital resource (food). Perhaps it is unjust that the whaling regime is not evenhanded: Norway has been able to secure its interests and has resumed commercial whaling without much penalty, while other parties, especially Iceland and Japan, have been less successful. Aboriginal communities that aren't granted IWC quotas suffer an injustice, and all aboriginal communities are under intense and demeaning scrutiny to ensure that they engage in only "subsistence" practices, which might violate the right to freedom from interference.[8] Perhaps it is a violation of international justice and law that the regime depends on unilateral threats of sanctions from the United States, which violates the right of states to set policies without coercion. Cultural justice and human rights might be ill served in aboriginal communities that have whaled for centuries but suddenly find themselves under pressure to stop. Perhaps reform can help undo these wrongs.

Many of the principles of justice and rights invoked by whaling stakeholders are not widely accepted or conflict directly with other principles and rights. A right (or even obligation) to harvest whales for food conflicts directly with the right (or obligation) to prevent the killing of whales. Neither right (nor obligation) is universally accepted—at best, they are candidates for customary law. It might be argued that the right of whales to live is superior if whales are threatened with extinction, but the debate over commercial whaling concerns species that are not threatened. The conflict over rights is more intense in aboriginal whaling, where some communities target already endangered species. In that case, agreed rights do exist but they conflict directly—endangered whales versus aboriginal freedom—and there is little guide for how to resolve them. Three types of "justices" are particularly well established in international society and thus might be advanced through reform of the whaling regime: (1) equal treatment of states under international law; (2) sovereign freedom from intervention and coercion, such as by U.S. economic power and sanctions; and (3) protection of aboriginal practices. The discussion here addresses each in turn.

If all states have equal standing under international law, perhaps it is unjust that Norway seems able to get much of what it wants—to resume whaling under an objection to the moratorium and satisfy its domestic market for whale meat—but Japan and Iceland have been stymied. The causes of these differences reflect different circumstances and strategies. Norway has behaved strategically, built a unified national position in support of resumed whaling, and carefully documented the sustainability of its minke harvest. Iceland has also documented the sustainability of a proposed harvest, but the lack of national unity—splintered in part by the threat of U.S. sanctions— has eroded the country's ability to act strategically. Moreover, since the end of the Cold War Iceland has been less able to link issues such as this to other matters of vital interest to the United States (e.g., NATO bases), whereas Norway's continued prominent role in world politics (e.g., the Oslo Middle East peace process) has better insulated Norway from U.S. pressure.[9]

Japan has tried hard, but failed, to act strategically through the IWC. Perhaps reforming the IWC could help Japan realize its interests more fully. Japanese bureaucrats failed to stop the adoption of the Antarctic whale sanctuary. They have also failed in their effort to have Japanese communities that have long depended on small-type shore-based whaling to be given the same rights as aboriginal communities, which would have allowed them to continue limited whaling for local consumption. Friedheim has convincingly documented that many of these Japanese failures were the result of poor strategies and tactics by Japanese negotiators within the IWC, exacerbated by divided positions of the Japanese government and perhaps anti-Japanese and racist attitudes within the IWC.[10] Japanese delegations have pursued a scientific strategy aimed at adopting sustainable management quotas, but since 1982 the IWC has been a highly politicized body for which science has been (increasingly) irrelevant. Reforming the IWC could present Japanese diplomats with a window of opportunity to better secure their interests, but so far they have failed to use the windows that are already open. Is it worth reforming the IWC just so some parties, especially Iceland and Japan, can better obtain what they have halfheartedly and incompetently pursued so far?

In short, in the IWC the causes of unequal outcomes reflect differences in national interests and capabilities; although advocates of international society may lament the fact, it is hardly surprising that different inputs yield different outcomes. Nor is it a fundamental principle of justice that outcomes should be identical. That minority stakeholders with contested interests don't get what they want is hardly unusual in any society and is commonplace in international decision making. If that violates a principle of justice because some do better than others, that also is commonplace in societies—especially in international society—where the rule of law is weak and interests dictate outcomes. The rule of law would be no stronger in a reformed IWC.

Perhaps extensive reliance upon U.S. economic power is unjust to the countries and industries that are the targets of sanctions—no state, in a law-bound society, should be able to impose its will on others. This argument is ambiguous and difficult to apply for two reasons. First, many pragmatists support the use of such power because it serves their interests, and, they argue, U.S. sanctions help to uphold international law. Without such sanctions, there would be practically no mechanism for enforcing international obligations and principles. Furthermore, the vast majority of IWC members oppose whaling, and the United States is merely helping to secure what they have decided should be law. That argument has perhaps some validity, but it ignores the prior question—why should the IWC, packed with like-minded states that engage in no whaling, be the arbiter of international law on whaling? Moreover, Japan and Norway have formally complied with the ICRW—their whaling has been either for the scientific study of whales or conducted under an objection to the moratorium on commercial whaling, both of which are procedurally allowed in the ICRW. Yet both have been the targets of U.S.-threatened sanctions. In sum, U.S. power broadly supports the aim of the IWC majority, but it is hardly a detailed support of international law.

The second and more important reason why U.S. sanctions do not violate strict principles of justice is that international affairs are not strictly governed by principles of justice. We may want them to be, but such idealistic thinking will cloud policy advice and could hurt those who feel aggrieved even further. Today, the policy question is not whether to pursue a just international society but whether to maintain or reform the IWC, or perhaps to start again with a new organization. The United States will use its power not only in the current IWC but also in the negotiations to create, and enforce, any alternative. Parties that feel aggrieved by U.S. power are more likely to be harmed even further if nonwhaling states dominate the process of forming the alternative. Loopholes that currently allow whalers to pursue their interests—notably the provisions to allow lethal scientific research and to object to IWC decisions—surely would be tightened (or eliminated). Such a regime would harm pro-whaling interests further, which would require even more U.S. power to coerce their participation. Pro-whaling nations could establish their own pro-whaling regime, but the failure of NAMMCO so far to become a serious alternative to the IWC—in part due to fear of U.S. sanctions against NAMMCO members—suggests that strategy won't work. W. T. Burke (chapter 1, this volume) makes it clear that international law does not ordain the IWC as the only forum; what stands in the way is not law but coercion.

The strongest case that the existing regime is unjust is probably in the treatment of aboriginal communities. For centuries, Arctic aboriginal communities have hunted whales. Excessive commercial whaling caused depletion of stocks, which has interfered with aboriginal whaling. Today, some whale preservationists argue that all whaling—including aboriginal—should stop, which would compound injustices further. Perhaps reform is needed to secure aboriginal rights.[11]

Reform is not needed to protect aboriginal rights in part because those rights are less trampled than initially appears. A quota for bowhead whales has been granted to aboriginal communities even when all other killing of this endangered species has been stopped. The United States, a leading opponent of commercial whaling, has consistently advocated quotas for aboriginal communities within its territory. Increasingly aboriginal rights are recognized and protected in international and national law, and in practice the IWC already protects these rights. What more can be done? Certainly these rights could be sanctified further through reform of the IWC, and failure do so is perhaps the greatest injustice of the current system. But it seems unlikely that these rights would be expanded in a regime created by whaling opponents, and they could be hurt further. The debate about whaling is principally about commercial whaling; aboriginal whaling is swept along, and the protection of aboriginal rights is best served by allowing the practice to occur but not focusing legal and diplomatic attention on it.

But two reforms to the handling of aboriginal rights might be needed, although both would be politically difficult for IWC members to adopt and implement. First, if the argument that avoiding extinction is a paramount goal of the IWC, then directing aboriginal whaling to nonendangered species will be needed. That is a sensitive business not least because particular species, methods, and locations are integral to aboriginal whaling practices. Second, even less feasible but nonetheless just, would be evenhanded application of the ICRW's exceptions for aboriginal whaling. As Milton Freeman shows (chapter 4, this volume), the present system of oversight is demeaning, the definitions of subsistence aboriginal whaling are internally inconsistent, and some coastal whaling communities (notably in Japan) might qualify but are shut out for political (perhaps racist) reasons. If these two reforms can be adopted quietly—which in practice means with the consent of both conservationist and preservationist stakeholders—then they deserve action, but neither is so crucial to merit the risk of sinking the fragile consensus on which the IWC survives.

In sum, the whaling regime tramples on some rights to be sure. It is probably at its worst in protecting aboriginal rights; however, in practice, the IWC has followed international norms and thus, in practice, actually values aboriginal rights quite highly. Nonetheless, arguments about rights are contested—some policy advocates claim that the rights of whales are superior. The question is not whether some rights are trampled, but what can be done in light of the contested nature of the issue and the need in the international system to reach legal agreements through political processes where power and coercion—whether one likes it or not—play a role. In that context, will any principles of justice be better served by substantially reforming—or even rewriting—the ICRW and the IWC's management practices? Could aboriginal communities possibly expect to achieve more under a new agreement, when the old agreement has allowed whaling of even endangered species? Can the targets of U.S. sanctions expect that the United States will make less use of its power during the negotiation of a new whaling agreement? If the new agreement is offensive to whaling interests and whalers exit, as Iceland did, will whale lovers in the United States allow the U.S. government to stand idly on the sidelines rather than force whalers to join? The answer to each of these questions is "no," and thus imperfect justice in the IWC is probably still better than the alternatives.

INTERNATIONAL LAW AND GOVERNANCE

The need to promote international law and governance might also be a compelling reason to reform the IWC. The practice of muddling through by ignoring legal texts may degrade the role of law in international society. It is also a bad precedent to openly flout science, the cornerstone of rational and sustainable governance. At minimum, good legal agreements should be transparent about their purposes and means, yet the IWC survives on unwritten understandings and veiled agendas. A civilized society, it might be claimed, resolves conflicts through law and principles, not politics and power.

International society governed by law is a noble aim, but it is also a public good that the IWC can do little to promote. Collectively, the mission to build international society based on law depends on building up the rule of law, which is degraded by each individual instance of farcical law. But the demand for that society, and thus the demand for law, depends fundamentally on the support of its members.[12] Regulation of whaling is a bad case for building up the role of law because it enjoys no such fundamental support. As argued, building a new agreement will result in either deadlock (and thus no agreement) or extensive reliance on U.S. power (and thus a coerced agreement). Neither is good for the rule of law. Yet the pragmatic needs of governance require that some legal instrument exist to protect against the resumption of unsustainable practices by the few remaining commercial whalers and to manage aboriginal whaling to the extent possible. In short, the ICRW is still needed; it serves a useful function in international law—as a backstop to provide political assurance that the "whale problem" is more or less under control. The alternatives are no better for the rule of law. Reforming the legal regime to advance the public good of international law is a mission for which political pressure is weak and thus bound to fail.

PARETO OPTIMAL OUTCOME?

The argument here has been that the imperfections of the whaling regime are less outrageous than initially appears, and importantly, I argue, the feasible alternatives would be worse than maintaining the uneasy status quo. The lack of a tight link between the preferences of the IWC's voting majority and actual behavior allows most parties to satisfy their interests. The result is a messy process—where de facto rules and practices are built up through unilateral interpretations and careful strategic action rather than strict adherence to IWC decisions and the rule of law. The present messy outcome may indeed be Pareto optimal—no party could do better without harming the others.

The interests of whalers are probably better served with the current regime than under plausible alternatives—indeed, pursuing reform of the IWC would be very risky for whaling states, who might fare much worse. Rather than lament the IWC's messy procedure, whalers should be grateful that it exists. The Irish proposal to allow the resumption of coastal commercial whaling, debated first in 1998 and quickly found stillborn, revealed the poor prospects for reforming the IWC. If that proposal lingers and brings new attention to whaling practices, new sanctions, and a rekindled militant protectionism among the nongovernmental organizations, whalers will find themselves longing for the good old days.

Even principles of justice are not completely ill-served by the current regime and might also fare worse under a reformed IWC. Achieving the ultimate objective of stopping whaling would require applying more power— sanctions and boycotts—that almost surely would trample rights further. Complete cessation of whaling would violate aboriginal rights. Stopping Norway's harvest could damage other aspects of international affairs and would require environmentalists to explain why they oppose the policy of sustainable whaling that was engineered by the former Norwegian prime minister Gro Harlem Brundtland, principal author of the most important political document on sustainable development—*Our Common Future*.[13] Stopping Japanese whaling completely would require more U.S. muscle, which would be illegal and also would unwisely test an already tense but vital commercial trading relationship.

Although whaling interests are relatively well served, are anti-whaling interests also satisfied by the regime? Surely they must view the one thousand whales that are killed per year as a failure. The ultimate objective of eliminating all whaling has not been met, but that has an advantage for pro-whaling public interest groups. Many nations have quit whaling since the 1970s, and perhaps never again will the whaling issue command such mass concern (and donations to public interest groups) as it did in the 1970s. But permanent cessation of whaling would forever deprive these groups of an issue and its attendant membership support. That the IWC continues to meet, debate whether to resume commercial whaling, and attract media attention keeps at least a small segment of the public worried about whale safety. Insofar as they are concerned about sustaining membership and budgets, the end of whaling does not serve their interests. Moreover, the complete end of whaling would require coercion, and that would harm the ability of these groups to be effective in other areas of international environmental law, where sanctions and coercion are also needed. Sanctions and other coercive techniques to enforce international environmental agreements might be illegal under international trade laws and are politically costly to implement. Probably it is worth reconciling environmental and trade conflicts on a more important issue for animal rights advocates and environmentalists (and the animals themselves), such as the protection of truly endangered species.

THE DANCE

Although the present situation may be practically Pareto optimal, it is not rigorously stable. Any of the main stakeholders—the whaling nations, the anti-whaling nations (notably the United States), aboriginal communities, and the public interest groups—might destroy the uneasy peace. Thus, insofar as this chapter offers any policy advice, it is not how to reform the IWC but rather how to survive without reform—to sustain the pragmatic, messy status quo while the norm against killing whales diffuses, demand for whale products declines, and whaling becomes even less profitable than it is today. In perhaps a decade or two, the only remaining (lethal) whaling activity may be in aboriginal communities. At that time, it should be easily possible to sustain a whaling regime that serves no purpose other than to manage aboriginal whaling, because awareness of the need for protection of aboriginal rights and cultures is rising worldwide, even as the norm against whaling solidifies. Perhaps the only conflict will be over whether and how to prevent aboriginal whaling of endangered species, such as the bowhead; starting down that path now would be a valuable undertaking for those interests that truly care about protection of endangered species. (However, that problem may also solve itself; some evidence suggests that the bowhead is becoming less endangered everywhere following the end of commercial whaling.)

Sustaining the uneasy peace won't be easy if any of the major parties finds its interests severely challenged. Thus the guiding principle for sustaining the IWC should be to avoid challenging actions. In practice, that means that the United States can continue to threaten sanctions but should also continue the practice of negotiating an outcome and not applying sanctions. Whaling nations must at least partially respond to U.S. concerns by changing their behavior. They should push for reform of the IWC but accept that reforms will be marginal and probably never will restore full-blown commercial whaling. Crucially, the major whaling nations should ensure that they work through the IWC rather than exiting. If Iceland were to resume commercial whaling outside the IWC and other members were also to leave, then the fiction that the IWC is the main whaling decision-making body would be difficult to sustain. If the IWC were to lose its pre-eminent role, the peace would unravel and most parties would suffer. With that objective in mind, IWC members should slow the pace of reform. Each iteration of "progress"—adopting again-and-again new systems for the management of whales—removes more of the fig leaf and makes crisis over the decision to resume commercial whaling more likely.

Will the parties exercise such restraint? They won't if they are unaware that avoiding reforms that expose their differences is the game that they should play. The risk of missteps is greatest in the whaling states, especially Japan, which incorrectly thinks that it will benefit from the formal resumption of commercial whaling. It appears that Japan still does not know how the game should be played. Japan's decision in 2000 to expand its scientific whaling to include charismatic Moby-Dick sperm whales earned scorn across the West and forced the U.S. government to renew threats of sanctions. Environmental and animal rights groups presumably also oppose this muddling-through strategy and also might push to overhaul the IWC. Luckily for them, and for progress on many other international issues, hard-core preservationists probably do not have the political support and power to force such a crisis.

IMPLICATIONS FOR INTERNATIONAL GOVERNANCE

To close, it should be underscored that whaling is a topic for a small, shrinking audience. However, the operation of the IWC and the debate about how to manage the whaling problem is one of generic importance because it concerns sustainability, justice, international law, and thus governance. One lesson from the whaling regime is that a principal means by which international law influences behavior has not been the imposition of rules (although coercive imposition has been a large part of the whaling story) but by helping to reinforce norms and diffuse values. The whaling problem is being solved principally because tastes are changing.

Another lesson is that although sustainability, justice, and international law have been imperfect in the whaling regime, they have not been irrelevant. Even though the parties fundamentally disagree about principles, they mime the need for sustainability rather than dissolve the regime. Aboriginal rights, which are the main issue of international justice at stake, have been recognized and protected. Special efforts have been made by pro- and anti-whaling forces not to radically flout the ICRW or established principles of international law. All parties have made special efforts to remain formally in compliance with their interpretation of international law. The United States has flexed its muscles, but it has been careful not to impose sanctions. For a case in which no agreement is possible, these are significant accomplishments—a sign that international norms and procedures matter.[14]

Finally, the regime survived *because* international law is weak—there is no perfect link between international legal principles and decisions and actual behavior. Scientific whaling and objections have allowed some whaling interests to be sustained. This incongruence between law and behavior is hardly unique to the whaling issue, or to international law. Even at the national level, some laws are not fully implemented but rather serve as symbols and goals. Such law is not necessarily empty but allows considerable deviation between standards and behavior—it allows incompatible interests to remain inside the same legal regime, and it engages stakeholders in processes.[15] Those processes also sometimes mask underlying differences, leading to messy outcomes such as in the IWC today. But in many regimes, including the IWC, the alternative (coercion or no law) could be worse for most if not all parties. This model—a regime built on faulty premises and sustained through tacit agreement that law should be sidestepped—is hardly the model for a society governed strictly by the rule of law. Nor is it a proper guide for how to achieve costly and detailed regulation, as may be required when solving problems like global warming. But in areas such as whaling—where parties are far apart and the problem is not very important—the model is better than the alternatives.

NOTES

1. Robert Friedheim, "Toward a Sustainable Whaling Regime: A Preparatory Paper," May 1997, unpublished.

2. J. L. McHugh, "The Role and History of the Whaling Commission," in *The Whale Problem: A Status Report*, ed. William E. Schevill (Cambridge: Harvard University Press, 1974); Johan Nicolay Tønnessen and Arne Odd Johnsen , *The History of Modern Whaling* (Berkeley and Los Angeles: University of California Press, 1982);

Patricia Birnie, *International Regulation of Whaling: From Conservation of Whaling to Conservation of Whales and Regulation of Whale-Watching*, 2 vols. (Dobbs Ferry, N.Y.: Oceana, 1985); M. J. Peterson, "Whalers, Cetologists, Environmentalists, and International Management of Whaling," *International Organization* 46, no. 1 (winter 1992): 149–53; Robert L. Friedheim, "Moderation in the Pursuit of Justice: Explaining Japan's Failure in the International Whaling Negotiations," *Ocean Development and International Law* 27 (1996): 349–78; Steinar Andresen, "The Making and Implementation of Whaling Policies: Does Participation Make a Difference?" in *The Implementation and Effectiveness of International Environmental Commitments: Theory and Practice*, ed. David G. Victor, Kal Raustiala, and Eugene B. Skolnikoff (Cambridge: MIT Press, 1998).

3. For example, Elinor Ostrom, *Governing the Commons: The Evolution of Institutions for Collective Action* (Cambridge: Cambridge University Press, 1990).

4. For example, Garrett Hardin, "The Tragedy of the Commons," *Science* 162 (1968): 1243–48; Arild Underdal, *The Politics of International Fisheries Management: The Case of the Northeast Atlantic* (Oslo: Universitetsforlaget, 1980).

5. Chapter 3, "Science and the IWC," by William Aron, presents some data on Antarctic catches. Note that the actual catch was typically slightly higher than the quota because of time delays between the closing of the season and the actual cessation of whaling and because of a normal level of noncompliance; the point is that catch and quotas correspond closely. Although it was easy to comply with IWC decisions and the ICRW, nonetheless there were some cases of noncompliance— Panama falsified its data, and recently it has been discovered that the Soviet Union also manipulated whaling statistics.

6. Robert L. Friedheim and Tsuneo Akaha, "Antarctic Resources and International Law: Japan, the United States, and the Future of Antarctica," *Ecology Law Quarterly* 16, no. 1 (1989): 119–54.

7. For more on Iceland and Norway, see Andresen, "Whaling Policies."

8. For an analysis of the cultural and legal issues, and the practice under the IWC, see chapter 4, "Is Money the Root of the Problem? Cultural Conflict in the IWC," by Milton Freeman, in this volume.

9. For comparisons of the Icelandic and Norwegian strategies, see Andresen, "Whaling Policies."

10. Friedheim, "Pursuit of Justice."

11. A case might also be made that small-type artisanal whaling deserves similar protection because, like aboriginal whaling, it is a traditional practice. That norm has not been as widely accepted in international law and practice, and thus it is no surprise that small-type whalers have fared worse in the whaling regime than have

aboriginal communities. Because small-type whaling rights are not sanctified, it is unlikely that countries could successfully push for their small-type whalers to be given special rights unless the country as a whole were willing to risk penalties imposed by other countries (e.g., U.S. sanctions)—Japan has tried, and failed. The differences in outcomes between nonaboriginal small-type whaling and aboriginal whaling is some evidence that principles of justice do influence law and practice, even at the international level.

12. Louis Henkin, *How Nations Behave: Law and Foreign Policy* (New York: Praeger, 1968).

13. World Commission on Environment and Development, *Our Common Future* (Oxford: Oxford University Press, 1987).

14. For more on the influence of international procedures and names see Marc A. Levy, "European Acid Rain: The Power of 'Tote-Board' Diplomacy," in *Institutions for the Earth: Sources of Effective International Environmental Protection,* ed. Peter M. Haas, Robert O. Keohane, and Marc A. Levy (Cambridge: MIT Press, 1993); Abram Chayes and Antonia Handler Chayes, *The New Sovereignty: Compliance with International Regulatory Agreements* (Cambridge: Harvard University Press, 1995).

15. For more on devices such as opt-out provisions and nonbinding instruments that allow for substantive law that accounts for national differences, see Kal Raustiala and David G. Victor, "Conclusions," in *The Implementation and Effectiveness of International Environmental Commitments: Theory and Practice,* ed. David G. Victor, Kal Raustiala, and Eugene B. Skolnikoff (Cambridge: MIT Press, 1998).

11 / Fixing the Whaling Regime

A Proposal

ROBERT L. FRIEDHEIM

PARETO OPTIMAL IT MAY BE, BUT SATISFACTORY IT IS NOT

The current preservation-based whaling regime probably is "Pareto optimal," as David Victor says (chapter 10 of this volume). That is, no party can improve its position without worsening the position of someone else. Since the preservation forces insist upon a complete cessation of whaling and the anti-preservation forces insist upon the right to catch a larger—albeit sustainable—number of whales, it is hard to envisage any other arrangement that would create "a different distribution of benefits among the affected parties" and not worsen the position of one of the contestants.[1]

I agree that Victor's assessment is insightful, but I cannot agree with his prescription—leave the situation alone and accept the current distribution of benefits. This is not merely a matter of personal preference; neither of the contending camps is willing to leave it alone. Therefore we must find a different solution with a different distribution of benefits. It is not just the practical division of the benefits that divides them. The stakeholders cannot leave well enough alone because they cannot agree on a commonly accepted definition of what is just and equitable in a regime for the future management of whales and whaling.

In sum, neither side is willing to accept as a permanent outcome the current regime, which includes (1) a formal ban on whaling through a moratorium and Southern Ocean sanctuary, (2) an informal and grudging tolerance of Japanese scientific whaling and Norwegian regional commercial whaling, (3) seemingly equally grudging acceptance of indigenous whaling, and (4) no provision for whaling rights for artisanal claimants. The preservationists are not willing to concede what the whalers now "have," and the whalers insist upon having more than what they have been allotted under the present regime. Both push toward improving their situations and force us to consider, if they continue, whether the International Whaling Commission (IWC) will remain a viable international resource management organization.

Unfortunately, events are undermining the opportunity to follow Victor's advice of leaving well enough alone. In IWC plenary meetings, the strategy of the like-minded coalition is obvious—use any and all means to get the Japanese to surrender their right to conduct scientific whaling, and accuse Norway of lying (but not officially) about the data they provided to establish the Revised Management Procedure (RMP). One approach is to insist that only research approved by the Scientific Committee would be considered scientific whaling. Lethal research, according to this argument, is no longer needed and therefore should not be authorized. But all the preservationists have achieved is a standoff. Whatever the opinion of the majority might be concerning whether scientific needs justify data gathering via lethal methods (as expressed through resolutions), Japan has a treaty right to continue conducting scientific research, presumably by methods of its choice. Moreover, while a working group of the Scientific Committee did not give Japan's lethal scientific research a ringing endorsement, it did say that "there were non-lethal methods available that could provide information about age structure (e.g., natural markings) but that logistics and the abundance of minke whales in the relevant Areas probably precluded their successful application."[2]

Norway corrected data they provided. Hallway gossip and news reports implied deliberate deception on Norway's part.[3] These actions just keep the pot boiling and make it more difficult for the parties to accept Victor's policy advice that all parties should learn not to rock the boat.

Both pro-whaling and anti-whaling forces *are* rocking the boat. What they do is fuel each other's paranoia and suspicion. The North Atlantic Marine Mammal Commission (NAMMCO) remains in the wings as a potential rival organization. There are strong pressures in NAMMCO to become an allocating agency. To this point, Norway, playing a more subtle game than the other members of this High North group, allows expertise to be created in the organization but prevents it from substituting for the IWC. If the anti-whaling majority in the IWC does not allow Norway and other whalers to retain the right to capture the number of whales the NAMMCO Scientific Committee calculates can be taken sustainably, NAMMCO could be transformed into an allocation agency. In March 1999, NAMMCO received a potential boost as a substitute for the IWC. Iceland's parliament, the Althingi, voted overwhelmingly to recommend that the government permit a resumption of commercial whaling, based on the scientific data collected by NAMMCO. In addition, the Foreign Minister proposed that Iceland join the Convention on International Trade in Endangered Species (CITES) in order to fight the listing of abundant whale species as "threatened with extinction."[4]

Pressure from underrepresented communities that wish to have their voices heard on the whaling issue also has increased. They have not conveniently disappeared (as preferred by some metropolitan-based anti-whaling forces), and in our more homogenized world, they have learned to take advantage of modern communications and transportation by forming a transnational educational, but politically aware, organization. At its first general meeting in March 1998, a World Council of Whalers was formed with delegates from nineteen countries participating. Its members insist that no people be deprived of their own means of subsistence. Its purpose is to provide a collective informed voice for whaling people, community-based as well as aboriginal subsistence, around the world. Members of this organization remind us that the problem we face is a matter not just of animal rights but of human rights.

There is also pressure from the flank. Some states are pushing in the conference of the parties of CITES to get minke whales reclassified from Appendix I (particularly strict trade regulation and authorized only in exceptional circumstances) to Appendix II (strict trade regulation). The Norwegian proposal for reclassification gained a majority (fifty-seven yea, fifty-one nay, six abstentions), but not a requisite majority. Moreover, limited trade of products of the other charismatic megafauna—the elephant—was restored for those states that had a good elephant conservation record. The restoration of limited trade in elephant products is one reason I would not include a strict trade ban in an IWC reform. Rather, it seems to make sense that if the limited elephant product regime is successful, it should provide a set of practices that the IWC can emulate. In any case, I expect that what has been happening in CITES will reverberate in the IWC.

Essentially I have shown is that both sides reject the criterion of "Pareto optimality" as the appropriate model for an acceptable outcome. By their actions participants in both camps clearly indicate that informally and intuitively they understand the costs and benefits of the current outcome but see Pareto optimality telling only half the story. Pareto optimality addresses the question of whether, between two parties, the distribution of benefits between them is efficient. It is the highest welfare position. But "it says nothing about the *justice* of the distribution of claims. That is, about the issue of distribution (an ethical problem) as contrasted with *efficiency*" (emphasis in the original).[5] While, as we have seen, it is extremely difficult to get people to resolve claims based on ethics, we must still try.

A MODEST PROPOSAL

Below is a proposal I developed designed to get peoples, governments and their delegates, and representatives from nongovernmental organizations (NGOs) thinking about how to negotiate a more satisfactory outcome than the present stalemate. I think it is a *modest* proposal. Since I do not believe a proposal with all the important points my colleagues made could be negotiable, I did not try to incorporate all their wisdom in my draft. As a result, some of them think the proposal *too modest.* I expect many whaling opponents, especially those for whom the sanctity of whale lives is a cardinal principle, will find the proposal *immodest.* But I do not expect to have the final word. I hope to stimulate action and would be pleased to see a serious negotiation begin even if all of my points are taken merely as a starting point.[6]

There are reasoned arguments and reasoning persons on the anti-whaling side of the issue. This book was written to appeal to them. Though some may conclude that we believe all opponents of sustainable whaling are fools, scoundrels, or mad persons, that is not our intent. Perhaps the recent bullying and bribing and other forms of uncivil behavior have been displayed more by supporters of a total whaling ban than by supporters of sustainable whaling; there is enough blame for the present impasse to go around. The whaling problem is a clash of rights and interests. Unfortunately it has been played out as a Manichaean struggle of light and darkness, good and evil, by the more ardent proponents of both sides. It is very difficult, therefore, to keep one's "cool," but we have tried, and I hope my proposal is viewed as an attempt to provide a positive input into the mix.

I am convinced that if national leaders developed the "will," it would be possible to put the IWC right, and indeed improve its procedures, increase its mandate concerning the management of whales and whaling worldwide, and become a model regime for demonstrating that living resource management regimes can be used to achieve sustainability. But as I noted in chapter 7, the issue is not "ripe" for a negotiated solution. Using the bargaining and international bargaining literatures, I tried to demonstrate what might be done to make the issue ripe. It is ugly. Many, including early readers of my manuscript, recoiled in dismay and asked, "who would take such advice?" The answer is *no one*. No one wants to pay the price. Consequently, the IWC remains in stasis.

How do we develop the will? Under the current circumstances of wide division, I have little faith that arguments developed to change "perspective" alone will do the job. Put another way, I am skeptical that all parties to this particular dispute will behave as "rational people" who will listen to the arguments and adjust their views accordingly so that an accommodation can be found. I suspect the problem will move toward crisis or irrelevance before the parties become willing to listen to ideas that might prevent the breakdown of the whaling regime. Even if breakdown is avoided, I doubt whether those stitching the IWC back together will find a better available path toward a long-run solution. In short, it is most likely that the stakeholders will take their best short-run options. But human beings *are* capable of learning, although it is difficult for them to change their perceptions to favor anything but their short-run interests.[7] Obviously, what is needed to provide the incentive for stakeholders to reassess their positions is a "triggering event." As best I can tell, we have none, but there might be early rumblings of a shift in approach to ocean management—at least in the United States. In February 1999 the National Oceanic and Atmospheric Administration issued a report proposing that some selected unmanageable California sea lions or Pacific harbor seals—totally protected by the Marine Mammal Protection Act of 1972—be "lethally removed" because they were preying on salmonids listed under the Endangered Species Act.[8] Perhaps this is the beginning of a willingness to manage an entire ecosystem rather than privileging single species.[9]

The IWC needs not be thrown out like the baby with the bathwater, and a new whaling regime or regimes need not be created. The IWC does not require extensive structural reform. This has became clear in my coauthors' reviews of the IWC's scientific, structural, and organizational capabilities. The IWC can be renewed as the premier world manager of whales and whaling if it can (1) develop and approve an effective Revised Management Scheme; (2) develop a management scheme for coastal whaling that will attract the many states that conduct coastal whaling into the IWC fold (to do this it is essential for the IWC to alter its approach to what constitutes "commercial" whaling); and (3) create a limited exception to the Southern Ocean sanctuary that guarantees that the exception will not lead to a resumption of a "whaling Olympics." The difficult part is not *what* needs to be done, but *how* to do it.

Knowledgeable readers will recognize that much of what I propose was touched upon by Michael Canny in his effort to end the hostility within the IWC, but I will treat the issues somewhat differently than it is said Canny does. For one, I do not propose to end scientific whaling. I think the issue will fade away if the other major issues relating to the take of whales for food are resolved. Moreover, I agree with my legal colleagues William Burke and Jon Jacobson that to terminate a right under the treaty would require a formal treaty amendment if the states concerned are to be legally bound by the change. For another, I also do not propose to end all trade in whale products. As the United States and the members of CITES are discovering, trade sanctions, especially bans, create their own problems. Controlling trade is a better and more flexible approach.

The linchpin of a serious effort to restore the IWC to proper functioning is the development and approval of a Revised Management Scheme (RMS). The "scheme" is the management program that should be in place before any resumption of whaling is authorized. It should (1) control entry into the "fishery" by authorizing *who* has the right to catch *what* and in what *numbers, where,* and *when;* (2) resolve whether the exploiter *pays* for an exclusive right to use a common pool resource owned by all peoples of the world; and (3) develop an implementation scheme that contains IWC plenary oversight, field inspection, market controls, certification, verification (including third-party audits), and enforcement rules including sanctions for violations so that observers can ensure that cheating does not become a problem.

DEVELOPING AN EFFECTIVE RMS

The condition precedent for development of an RMS is the adoption of an RMP or Revised Management Procedure. After much internal turmoil, suspicions that approval of an RMP was being held up for political reasons, and the resignation of the chairman of the Scientific Committee, the RMP was accepted and endorsed in 1994.[10]

The core of the RMP is a mathematical algorithm that is used to estimate stock size and is adjusted so that the impact of any change, including human consumptive uses, can be estimated with reasonable certainty. It now includes not only removals resulting from commercial whaling but removals from other anthropogenic actions (e.g., incidental catches, catches under scientific permits, aboriginal subsistence whaling).[11] While essential, it alone is not sufficient for restoration of whaling. Too often in the debates surrounding the RMP those forces proposing the restoration of whaling argued that all that was necessary was approval of the RMP. Some of the concern expressed by the more pragmatic proponents of a total whaling ban was based on the perception that those in favor of a resumption of whaling were not taking the RMS seriously early in the moratorium years.

The pro-whaling forces did learn their lesson. Indeed, Norway submitted a draft amendment to incorporate the RMS into the Schedule. The draft included a proposed program of supervision and control, the only remaining component of the RMS that needs to be completed and approved. The program was summarily dismissed as inadequate, with no recommendations concerning what should be done to make it acceptable. Norway withdrew it. Subsequently, Japan submitted a paper to the Working Group on the Revised Management Scheme concerning an inspection scheme but has not submitted an amendment to the Schedule.[12] Assuming that any future proposals would be handled with, at best, faint praise, or as roughly as proposals for inspection and control have been handled so far, those who wish to whale have little incentive to try again. They must, or stasis will continue. But reasonable people in the pro-whaling camp need help from reasonable people in the anti-whaling camp, at least from those whose opposition to the resumption of whaling is on pragmatic grounds (i.e., people who fear that resumption of whaling could threaten the survival and stocks of whales). If the whaling proponents must propose, the opponents of whaling must critique and suggest (if they think whaling based on a sustainability standard can be maintained). It is a daunting task for several reasons. For one, drafting a legally, politically, and environmentally acceptable proposal is a formidable task in draftsmanship. There are many traps because of the second reason: the world community has not been notably successful in creating effective international natural resource management regimes for open-access resources. Closing or at least controlling access has been particularly difficult. As a result, there is deep skepticism among some experienced resource analysts and managers that it can be done.[13] On the other hand, there are those, including in the ocean science and policy communities, who believe that better—indeed adaptive style management—practices can be developed to allow exploitation within a sustainability standard, and I stand with them.[14]

Controlling entry is one of the most valuable management tools available to the IWC—being able to say *who* can take *what, when, where,* and *how* (or using what equipment). A right or duty to limit entry or allocate a valuable resource to a particular claimant (the *who*) is a right or duty denied to many other international resource management organizations. Though the International Convention for the Regulation of Whaling (ICRW) explicitly denies the IWC a right to allocate via assigning quotas to state claimants, it has done so overtly for indigenous catches over many years.[15] In addition, in the past, for "commercial" whaling, members have agreed among themselves to limit entry and allocate the resource, even though there was no legal obligation to do so. Considering the difficulty other international resource agencies have had in acquiring any right to allocate, I'm struck by the seemingly cavalier attitude of whaling opponents to the possible breakup of the IWC and loss of an international agency with such powers. The world community is struggling toward a right to allocate and exclude nonpermitted claimants in some newer fisheries agreements such as the Straddling Stocks Agreement and the North Pacific Pollock Agreement.[16] Instead of taking actions that might encourage the loss of an allocating agency, the world community should encourage its growth.

If selective whaling were restored, presumably the basis of allocation decisions would be the RMP. But seldom discussed in the IWC is a key assumption of the RMP—that human beings may consume whales. After all, why worry about analyzing the consequences of takings unless takings are allowed? Although largely unstated, this was the substantive reason (in addition to the strategic bargaining reason) that it took so much effort to get the RMP approved. If IWC members are to build on the RMP and restore some whaling, then the RMS must be used to supplement the RMP by being incorporated into the Schedule. It must be used to build confidence that any consumptive activities do not go beyond sustainability. This may be done through the application of the precautionary principle.

The precautionary principle also has been little invoked in IWC proceedings, as least as far as my perusal of IWC documents shows. But the RMP is very precautionary. This means that, while little invoked in the IWC, the RMP is fully consistent with the current trend in international management.

Indeed it might be claimed that the Scientific Committee of the I W C is a leader in applying the principle to a real management problem. The precautionary principle has been central to international interactions on environmental issues since the Brundtland Report and the Rio Earth Summit (Principle 15 of the R I O Declaration).[17] Although there is no "cookbook" formula for the precautionary principle, some ideas are emerging that are viewed by many observers as requirements for invoking the principle in practice. Among them are "best scientific evidence," and "sound statistical evidence."[18] Obviously, an R M S based on an R M P could qualify on these criteria. But such an R M S would also require that management be conducted with recognition of uncertainty. That might mean establishing quotas at some point below what previous stock models predict as "safe." That is built into the R M P. It might be claimed that the Scientific Committee could propose quotas consistent with their precautionary "reference points" as required in the Straddling Stocks Agreement,[19] or quotas based on "pessimistic models." Certainly the target quotas allowed under the R M P would be well below maximum sustainable yield (the old M S Y!).

But the Scientific Committee might take on other tasks associated with the precautionary approach such as stepwise development with impact monitoring rather than an immediate full resumption of exploitative activity. It would also help to have in place a management system that can gather new information and adapt quickly to changes in the information base. Some of the oversight might also come from participation in decision making by those who are nonconsumptive users. Identification of "brakes" to avoid explosive development is another related idea, as is development of multispecies management procedures. Finally, a well-developed system of monitoring and controlling implementation of whalers and whaling must be in place (as will be discussed below).

The precautionary principle requires the development of a risk-averse, not a risk-free approach, and perhaps one whose financial costs might make the benefits seem not worthwhile in the eyes of potential future whalers. But they must consider such measures as the price of getting whaling restored and try to make them workable. It is possible to get the costs of managing world whaling down to an acceptable level and still have in place a control system in which all stakeholders have confidence. This might mean the gradual restoring of limited whaling rights rather than full immediate restoration. It might mean allowing some inspection efforts be in place on national territory (raising questions of sovereignty). It would mean greater costs, since part of the precautionary principle is that the user pays. It would probably mean that N G Os, some not inclined to be "reasonable," would have to be part of inspection and verification efforts so that it becomes clear that exploitation of whales for consumptive purposes is not getting out of control.[20] There must be assurance that another "whaling Olympics" cannot happen.

WHO SHOULD PAY? HOW MUCH? AND TO WHOM?

It is important to get the costs of whale use correct. By use I mean not only consumptive use but all uses. At this point some readers may opine that I am getting excessively arcane and theoretical. Perhaps, but the central problem underlying the overutilization of open-access or common pool resources is that we treat them as essentially "free goods." The only private costs of using whales are the catching and observing efforts. "Nature" provides the habitat favorable for whales to be born, adequate food for their sustenance, clean waters through which they can roam, and so on. In order for humans to recognize that what they do has consequences, they must internalize the costs of the goods and services they take so freely. This means putting a cost on use. "User pays" is a broader version of the "polluter pays" principle. All members of the community should not pay for activities that may be subsidized to benefit a few.[21] Rather, all whale "users"—lethal or nonlethal—might pay into a special fund (a "World Whale Fund," perhaps managed by the I W C) that could be used to deal with whale issues not currently well addressed, such as habitat protection, or to provide incentives for poorer states with whales off their coasts to bring their whaling activities under I W C direction.

In a sense, one way the cost of whales currently is internalized is via dues paid to the I W C, based on the number of whales caught. In recent years, the anti-whaling forces have injected into I W C debate the claim that whale watching is economically more valuable than lethal takings; whales are more valuable alive than dead. In short, whales have "existence values." The debate has an Alice-in-Wonderland quality, since many anti-whaling N G Os argue simultaneously that it is unethical for humans to benefit by using whales, that is, that there should be no pragmatic criterion for judging the question of whale killing, while claiming that humans can benefit financially by keeping whales alive.[22] Usually these N G O representatives never show that one activity competes with the other in practice, that lethal taking will somehow diminish the experience of those watching whales. It seems to me that most whales watched are species that are not threatened with lethal taking.[23] Most whaling takes place in remote regions where there are few ecotourists. But if whaling opponents are sincere, they too should pay for the uses they claim. To acknowledge that they value the resource, states opposed to whaling should pay a proper share of IWC expenses.

In any case, the debate as it is now conducted is not fruitful. But it does indicate that some competent natural resource economists should examine how to properly cost the use of whales so that the burdens are not shifted to others.

AN EFFECTIVE IMPLEMENTATION SCHEME

The heart of an RMS would be institutions, rules, and resources for ensuring that (1) the RMP is adequate for predicting the effects of takings on the stocks, (2) accurate data necessary for making sound management decisions are provided to the IWC in a timely manner, and (3) whalers do not exploit in a manner inconsistent with the rules (humane killing in particular) or take more than has been allocated. These functions must be "cost-effective," that is, the costs should not exceed the benefits. Achieving such an institutional development is not as difficult as whaling opponents allege, nor as onerous as whaling proponents fear.

A permanent Implementation Subcommittee (whose mandate would be broader than an "Infractions Subcommittee") of the Technical Committee could be created to supervise operation of an implementation scheme. Obviously, the work of the committee and subcommittee would be subject to plenary oversight. The work of the Implementation Committee should be transparent, with all proceedings and data available openly. A field inspection corps must be created that uses nationals of both whaling and non-whaling nations. Certainly, when feasible these inspectors must go to sea with the whalers. Further, only vessels in good standing on a Register of Whaling Vessels would be issued permits. Inspections at landing sites, as called for in the original Schedule, would also be needed. Monitoring should be made easier by requiring the use of new technology, such as transponders.

A system of third-party audits, similar to what is being proposed to verify forest-based carbon offsets,[24] could be developed to ensure that the exploitation of a common pool resource remains within the rules. Auditors might have to penetrate national boundaries and audit, say, domestic fish markets to determine whether what is on sale has been caught legally. They should be in a position to inspect fishing vessels and should be able to cite those who violate rules. Perhaps cooperating fishing companies and markets who have passed inspection can be awarded a "legally caught" stamp similar to the "dolphin safe" logo on cans of tuna. Such an indicator of its legitimacy might be based on DNA testing.

Sanctions and punishment are always a difficult issue in the management of wildlife, especially beyond national borders. Nation-states are very reluctant to give up the right to punish their own nationals. While it is worth exploring the creation of transboundary sanctions, I would initially leave direct punishment for violations to the national government sponsoring the whaler. The sponsoring government would have an incentive to keep its whalers under control if a new Implementation Committee or Subcommittee would have the power to request that the plenary reduce or eliminate a quota for cause. In any case, the power of environmental groups to stage boycotts remains a powerful disincentive for some governments to cheat.

It is technically difficult to develop an effective institution and powers. Not only must the institution's managers have appropriate tools for dealing with the issue as we view it today, but also the institution must be flexible enough to promote new norms as it learns what must be done. The problem we face is not that many of the issues, and many of the control mechanisms, are unknown; indeed, they have been extensively debated within a working group of the Technical Committee. But the divisions were so wide that "very little common ground seemed to exist," and therefore the working group could not "propose amendments to the Schedule."[25] A new procedure for developing a plan or design is needed. It cannot be developed on the basis of debate in committee or plenary session. Committee and plenary consideration can be used to amend, judge, and approve or disapprove, but not create a plan. I propose creation of a committee of "Wise Persons" similar to that created in 1960 outside the ordinary agenda of the Scientific Committee to assess the Antarctic catch limit so that it could be reconciled with scientific findings. The committee should be authorized to recruit a consulting staff of experienced regime scholars, resource managers, and "reasonable" stakeholders (and the Wise Ones should be left to judge who is reasonable) who would put together a draft plan, a form of single negotiating text (SNT). The SNT would be subject to debate and amendment. What comes out of the consideration process, after approval by the IWC plenary would be promulgated as an amendment to the Schedule.

Surely a technically sound plan can be assembled. Whether there is sufficient "political will" to create a committee of Wise Persons and to adopt its work are more in doubt. Obviously, all parties to the controversy will have to compromise, something they have so far been unwilling to do. Both sides must recognize that times have changed: whalers must recognize that whaling can never be pursued as it was in "the good old days"; whaling opponents must recognize that the good old days are not likely to come back, partly because of their scrutiny. The existence of environmental NGOs with resources and expert staffs means that any future whaling must occur under transparent conditions. That is a fact of current international life.

Another fact of current international life is that, even if the claim that only the nation-state's writ could run within its borders were ever true in the past (a doubtful claim), it is less true today. The process of globalization has demonstrated clearly that contemporary nation-states have great trouble controlling transnational business, transnational pollution, and transnational communication and ideas within their borders![26] Some of the controls needed to assure a smoothly functioning implementation regime (e.g., inspection of fish markets) would impinge upon traditional notions of sovereignty. So be it.[27] When guerrilla DNA scientists can set up shop in a hotel room and test samples (even if such tests may not be credible because they are not subject to peer review, did not retain samples, or lack Scientific Committee oversight, etc.), scrutiny of domestic activities cannot be avoided. While whaling advocates must concede this, whaling opponents must also recognize the degree to which their own actions have transformed the international system and made a return to the bad old days unlikely.

Are the costs of a tight, well-developed implementation scheme too high to be feasible? Certainly they would be if condition were piled on condition for the purpose of stopping the process. But for carefully considered measures, although they might seem costly at first, the costs could be brought into line. A good part of that first response from whaling proponents is more a judgment that "controls [would be] insultingly onerous" than that the monetary costs would remain out of line.[28] When examined in detail, the costs, I suspect, would prove to be more moderate than first estimated. When the United States accepted the idea of a two-hundred-mile Exclusive Economic Zone (EEZ), the U.S. Coast Guard estimated that the costs of implementation and compliance would run into the billions. But in fact the cost has been nothing like that. More recently, the costs of reducing acid rain through market mechanisms are becoming increasingly reasonable and therefore feasible.

MANAGING COASTAL WHALING

It is important to create a management scheme for controlling whaling by dependent communities in the two-hundred-mile EEZ, and perhaps beyond. There are two compelling reasons for solving the problem of coastal whaling—the restoration of the IWC as a relevant force in coastal whaling and justice. One cannot be separated from the other.

The legal situation concerning whaling within the two-hundred-mile EEZ is somewhat murky. Does the IWC have jurisdiction over whaling within two-hundred-mile EEZs when the UN Convention on the Law of the Sea (UNCLOS) grants coastal states a sovereign right to explore and exploit the EEZ's natural resources (Article 56)? Article 65 of UNCLOS, the only provision to deal with whales, conveys responsibility for whales to international organizations (please note the plural), but it is remarkably silent about where the responsibilities of coastal states end and those of international organizations begin, implying possible overlapping jurisdiction in the EEZ. But while this question is important to international lawyers, we should not be bogged down by it.

In the world beyond the IWC, most coastal states act as if they have the right to control *all* exploitative activities within the zone. Those who do not forbid whaling or do not have an ideological reason for joining the IWC have also assiduously avoided membership in the IWC. Over one hundred states have whales or other smaller cetaceans within their coastal zone. The citizens of a number of these states take whales and porpoises. If there is any regulation or control over whaling, it is, and likely will remain, solely national unless the actions of the IWC indicate to many states that allow the take of whales in their EEZs that they will benefit by joining (such as having some of their management costs subsidized by a "World Whale Fund," which I previously described). Under current circumstances, there is little incentive to join the IWC. Not only are there no financial benefits, but observant nonmember diplomats watch members who wish to gain authorization for smallholder whaling being roughly handled, and they anticipate that the same would be their fate if their countries were to join. In short, the IWC's mandate will remain limited unless or until it finds a way to provide benefits to coastal states.

To be sure, if indiscriminate taking of cetaceans were to occur within the EEZs, some stocks of whales and porpoises would face disaster. That is not happening now, but there is good reason to try to bring the taking of marine mammals in the EEZ under IWC control, as many developing states modernize, acquire more effective exploitation tools, and pollute the coastal ocean, threatening the habitat of whales and other cetaceans.

Most states that allow whaling but are not members of the IWC are "middle developing states," or what were formerly called "less-developed states." Many are struggling to establish the basis of a modern economy, human rights, and effective government based upon both their inherited values and "modern" notions of economy and governance. They look in bemusement or confusion at the IWC notion of "aboriginal." Many nationalities can trace their occupation of their lands and waters back thousands of years, but under the IWC definition, they are not aboriginal or indigenous people. The definition of "aboriginal" in the IWC is a historical accident. It refers to minority populations in remote regions of states controlled by Caucasian majorities. It allows "aboriginal" people a "privilege"(?) of taking wildlife because of an early-twentieth-century romantic notion of "the noble savage," who by definition has not been corrupted by a market economy. Little wonder that scholars whose work is human-related—anthropologists, sociologists, and cultural geographers—get very upset at the definition of aboriginal and how it is interpreted.[29] It is very difficult to find a human group that has not engaged in exchange with another human group, even in very remote locations. For example, "Columbus himself caught a glimpse of Amerindian commerce when he intercepted a very large, fully loaded merchant canoe off the coast of present day Honduras."[30] There was no "Golden Age," and most people have engaged in some form of commerce. The current interpretation of *aboriginal* does not work well for even those who are privileged under it, and obviously it works less well so for those who are not be eligible under it. But they would be able to whale under an interpretation of a two-hundred-mile EEZ giving them a sovereign right to take whales within that zone.

The world needs a solution to the problem of coastal whaling by smallholders, whether they are considered aboriginal under current IWC practice or not. For practical reasons, it may not be sensible to take on the task of broadly "reforming" the IWC definition. Those who fit within the current definition might feel disadvantaged by the manner in which a revision would be implemented. Asking them to suggest a better approach to the management of aboriginal whaling would initiate a dialog in which their voice would be heard, and in which they could listen to others. But other subsistence whalers and artisanal whalers also should be accommodated. It is possible to do this without it becoming the wedge that causes the resumption of large-scale, uncontrolled commercial whaling, even if some of these people exchange or sell whale products. To require absolutely no exchange or commerce as a condition for the right to whale would be to establish a standard no human community could meet. But a reasonable and functional standard for those who would be eligible under an artisanal definition is available.

A rigorous definition of who might be eligible was offered several years ago by a team of social scientists. Two of the present authors were members. It would require that whaling be conducted by socially defined groups, that whaling involve practices that are socially reproducible over time, that permissible whaling practices be valued along a number of dimensions (not just economic), and that it be biologically sustainable.[31] Only a limited number of coastal villagers around the world could meet the test, but perhaps enough to convince governments of developing states to trust the IWC to help them manage their inshore living resources.

Acceptance of a broader definition of artisanal eligibility rules would also solve a problem of justice. I cannot accept the notion that developed states controlling IWC decisions have the prerogative to forbid a traditional right to an artisanal whaler because "we cannot restore every traditional culture" or because the developed can punish others by claiming that it really is not punishment since those displaced can be compensated by being allowed to assume new economic and social roles.[32] It is not the business of affluent, and perhaps guilty, developed states to destroy or restore the rights or livelihoods of others. As long as aboriginal or artisanal whalers do not diminish the ability of the world community or specific other states to enjoy whales in some form (what they do must be sustainable), major states should not impose metropolitan values upon them.

But we need to go beyond finding a definition of permissible artisanal whaling. We need to find a scheme that would attract to the IWC more states who fear the loss of rights in their two-hundred-mile EEZ if they join. We need to find a way to bring them into the IWC fold by creating a permissible whaling scheme for the EEZ. John Knauss, former U.S. commissioner to the IWC, proposed such a solution.[33] Michael Canny, the Irish commissioner, followed it up. We should take these proposals seriously and work out a detailed plan. If we succeed here, perhaps we should cast the net wider and find a way to link coastal states into regional whale or cetacean organizations, and link those regional organizations into arms of the IWC rather than rivals. If the task of worldwide supervision of implementation is formidable, perhaps it can be made easier by effective use of organizations closer to the ocean areas of concern.

CREATING A LIMITED EXCEPTION TO A SOUTHERN OCEAN SANCTUARY

Creating a limited exception to allow for some commercial whaling in the Southern Ocean might be the most formidable problem the commission must face. If the IWC majority has any pragmatism left, it will be forced someday to develop an RMS and find an accommodation for most coastal artisanal whalers. Unless they harbor a death wish for the commission, the anti-whaling forces will have to do something to prevent defections from the commission, development of alternate international—most likely regional—arrangements, and the continuance of unregulated coastal whaling among nonmembers. But the Southern Ocean is a different kettle of fish—and whales.

Dealing adequately with the question of whether creating a sanctuary is the appropriate policy for preventing a return to previous harmful practices in the Southern Ocean is, in some respects, different from dealing with nearshore problems. For the foreseeable future the only state that is discommoded by a Southern Ocean sanctuary is Japan. How other states, NGOs, and individuals perceive Japan's role in modern ocean history is critical to resolving this problem. Feelings, perceptions, and "historical memory" play as much a role as strategic calculation in the way anti-whaling representatives and NGOs have approached the situation. As a result, it is easy for them to forget that Japan conceded much of what they wanted—a sanctuary in which all whales save minke whales would enjoy long-term protection.[34] Japan, in effect, said that it would not target blues, humpbacks, rights, sperms, or any other whales in any limited hunt they would stage in the Southern Ocean, that there would not be a return to the "whaling Olympics." Moreover, any hunt would be closely regulated, especially if a tightly written RMS were available. In sum, what is needed is to limit the scope of the sanctuary, not eliminate it.

But that is not good enough for those who are committed to ending commercial whaling forever. It is hard to say whether many anti-whaling governments' and groups' devotion to an ideal leads them into quite anti-Japanese sentiments and measures, or their hostility toward Japan leads them to support a measure that backs Japan into a corner. Being the only major state left that wishes to engage in relatively large-scale commercial whaling in regions remote from its home waters makes Japan a natural target. Those whose position is based solely on their own moral considerations obviously cannot compromise, and therefore a substantial but not complete victory is not good enough. They do not have to compromise because, along with allies whose actions betray a mix of self-interest and idealism, they can form a majority to reduce Japan's presence on the ocean.[35] In sum, Japan has no natural allies on the question of access to whaling in the Southern Ocean.

It is easy for representatives of states and cultures that are not ocean-oriented to misunderstand Japanese behavior on the oceans. Japanese see themselves as an ocean-dependent people, and being an island country, Japan must always be concerned about food security.[36] Since the Second World War Japan has fished the seas hard to provide animal protein. For much of this time, although the United States Occupation Authorities encouraged the Japanese to return to fishing and whaling to help lighten the burden of feeding the Japanese population, the theme of U.S. Pacific fisheries policy has been to push the Japanese back, to reduce their take through measures like abstention lines, end of drift net fisheries, and creation of a two-hundred-mile EEZ. To force Japan out of whaling seems to Japanese another measure to reduce their independence and create a sense of dependence on outside food sources. Little wonder they have been stubborn.

But, at the same time, Japan has always bargained sincerely (witness the fact that they did not vote strategically on American requests for an aboriginal quota). When they promised to do something, they did it. In short, if they agree, they have a good implementation record. Lately, U.S.-Japan fisheries cooperation has been good; the two states worked cooperatively to eliminate the drift net fishery. That was possible not by trying to eliminate Japan entirely from the enjoyment of ocean products but by finding a way to assure them that they will get the ocean products their consumers prefer, while getting them to promise to obey rules that reduce the scope—or, as seen by some, the destructiveness—of their ocean activities. Indeed, while Japanese high seas salmon fishing was phased out, "Japanese firms cut their losses through direct foreign investment," by purchasing Alaskan packing and processing facilities.[37] This arrangement allowed Japanese fishing companies to supply salmon to Japanese consumers. The adjustment was painful, and it took some years to complete. Japanese former fishing, now fish trading, companies had to absorb the costs of terminating the livelihood of some of their fishermen.[38]

Since Japanese trading companies (which have eliminated their fishing divisions and scrapped their distant-water fleets) are vertically integrated, it is possible to rely upon any promise the Japanese government would make in relation to Southern Ocean whaling. The very close relationship between the government and the major Japanese fish-trading and processing companies guarantees that if the government agrees to a Southern Ocean regime, which limits the scope of Japanese activities, it will be with their consent and that of the Japan Fisheries Association.[39] The Japanese government will be able to implement an effective control scheme for Japanese fish markets. Even if the conditions are burdensome, if they can salvage some economic interest and be able to convince the Japanese public that they can still consume culturally approved ocean products, I think it is possible to have a limited and effectively regulated Southern Ocean minke-whaling fishery.

An exception to the Southern Ocean sanctuary to allow a limited minke catch could be created by reviving and accepting Japan's original amendment. It could, influenced by the precautionary principle, start small—say four hundred whales might be the appropriate starting number, or the number of whales taken by Japan to pay for the IWC scientific research program in the Southern Ocean largely funded by Japan—thereby eliminating the quarrel over "scientific whaling." Before it is implemented, all of the machinery required by an approved RMS included in the Schedule would have to be in place. Increases in quota numbers would depend upon how well the control plan was working, and the total finally allowed catch would be based on the effect on the ecosystem rather than the demands of the Japanese market. And Japan and any other user would have to pay a user's fee into an IWC-managed World Whale Fund for taking a resource common to all of us. It should be a modest fee. The purpose here is a symbolic gesture to indicate that the IWC is taking a lead in properly pricing extractions from the world's natural capital. If and when economists have a better framework for pricing, and other natural resource intergovernmental organizations also begin to charge for access to the resource, if appropriate, the price may rise.

Initially the Japanese are likely to be cautious concerning any plan for incrementally restoring limited commercial whaling in the Southern Ocean. They remember well that they acquiesced in the U.S.-sponsored trade-off under which Japan dropped its legal objection to the moratorium in return for a favored fishing position in the U.S. EEZ, only to have the U.S. renege on their fishing rights. They were left "holding the bag." But in this case they do not have to renounce their right to process whales caught for scientific purposes, but rather merely avoid stating a number for scientific whaling. If an initial quota is requested and granted, and the conditions on increasing the quota to some reference point are fulfilled, and a majority in the IWC reneges, they can keep their initial quota and reintroduce a scientific quota. At least they will have some leverage to keep the majority negotiating honestly.

Acceptance of a Japanese request for a quota would also solve another important problem. According to the ICRW, the commission does not have the right to "allocate specific quotas."[40] Of course, it would be legally optimal if member states amended the ICRW to give the commission the authority to award quotas. That is very unlikely. Somehow formal quotas were put in place for indigenous takes, and informal quotas were established for commercial takes. If Japan asked for a quota, and it was granted, perhaps the formal legal problem would not arise and therefore the legal issue would not become a problem. It would be a practical issue if Japan were the only claimant. Only if a competitor arose to demand a portion of the catch would a formal or, as in the past, informal decision need to be made to divide the quota. But establishing that a quota can be worked out through the IWC would be a useful precedent, even though it might not bind future claimants.

Failure to resolve the Southern Ocean sanctuary problem would hang heavy on the IWC, even if its other problems are resolved. While I think solving the coastal whaling problem should take priority over the Southern Ocean problem, obviously most or all of what I propose will be vehemently opposed by the most ardent anti-whaling cadres and governments. Perhaps this simple plan, even if not condemned as morally wrong, would be seen as inadequate.[41] But I must ask those opposed to reopening limited commercial whaling—is it in the long-term interest of whales and proper management of the ecosystem to try to humiliate and humble Japan? Japan will act like a "bootlegger" and will either remain defiant or look for a substitute.[42] Bootleggers often arise to provide a good or service when Baptists claim they have the moral right to enforce a total prohibition that would stop people from enjoying what is to be prohibited.[43] Perhaps the Baptists can create a new Jerusalem in the ocean, but I doubt it. If I am correct, but my proposal is not acceptable, I challenge those who wish to find a negotiated solution to develop a better plan, one that leaves something available to Japan so that its representatives and people can come out of the whaling contretemps with their sense of self intact, and the world come out of it without a loss of human liberty.[44]

The outcome will depend to a considerable extent on the willingness of reasonable people on both sides to accept the risk of resuming some whaling. A complete cessation of whaling means total victory for the forces of a post-modern sensibility and unwillingness to accept difference. There must be some give and disposition on both sides to take a risk, not an unreasonable risk, but a risk nevertheless. No one wants to resume unrestricted whaling, but would resumption of carefully controlled sustainable whaling eventually lead to a breakdown of the system, a restoration of unsustainable whaling and unmitigated disaster? One could not claim this is impossible. On the other hand, those who would restore whaling also must take risks—that whaling will never again be anything but a small enterprise, that they will always be scrutinized, monitored, and perhaps condemned, and that as their own people become more post-modern in values, whaling may fade away and the rights they fought so hard to protect will become irrelevant.

Richard Feynman, the Nobel laureate physicist, learned an important lesson from a Buddhist parable: "To every man is given the key to the gates of heaven: the same key opens the gates of hell."[45] The restoration of whaling may open the gates of hell, but the same key may lead us, with a restoration of sustainable whaling, to the heaven of a better regime for managing ocean resources.

NOTES

1. William J. Baumol and Wallace E. Oates, *The Theory of Environmental Policy* (Englewood Cliffs, N.J.: Prentice Hall, 1975), 188.

2. Appendix H, Report of the Scientific Committee, *Proceedings of the 49th Meeting of the IWC*, 1997 (SC/49).

3. Alison Motluk, "Norway's Wrong Numbers Fuel War on Whaling," *New Scientist* (1995): 1979.

4. *HNWNews*, 11 March 1999.

5. Allen V. Kneese, *Economics and the Environment* (Middlesex, England: Penguin, 1977), 20.

6. Our hopes in writing this volume are well expressed by the editors of another volume: "what is needed is a change of will on the part of the peoples of the world and their national leaders. In any event, studies such as the present book cannot accomplish this. They may, however, make some modest contribution towards changed perspective, and to identify some practical steps that could help effectuate it. Our purpose here is to operationalize, not to propagandize." John Lawrence Hargrove and Anthony D'Amato, "An Overview of the Problem," in *Who Protects the Ocean? Environment and the Development of the Law of the Sea*, ed. John Lawrence Hargrove (St. Paul, Minn.: West, 1975), 3. It is ironic that one of the authors of the quote, Anthony D'Amato, is one of the chief spokespersons of the whale preservation movement. For the record, Robert Friedheim was also a contributor to this volume.

7. Kenneth E. Boulding, "What Do We Want to Sustain? Environmentalism and Human Evaluations," in *Ecological Economics*, ed. Robert Costanza (New York: Columbia University Press, 1991), 22.

8. U.S. Department of Commerce, NOAA, NMFS, *Report to Congress: Impacts of California Sea Lions and Pacific Harbor Seals on Salmonids and West Coast Ecosystem*, 10 February 1999.

9. Perhaps rather than a serious reconsideration, it will merely lead to a new struggle. As might be expected, some in the environmental community responded immediately and did not address the issue, but accused the government of "scapegoating" seals and sea lions. See http://www. earthisland.org/news/news_pinniped1.html.

10. IWC Resolution 1994–95 in Chairman's Report, *Proceedings of the 46th Meeting of the IWC*, Puerto Vallarta, Mexico, 23–27 May 1994. Also see Final Press Release, *Proceedings of the 50th Meeting of the IWC*, Muscat, Oman, 20 May 1998.

11. Final Press Release, *Proceedings of the 50th Meeting of the IWC*, 20 May 1998, 1.

12. Chairman's Report, *Proceedings of the 49th Meeting of the IWC*, 20–24 October 1997, 24.

13. Donald Ludwig, Ray Hilborn, Carol Walters, "Uncertainty, Resource Exploitation, and Conservation: Lessons from History," *Science* 260 (2 April 1993): 17ff.

14. Louis W. Botsford, Juan Carlos Castilla, Charles H. Peterson, "The Management of Fisheries and Marine Ecosystems," *Science* 277 (25 July 1997): 508; Biliana Cicin-Sain and Robert W. Knecht, *Integrated Coastal and Ocean Management: Concepts and Practices* (Washington, D.C.: Island Press, 1998), 172–73.

15. Article V(2)(c), *International Convention for the Regulation of Whaling*, 2 December 1946, 62 Stat. 1716, 161 U.N.T.S. 74 [hereinafter ICRW].

16. United Nations, *Agreement for the Implementation of the United Nations Convention on the Law of the Sea of 10 December 1982 Relating to the Conservation and Management of Straddling Fish Stocks and Highly Migratory Fish Stocks* (A/CONF.164/37), 8 September 1995 [hereafter the Straddling Stocks Convention]. *Convention on the Conservation and Management of Pollack Resources in the Central Bering Sea, 16 June 1994,* reprinted in United Nations, *Law of the Sea: Bulletin No. 27* (New York: United Nations, 1995) [hereafter the Central Bering Sea Convention].

17. World Commission on Environment and Development, *Our Common Future* (New York: Oxford University Press, 1987); Peter M. Haas, Marc A. Levy, and Edward A. Parson, "Appraising the Earth Summit: How Should We Judge UNCED's Success?" *Environment* 34, no. 8 (October 1992): 6–15.

18. S. M. Garcia, "The Precautionary Principle: Its Implications in Capture Fisheries Management," *Ocean and Coastal Management* 22 (1994): 99–125.

19. Article 6 (3)(b), Straddling Stocks Convention.

20. One reason some supporters of preservation take their position is that inspection is not a new idea, but one that failed previously. After all, the first "Schedule" developed by the IWC called for inspectors on factory ships and land stations. They did not prevent the "whaling Olympics." Article 1(a)(b), Schedule, ICRW.

21. Christopher Stone, "Too Many Fishing Boats, Too Few Fish: Can Trade Laws Trim Subsidies and Restore the Balance in Global Fisheries?" *Ecology Law Quarterly* 24, no. 3 (1997): 505–44; Christopher Stone, "Can the Oceans Be Harbored? A Four-Step Plan for the Twenty-first Century" (paper prepared for a conference, "Toward the International Protection of the Oceans: From Rules to Compliance," Lisbon, 17–19 September 1998). Also see David Malin Roodman, *The Natural Wealth of Nations* (New York: Norton, 1998).

22. See, for example, Kate O'Connell, "The 1997 IWC Meeting and Beyond: The Turmoil Continues," *Whales Alive* 7, no. 1 (January 1998).

23. With perhaps the exception of eastern Pacific gray whales that might in theory be caught by the Russian Chukotkans and the American Makah people in their authorized hunts. (The Makah quota of five has been deducted from the Russian quota.) But even this externality would be limited since together they have been authorized to take annually 140 whales from a stock of about 23,000. The actual Chukotkan catch in 1997 was seventy-nine whales. The Makah have been able to capture only one of the five gray whales they have been allocated.

24. Roger A. Sedjo, "Harvesting the Benefits of Carbon 'Sinks'" *Resources* 133 (fall 1998): 12.

25. Chairman's Report, *Proceedings of the 47th Meeting of the IWC*, 29 May–2 June 1995, 19.

26. "the Westphalian principle of nonintervention in internal affairs has been eroded by interventions in the name of dispute resolution, economic stability, and human rights." James G. March and Johan Olsen, "The Institutional Dynamics of International Political Orders," *International Organization* 52, no. 4 (autumn 1998): 946.

27. But I recognize that if the trend of events continues, the state's effectiveness as a civil association might erode significantly and create a crisis of legitimacy. Philip G. Cerny, "Globalization and the Changing Logic of Collective Action," *International Organization* 49, no. 4 (autumn 1995): 618–19. Also see John Gerard Ruggie, "Territoriality and Beyond: Problematizing Modernity in International Relations," *International Organization* 47, no. 1 (winter 1993): 139–74.

28. "Annual Meeting of the IWC: Lots of Bad Blood; Nothing Achieved" *HN-News*, 11 July 1998.

29. Martin W. Lewis, *Green Delusions: An Environmentalist Critique of Radical Environmentalism* (Durham: Duke University Press, 1992), 242–51.

30. B. L. Turner II and Karl W. Butzer, "The Columbian Encounter and Land-Use Change," *Environment* 34, no. 8 (October 1992): 40. Also see Jared Diamond, "The Golden Age That Never Was," in *The Third Chimpanzee* (New York: Harper, 1993), 317–38.

31. O. R. Young et al., "Subsistence, Sustainability, and Sea Mammals: Reconstructing the International Whaling Regime," *Ocean and Coastal Management* 23 (1994): 117–27.

32. Harry N. Scheiber, "Historical Memory, Cultural Claims, and Environmental Ethics in the Jurisprudence of Whaling Regulation," *Ocean and Coastal Management* 38, no. 1 (1998): 35.

33. John A. Knauss, "The International Whaling Commission: Its Past and Possible Future," *Ocean Development and International Law* 28, no. 1 (1997): 79–99.

34. Government of Japan, "Proposed Consideration for Sanctuary in the Southern Ocean," *Report of the 46th Meeting of the IWC*, Puerto Vallarta, Mexico, May 1994 (IWC/46/35 Rev.1).

35. For those who claim that "Japan bears a special responsibility to hold its hand on the whaling question" (Scheiber, "Historical Memory," 32), I would respond that the United States came to the table with less than clean hands on many ocean issues. The political forces in Alaska and the Pacific Northwest who were behind the many efforts to limit Japanese fishing did so often for self-interested as well disinterested concern for ecosystem survival. See William T. Burke, Mark Freeberg, Edward L.

Miles, "United Nations Resolutions on Driftnet Fishing: An Unsustainable Precedent for High Seas and Coastal Fisheries Management," *Ocean Development and International Law* 25 (1994): 127–86.

36. See Reiko Niimi, "The Problem of Food Security," in *Japan's Economic Security,* ed. Nobutoshi Akao (New York: St. Martin's, 1983). Also see "The Kyoto Declaration and Plan of Action on the Sustainable Contribution of Fisheries to Food Security," Document 4–8, International Conference on the Sustainable Contribution of Fisheries to Food Security, FAO, 4–9 December 1995.

37. Virginia Walsh, "Eliminating Driftnets from the North Pacific Ocean: U.S.-Japanese Cooperation in the International North Pacific Fisheries Commission, 1953–1993," *Ocean Development and International Law* 29, no. 4 (1998): 317.

38. Olav Schram Stokke, "Transnational Fishing: Japan's Changing Strategy," *Marine Policy* (July 1991): 231–43.

39. Tsuneo Akaha, *Japan in Global Ocean Politics* (Honolulu: University of Hawaii Press, 1985), 21.

40. Article V(2)(c), ICRW.

41. While Stephen Kellert claims that there exist "a common understanding of nature characteristic of all people independent of tradition, society, and geography," there are important differences in Western and Japanese attitudes toward nature. We should respect these differences. Stephen R. Kellert, "Concepts of Nature East and West," in *Reinventing Nature? Responses to Postmodern Deconstruction*, ed. Michael E. Soule and Gary Lease (Washington, D.C.: Island Press, 1995), 103.

42. Elizabeth R. DeSombre, "Baptists and Bootleggers for the Environment: The Origins of United States Unilateral Sanctions," *Journal of Environment and Development* 4, no. 1 (winter 1995): 53–75.

43. This point has been demonstrated in behavioral research relating to environmental issues, for example, "The targeted actors may also resist a policy or program after it has been enacted, thus rendering it ineffective or even counterproductive." Gerald T. Gardner and Paul C. Stern, *Environmental Problems and Human Behavior* (Boston: Allyn & Bacon, 1996), 169.

44. "Humankind depends for its liberty on variety and difference." Benjamin R. Barber, *Jihad vs. McWorld: How Globalism and Tribalism Are Reshaping the World* (New York: Ballantine, 1995), 296.

45. Richard Feynman, *"What Do You Care What Other People Think?" Further Adventures of a Curious Character* (New York: Bantam, 1989), 241.

BIBLIOGRAPHY

Adede, Andronico O. *International Environmental Law Digest*. Amsterdam: Elsevier, 1993.

Aftenposten (Oslo, Norway), 5 April 1925.

Akaha, Tsuneo. *Japan in Global Ocean Politics*. Honolulu: University of Hawaii Press, 1985.

Akimichi, Tomoya, Harumi Befu, Stephen R. Braund, Helen Hardacre, Arne Kalland, Brian D. Moeran, Pamela J. Asquith, Theodore C. Bester, Milton M. R. Freeman, Masami Iwasaki, Lenore Manderson, Junichi Takahashi. *Small-Type Coastal Whaling in Japan*. Edmonton, Alberta: Boreal Institute for Northern Studies, 1988.

Alaska Eskimo Whaling Commission (AEWC). *1995 Tri-Annual Alaska Eskimo Whaling Captains Convention*. Barrow, Ala.: AEWC, 13–15 February 1995.

Allen, K. Radway. *Conservation and Management of Whales*. Seattle: University of Washington Press, 1980.

Anaya, S. James. *Indigenous Peoples in International Law*. New York: Oxford University Press, 1996.

Andresen, Steinar. "Science and Politics in the International Management of Whales." *Marine Policy* 13 (1989): 99–117.

———. "International Verification in Practice: A Brief Account of Experiences from Relevant International Cooperative Measures." In *Achieving Environmental Goals*, ed. Erik Lykke. London: Belhaven Press, 1992.

———. "NAMMCO, IWC, and the Nordic Countries." In *Whaling in the North Atlantic: Economic and Political Perspectives*, ed. Gudrun Petursdottir. Reykjavík:

Fisheries Research Institute, University of Iceland, University of Iceland Press, 1997.

———. "The Making and Implementation of Whaling Policies: Does Participation Make a Difference?" In *Implementation and Effectiveness of International Environmental Regimes: Confronting Theory with Practice,* ed. David G. Victor, Kal Raustiala, and Eugene B. Skolnikoff. Cambridge: MIT Press, 1998.

———. "The Whaling Regime: The International Convention for the Regulation of Whaling (ICRW) and the International Whaling Commission." In *Science and International Environmental Regimes: Integrity and Involvement,* ed. Arild Underdal, Steinar Andresen, Tora Skodvin, and Jørgen Wettetstad. Manchester, England: Manchester University Press, 1998.

———. "The International Whaling Commission: The Failure to Manage Whales Effectively." In *Explaining Regime Effectiveness: Confronting Theory with Evidence,* ed. Edward Miles et al. Cambridge: MIT Press, 1999.

Andresen, Steinar, and Jørgen Wettestad. "International Resource Cooperation and the Greenhouse Problem." *Global Environmental Change* (December 1992): 277–91.

Arhem, Kaj. "The Cosmic Food Web: Human-Nature Relatedness in the Northwest Amazon." In *Nature and Society: Anthropological Perspectives,* ed. Phillipe Descola, and Gisli Palsson, 185–204. London: Routledge, 1996.

Aron, William. "The Commons Revisited: Thoughts on Marine Mammal Management." *Coastal Management* 16 (1988): 99–110.

Aron, William, William Burke, and Milton Freeman. "Flouting the Convention." *Atlantic Monthly,* May 1999, 22–29.

Arrow, Kenneth, B. Bolin, R. Costanza, P. Dasgupta, C. Folke, C. S. Hollings, B.-O. Jansson, S. Levin, K.-G. Mäler, C. Perrings, and D. Pimenthal. "Economic Growth, Carrying Capacity, and the Environment." *Science* 268 (1995): 520–21.

Ashkenazi, Michael. "From Tachi Soba to Naori: Cultural Implications of the Japanese Meal." *Social Science Information* 30, no. 2 (1991): 287–304.

Ashkenazi, Michael, and Jeanne Jacob. *Summary of Whalemeat as a Component of the Changing Japanese Diet in Hokkaido* (IWC/44/SEST2). Cambridge, England: IWC, 1992.

Ausbel, Jesse, and David Victor. "Verification of International Environmental Agreements." *Annual Review of Energy and the Environment* 17 (1992): 1–43.

Axelrod, Robert. *Conflict of Interest: A Theory of Divergent Goals with Applications to Politics.* Chicago: Markham, 1969.

Bailey, Ronald "Prologue: Environmentalism for the Twenty-first Century." In *The True State of the Planet,* ed. Ronald Bailey. New York: Free Press, 1995.

Baranzini, Andrea, and Gonzague Pillet. "The Physical and Biological Environment-the Sociobiology of Sustainable Development." In *Economy, Environment, and Technology,* ed. Beat Bürgenmeier. Armonk, N.Y.: M. E. Sharpe, 1994.

Barber, Benjamin R. *Jihad vs. McWorld: How Globalism and Tribalism Are Reshaping the World.* New York: Ballantine, 1995.

Barkin, J. Samuel, and George E. Shambaugh, eds. *Anarchy and the Environment.* Albany: State University of New York Press, 1999.

Barrett, Hazel R. *The Marketing of Foodstuffs in the Gambia, 1400–1980: A Geographical Analysis.* Aldershot, England: Avebury, 1988.

Barsh, Russel L. "The Substitution of Cattle for Bison on the Great Plains." In *The Struggle for the Land,* ed. Paul A. Olson. Lincoln: University of Nebraska Press, 1990.

———. "Indigenous Peoples' Role in Achieving Sustainability." In *Green Globe Yearbook 1992,.* Oxford: Fridtjof Nansen Institute, Oxford University Press, 1992.

———. "Canada's Aboriginal Peoples: Social Integration or Disintegration?" *Canadian Journal of Native Studies* 14, no. 1 (1994): 1–46.

———. "Fire on the Land." *Alternatives* 23, no. 4 (1997): 36–40.

———. "Chronic Health Effects of Dispossession and Dietary Change: Lessons from North American Hunter-Gatherers." *Medical Anthropology* 18, no. 1 (1998): 1–27.

———. "Taking Indigenous Science Seriously." In *Biodiversity in Canada: An Introduction to Environmental Studies,* ed. Stephen Bocking. Toronto: Broadview Press, 1999.

Barston, R. P., and Patricia Birnie, eds. *The Maritime Dimension.* London: Allen and Unwin, 1980.

Bartelson, Jens. "The Trial of Judgment: A Note on Kant and the Paradoxes of Internationalism." *International Studies Quarterly* 39, no. 2 (1995): 255–79.

Bartos, Otomar J. *Simple Models of Group Behavior.* New York: Columbia University Press, 1967.

Baumol, William J., and Wallace E. Oates. *The Theory of Environmental Policy.* Englewood Cliffs, N.J.: Prentice Hall, 1975.

Bennett, J. W. "Food and Social Status in a Rural Society." *American Sociological Review* 8 (1943): 561–69.

Berger, Joel, and Carol Cunningham. "Active Intervention and Conservation: Africa's Pachyderm Problem." *Science* 263 (1994): 1241.

Bergesen, Helge Ole, and Georg Parman, eds. *Green Globe Yearbook of International Cooperation on Environment and Development.* Oxford: Fridtjof Nansen Institute, Oxford University Press, 1997.

Berkes, Filkert, D. Feeny, B. J. McCay, and J. M. Acheson. "The Benefits of the Commons." *Nature* 340 (13 July 1989): 91–93.

Bernauer, T. "The Effect of International Environmental Institutions, How We Might Learn More." *International Organization* 49 (spring 1995): 351–77.

Berton, Peter, Hiroshi Kimura, and I. William Zartman, eds. *International Negotiation: Actors, Structure, Process, Values.* New York: St. Martin's, 1999.

Bestor, Theodore C. *Socio-Economic Implications of a Zero Catch Limit on Distribution Channels and Related Activities in Hokkaido and Miyagi Prefectures, Japan* (IWC/41/SE1). Cambridge, England: IWC, 1989.

Birnie, Patricia. "The Role of Developing Countries in Nudging the International Whaling Commission from Regulating Whaling to Encouraging Nonconsumptive Uses of Whales." *Ecology Law Quarterly* 12 (1984): 962.

———. *International Regulation of Whaling: From Conservation of Whaling to Conservation of Whales and Regulation of Whale-Watching.* 2 vols. Dobbs Ferry, N.Y.: Oceana, 1985.

———. "UNCED and Marine Mammals." *Marine Policy* 17 (1993): 501, 504.

———. "Opinion on the Legality of the Designation of the Southern Ocean Whale Sanctuary by the International Whaling Commission." *Proceedings of the 47th Meeting of the International Whaling Commission,* 1995.

———. "Are Twentieth-Century Marine Conservation Conventions Adaptable to Twenty-first-Century Goals and Principles?" *International Journal of Marine and Coastal Law* 12 (1997): 307, 438.

Bock, Paula. "The Accidental Whale." *Pacific Magazine (Seattle Times),* 26 November 1996.

Bodansky, Dan. "The United Nations Framework Convention on Climate Change: A Commentary." *Yale Journal of International Law* 18, no. 2 (1993): 451–558.

Bodin, Jean. *On Sovereignty.* Ed. Julian Franklin. Cambridge: Cambridge University Press, 1992.

Bohannan, Paul. "The Impact of Money on an African Subsistence Economy." *Journal of Economic History* 19, no. 4 (1959): 491–503.

Bongaarts, John. "Population Policy Options in the Developing World." *Science* 263 (11 February 1994): 771–76.

———. "Can the Growing Human Population Feed Itself?" *Scientific American,* March 1994, 36–42.

Borgese, Elizabeth Mann, Norton Ginsburg, and Joseph R. Morgan. *Ocean Yearbook 11.* Chicago: University of Chicago Press, 1994.

Borton, Hugh. *Japan's Modern Century.* New York: Roland, 1955.

Bibliography

Botkin, Daniel. *Discordant Harmonies: A New Ecology for the Twenty-first Century*. New York: Oxford University Press, 1994.

Boulding, Kenneth E. *Ecodynamics*. Beverly Hills: Sage, 1978.

———. "What Do We Want to Sustain? Environmentalism and Human Evaluations." In *Ecological Economics*, ed. Robert Costanza, 22–31. New York: Columbia University Press, 1991.

Bowett, D. W. *The Law of International Institutions*. London: Stevens, 1982.

Boyden, Stephen V. *Biohistory: The Interplay between Human Society and the Biosphere*. Paris: UNESCO and Parthenon, 1992.

Boyle, A. E. *Environmental Regulation and Economic Growth*. Oxford: Clarendon, 1994.

Braund, Stephen R., Milton M. R. Freeman, and M. Iwasaki. *Contemporary Sociocultural Characteristics of Japanese Small-Type Whaling* (IWC/41/STW1). Cambridge, England: IWC, 1989.

Braund, Stephen R., J. Takahashi, J. A. Kruse, and Milton Freeman. *Quantification of Local Need for Minke Whale Meat for the Ayukawa-Based Minke Whale Fishery* (TC/42/SEST8). Cambridge, England: IWC, 1990.

Brookfield, H. C. *Colonialism, Development, and Independence: The Case of the Melanesian Islands*. Cambridge: Cambridge University Press, 1972.

Brown, Amanda. "Campaigners to Urge Irish to Drop Whaling Proposals." Web page.

———. "Tourism Threat over Whales Vote-Buying Claim." *Press Association Newsfile*, 30 June 1992.

Brown, Lester R. *Who Will Feed China?* New York: Norton, 1995.

Brown, Paul. "Playing Football With the Whales." *Guardian*, 1 May 1993, 26.

Bryden, M. M., and P. Corkeron. "Intelligence." In *Whales, Dolphins, and Porpoises*, ed. R. Harrison and M. M. Bryden, 160–65. New York: Facts on File, 1988.

Brydon, Anne. "Whale-Siting: Spatiality in Icelandic Nationalism." In *Images of Contemporary Iceland: Everyday Lives and Global Contexts*, ed. G. Palsson, and P. Durrenberger, 25–45. Iowa City: University of Iowa Press, 1996.

Burger, Anna. *The Agriculture of the World*. Aldershot, England: Avebury, 1994.

Burke, William T. "Aspects of Internal Decision-Making Processes in Intergovernmental Fishery Commissions." *Washington Law Review* 43 (1967): 115, 140–42.

———. "UNCED and the Oceans." *Marine Policy* 17 (1993): 519–33.

———. "Memorandum of Opinion on the Legality of the Designation of the Southern Ocean Sanctuary by the IWC." *Ocean Development and International Law* 27 (1996): 315–26.

———. "Legal Aspects of the IWC Decision on the Southern Ocean Sanctuary." *Ocean Development and International Law* 28 (1997): 311–54.

Burke, William T., Mark Freeberg, and Edward L. Miles. "United Nations Resolutions on Driftnet Fishing: An Unsustainable Precedent for High Seas and Coastal Fisheries Management." *Ocean Development and International Law* 25 (1994): 127–86.

Butterworth, D. S. "Science and Sentimentality." *Nature* 357 (18 June 1992): 532–34.

Caldwell, Lynton Keith. *International Environmental Policy,* 2d ed. Durham, N.C.: Duke University Press, 1990.

Cameron, James. *Opinion of Law* (IWC/47/OS). Cambridge, England: IWC, 1995.

Cannon, Terry. "Indigenous Peoples and Food Entitlement Losses Under the Impact of Externally-Induced Change." *GeoJournal* 35, no. 2 (1995): 137–50.

Cantor, Leonard. *The Changing English Countryside, 1400–1700.* London: Routledge and Kegan Paul, 1987.

Capie, Forrest, and Richard Perren. "The British Market for Meat, 1850–1914." *Agricultural History* 54, no. 4 (1980): 502–15.

Caron, David D. "International Sanctions, Ocean Management, and the Law of the Sea: A Study of Denial of Access to Fisheries." *Ecology Law Quarterly* 16, no. 1 (1989): 311–54.

———. "The IWC and the North Atlantic Marine Mammal Commission: The Institutional Risks of Coercion in Consensual Structures." *American Journal of International Law* 98, no. 1 (1995): 154–73.

Carter, James Earl. "Message from the President of the United States." *Public Papers of the President of the United States.* Washington, D.C.: GPO, 1979.

Castillero-Calvo, Alfredo. "Niveles de vida y cambios de dieta a fines del período colonial en Amèrica." *Anuario de Estudios Americanos* 44 (1987): 427–76.

Caulfield, Richard A. *Qeqertarsuarmi Arfanniarneq: Greenlandic Inuit Whaling in Qeqertarsuaq Kommune, West Greenland* (TC/43/AS4). Cambridge, England: IWC, 1991.

———. *Whaling and Sustainability in Greenland* (IWC/46/AS1). Cambridge, England: IWC, 1994.

———. *Greenlanders, Whales, and Whaling: Sustainability and Self-Determination in the Arctic.* Hanover, N.H.: University Press of New England, 1997.

Cerny, Philip G. "Globalization and the Changing Logic of Collective Action." *International Organization* 49, no. 4 (1995): 595–625.

Charnovitz, Steve. "Encouraging Environmental Cooperation through the Pelly Amendment." *Journal of Environment and Development* 3, no. 1 (winter 1994): 11.

Chayes, Abram, and Antonia Handler Chayes. *The New Sovereignty: Compliance with International Regulatory Agreements.* Cambridge: Harvard University Press, 1995.

Christy, Francis, Jr., and Anthony Scott. *The Common Wealth in Ocean Fisheries.* Baltimore: Johns Hopkins University Press, 1965.

Clark, Colin W. "A Delayed Recruitment Model of Population Dynamics, with an Application to Baleen Whale Populations." *Journal of Mathematical Biology* 3 (1976): 381–91.

———. *Mathematical Bionomics: The Optimal Management of Renewable Resources.* New York: Wiley Interscience, 1976.

———. "Economic Aspects of Renewable Resource Exploitation as Applied to Marine Mammals." *FAO Fish Ser.* 3, no. 5 (1981).

Clingan, Thomas A., Jr. "Issues Involved in a Broadened View of International Management of Whales." In *Report: International Legal Workshop, Sixth Annual Whaling Symposium,* ed. Robert L. Friedheim. Tokyo: Institute of Cetacean Research, 1996.

Cohen, Joel E. "Population Growth and Earth's Human Carrying Capacity." *Science* 269 (21 July 1995): 341–46.

Collins, E. J. T. "Why Wheat? Choice of Food Grains in the Nineteenth and Twentieth Centuries." *Journal of European Economic History* 22, no. 1 (1993): 7–38.

Collins, Jane L. "Smallholder Settlement of Tropical South America: The Social Causes of Ecological Disaster." *Human Organization* 45, no. 1 (1986): 1–10.

Conrad, Jon, and Trond Bjørndal. "On the Resumption of Commercial Whaling: The Case of the Minke Whale in the Northeast Atlantic." *Arctic* 46, no. 2 (1993): 164–71.

Convention on International Trade in Endangered Species (CITES). "Proposal from Norway to Transfer Minke Whale (*Balaenoptera acutorostrata*) from Appendix I to Appendix II: Draft for Consultation with Range States." *CITES, 19th Meeting.*

Costanza, Robert, F. Andrade, P. Antunes, M. van den Belt, D. Boersma, D. F. Boesch, F. Catarino, J. Hanna, K. Linberg, B. Low, M. Molitor, J. G. Pereira, S. Rayner, R. Santos, J. Wilson, and M. Young. "Principles for Sustainable Governance of the Oceans." *Science* 281 (10 July 1998): 198–99.

Cowhey, Peter. "Elect Locally-Order Globally: Domestic Politics and Multilateral Cooperation." In *Multilateralism Matters,* ed. John Gerard Ruggie. New York: Columbia University Press, 1993.

Craig, Gordon A., and Alexander L. George. *Force and Statescraft,* 3d ed. New York: Oxford University Press, 1995.

Crawford, James. "Negotiating Global Security Threats in a World of Nation States: Issues and Problems of Sovereignty." *American Behavioral Scientist* 38, no. 6 (May 1995): 867–88.

Cronon, William. *Uncommon Ground: Rethinking the Human Place in Nature*. New York: Norton, 1995.

Crosby, Alfred W. *Ecological Imperialism: The Biological Expansion of Europe, 900–1900*. Cambridge: Cambridge University Press, 1986.

Crutchfield, James, and Giulio Pontecorvo. *The Pacific Salmon Fisheries: A Study in Irrational Conservation*. Baltimore: Johns Hopkins University Press, 1990.

Cunliffe, Barry. *Greeks, Romans, and Barbarians: Spheres of Interaction*. London: Guild, 1988.

Dales, John. *Pollution, Property, and Prices*. Toronto: University of Toronto Press, 1968.

D'Amato, Anthony. "Agora: What Obligation Does Our Generation Owe to the Next? An Approach to Environmental Responsibility." *American Journal of International Law* 84, no. 1 (January 1990): 190–198.

D'Amato, Anthony, and Sudhir K. Chopra. "Whales: Their Emerging Right to Life." *American Journal of International Law* 85, no. 1 (January 1991): 21–62.

Dawkins, Marian Stamp. *Through Our Eyes Only? The Search for Animal Consciousness*. Oxford: Oxford University Press, 1998.

Day, David. *The Whale War*. San Francisco: Sierra Club Books, 1987.

de Graaf, J. *The Economics of Coffee*. Waginingen, Netherlands: Pudoc, 1983.

de Klemm, Cyrille, and Claire Shine. *Biological Diversity, Conservation, and the Law*. Gland, Switzerland: IUCN, 1993.

Derr, Mark. "To Whale or Not to Whale." *Atlantic* (1997): 22–26.

Descola, Philippe. *In the Society of Nature: A Native Ecology in Amazonia*. Cambridge: Cambridge University Press, 1994.

DeSombre, Elizabeth R. "Baptists and Bootleggers for the Environment: The Origins of United States Unilateral Sanctions." *Journal of Environment and Development* 4, no. 1 (winter 1995): 53–75.

———. *Domestic Sources of International Environmental Policy: Industry, Environmentalists, and U.S. Power*. Cambridge: MIT Press, 2000.

Deudney, Daniel H., and Richard A. Matthew, eds. *Contested Grounds: Security and Conflict in the New Environmental Politics*. Albany: State University of New York Press, 1999.

Devall, Bill, and George Sessions. *Deep Ecology: Living As If Nature Mattered*. Salt Lake City, Utah: Peregrine Smith, 1985.

Diamond, Jared. *The Third Chimpanzee*. New York: Harper, 1993.

———. *Guns, Germs, and Steel*. New York: Norton, 1997.

Donovan, G. P., ed. *Aboriginal/Subsistence Whaling (with Special Reference to the Alaska and Greenland Fisheries)*. Reports of the International Whaling Commission, special issue 4. Cambridge, England: IWC, 1982.

Bibliography

Douglas, Mary. *In the Active Voice*. London: Routledge and Kegan Paul, 1982.

Douglas, Mary, ed. *Food in the Social Order: Studies of Food and Festivities in Three American Communities*. New York: Russell Sage Foundation, 1984.

Dryzek, John S. *The Politics of the Earth: Environmental Discourses*. New York: Oxford University Press, 1997.

Dunning, W. A. *A History of Political Theories: From Luther to Montesquieu*. New York: Macmillan, 1905.

Edelman, Murray. *The Symbolic Uses of Politics*. Urbana: University of Illinois Press, 1967.

Edwards, Stephen R. "Conserving Biodiversity: Resources for Our Future." In *The True State of the Planet*, ed. Ronald Bailey. New York: Free Press, 1995.

Eichstaedt, Richard Kirk. "'Save the Whales' v. 'Save the Makah': The Makah and the Struggle for Native Whaling." *Animal Law* 4, no. 1 (1998): 145–71.

Eide, Asbjorn. *Report on the Right to Adequate Food as a Human Right, Submitted by Mr. Asbjorn Eide, Special Rapporteur*. 1987. UN Doc. E/CN.4/Sub.2/23.

Evans, Peter B., Harold K. Jacobson, and Robert D. Putnam. *Double-Edged Diplomacy: International Bargaining and Domestic Politics*. Berkeley and Los Angeles: University of California Press, 1993.

Farb, Peter, and George Armelagos. *Consuming Passions: The Anthropology of Eating*. Boston: Houghton Mifflin, 1980.

Fenwick, Charles G. *International Law*, 4th ed. New York: Appleton-Century-Crofts, 1965.

Ferguson, R. Brian. "Game Wars? Ecology and Conflict in Amazonia." *Journal of Anthropological Research* 45, no. 2 (1989): 179–206.

Feynman, Richard P. *What Do You Care What Other People Think? Further Adventures of a Curious Character*. New York: Bantam, 1989.

Fine, Ben, Michael Heasman, and Judith Wright. *Consumption in the Age of Affluence: The World of Food*. London: Routledge, Chapman and Hall, 1996.

Finel, Bernard I., and Kristin M. Lord. "The Surprising Logic of Transparency." *International Studies Quarterly* 43, no. 2 (1999): 315–40.

Fineman, Mark. "Dominica's Support of Whaling Is No Fluke." *Los Angeles Times*, 9 December 1997, A1.

Finkelstein, Lawrence. "What Is Global Governance?" *Global Governance* 1 (1995): 367–71.

Finnamore, Martha, and Kathryn Sikkink. "International Norm Dynamics and Political Change." *International Organization* 52, no. 4 (autumn 1998): 881–90.

Floit, Catherine. "Reconsidering Freedom of the High Seas: Protection of Living Marine Resources on the High Seas." In *Freedom for the Seas in the Twenty-first Century*, ed. Jon M. Van Dyke et al. Washington, D.C.: Island Press, 1993.

Frank, Richard. "The Paradox of the American View on Utilization of Marine Mammals." *ISANA* 6 (May 1992): 11–13.

Freeman, Milton M. R. "Tradition and Change: Problems and Persistence in the Inuit Diet." In *Coping with Uncertainty in Food Supply,* ed. I. de Garine and G. A. Harrison, 150–169. Oxford: Oxford University Press, 1988.

———. "A Commentary on Political Issues with Regard to Contemporary Whaling." *North Atlantic Studies* 2, no. 1–2 (1990): 106–16.

———. "The Historical Legacy of Industrial Whaling and Current Problems in Japan's Coastal Fisheries." In *Ocean Resources: Industries and Rivalries Since 1800,* ed. Harry N. Scheiber. Berkeley: Center for the Study of Law and Society, 1990.

———. "Energy, Food Security, and A.D. 2040: The Case for Sustainable Utilization of Whale Stocks." *Resource Management and Optimization* 8, no. 3–4 (1991): 235–44.

———. "The International Whaling Commission, Small-Type Whaling, and Coming to Terms with Subsistence." *Human Organization* 52, no. 3 (1993): 243–51.

———. "Identity, Health, and Social Order: Inuit Dietary Traditions in a Changing World." In *Human Ecology and Health.* ed. M. L. Foller and L. O. Hansson. Goteborg: Goteborg University Press, 1996.

———. "Issues Affecting Subsistence Security in Arctic Societies." *Arctic Anthropology* 34, no. 1 (1997): 7–17.

Freeman, Milton M. R., et al. *Tradition and Change: Problems and Persistence in the Inuit Diet,* ed. I. de Garine and G. A. Harrison. Oxford: Oxford University Press, 1988.

———. "Science and Trans-science in the Whaling Debate." In *Elephants and Whales: Resources for Whom?* ed. Milton M. R. Freeman and Urs P. Kreuter, 143–58. Basel, Switzerland: Gordon and Breach, 1994.

———. *Inuit, Whaling, and Sustainability.* Walnut Creek, Calif.: AltaMira Press, 1998.

———. "Japanese Community-Based Whaling, International Protest, and the New Environmentalism." In *Japan at the Crossroads: Hot Issues for the Twenty-first Century,* ed. David Myers and Kotaku Ishido, 13–31. Tokyo: Seibundo, 1998.

Freeman, Milton M. R., L. S. Bogoslovskaya, Richard A. Caulfield, I. Egede, I. Krupnik, and M. G. Stevenson. *Inuit, Whales, and Sustainability.* Walnut Creek, Calif.: Altamira, 1998.

Freeman, Milton M. R., and Stephen R. Kellert. "International Attitudes to Whales, Whaling, and the Use of Whale Products: A Six-Country Survey." In *Elephants and Whales: Resources for Whom?* ed. Milton M. R. Freeman, and Urs P. Kreuter, 293–315. Basel, Switzerland: Gordon and Breach, 1994.

Freeman, Milton M. R., and U. P. Kreuter, eds. *Elephants and Whales: Resources for Whom?* Basel, Switzerland: Gordon and Breach, 1994.

Fridjonnson, Thordur. *Whaling and the Icelandic Economy.* Reykjavík: Conference on Whaling in the North Atlantic, 1997.

Friedheim, Robert L. "International Organizations and the Uses of the Ocean." In *Multinational Cooperation,* ed. Robert Jordan. New York: Oxford University Press, 1972.

———. *Negotiating the New Ocean Regime.* Columbia: University of South Carolina Press, 1993.

———. "Moderation in the Pursuit of Justice: Explaining Japan's Failure in the International Whaling Negotiations." *Ocean Development and International Law* 27 (1996): 349–78.

———. "Fostering a Negotiated Outcome in the IWC." In *Whaling in the North Atlantic: Economic and Political Perspectives,* ed. Gudrun Petursdottir, 135–57. Reykjavík: Fisheries Research Institute, University of Iceland, Iceland University Press, 1997.

Friedheim, Robert L., and Tsuneo Akaha. "Antarctic Resources and International Law: Japan, the United States, and the Future of Antarctica." *Ecology Law Quarterly* 16, no. 1 (1989): 139.

Galloway, J. H. *The Sugar Cane Industry: An Historical Geography from Its Origins to 1914.* Cambridge: Cambridge University Press, 1989.

Gambell, Ray. "The Management of Whales and Whaling." *Arctic* 46, no. 2 (1993): 97–108.

———. "The International Whaling Commission Today." In *Whaling in the North Atlantic: Economic and Political Perspectives,* ed. Gudrun Petursdottir. Reykjavík: Fisheries Research Institute, University of Iceland, Iceland University Press, 1997.

Garcia, S. M. "The Precautionary Principle: Its Implications in Capture Fisheries Management." *Ocean and Coastal Management* 22, no. 2 (1994): 99–125.

Gardner, Gerald T., and Paul C. Stern. *Environmental Problems and Human Behavior.* Boston: Allyn and Bacon, 1996.

Garnsey, Peter. *Famine and Food Supply in the Greco-Roman World: Response to Risk and Crisis.* Cambridge: Cambridge University Press, 1988.

Gaskin, David E. *The Ecology of Whales and Dolphins.* London: Heinemann, 1982.

George, Susan. *Ill Fares the Land: Essays on Food, Hunger, and Power.* Washington, D.C.: Institute for Policy Studies, 1984.

Gettleman, Jeffrey. "Drawn to the Sea, to Tradition, to Danger." *Los Angeles Times,* 29 August 1999, A26.

Gibbs, Walter. "Journal: Norwegian Whalers Say the Wind Is Turning in Their Favor." *New York Times,* 23 July 1997.

Goldberg, Carey "Downsizing Activism: Greenpeace Is Cutting Back." *New York Times,* 16 September 1997, A1.

Goldblat, Jozef, ed. *Maritime Security: The Building of Confidence.* New York: United Nations, 1992.

Graburn, Nelson H. H. "Whaling Towns and Tourism: Possibilities for Development of Tourism in the Former [*sic*] Whaling Towns: Taiji, Wada, and Ayukawa," Report submitted by the government of Japan. *Proceedings of the 42d Annual Meeting of the IWC.* Cambridge, England: IWC, 1990.

Greenpeace. "Whales." *Greenpeace Fact Sheet.* Greenpeace International via Greenbase, 1991.

Gross, Paul R., and Norman Levitt. *Higher Superstition: The Academic Left and Its Quarrels with Science.* Baltimore: Johns Hopkins University Press, 1994.

Gulland, John A. *The Management of Marine Fisheries.* Seattle: University of Washington Press, 1974.

———. "The End of Whaling?" *New Scientist* (29 October 1988): 42–47.

Haas, Peter. "Do Regimes Matter? Epistemic Communities and Mediterranean Pollution Control." *International Organization* 43, no. 4 (summer 1989): 377–404.

———. *Saving the Mediterranean: The Politics of International Environmental Cooperation.* New York: Columbia University Press, 1990.

———. "Introduction: Epistemic Communities and International Policy Coordination." *International Organization* 46, no. 1 (winter 1992): 1–36.

Hallam, Elizabeth M. *The Domesday Book through Nine Centuries.* London: Thames and Hudson, 1986.

Hallam, H. E., ed. *The Agrarian History of England and Wales.* Vol. 2. Cambridge: Cambridge University Press, 1988.

Hammond, Phillip. Letter of resignation to Ray Gambell, secretary of the IWC, 26 May 1993.

Hampson, Fen Osler, with Michael Hart. *Multilateral Negotiations: Lesson from Arms Control, Trade, and the Environment.* Baltimore: Johns Hopkins University Press, 1995.

Hanna, Susan S., Carl Folke, and Karl-Goran Maler, eds. *Rights to Nature.* Washington, D.C.: Island Press, 1996.

Hardin, Garrett. "The Tragedy of the Commons." *Science* 162 (1968): 1243–48.

Hargrove, John Lawrence, ed. *Who Protects the Ocean? Environment and the Development of the Law of the Sea.* St. Paul, Minn.: West, 1975.

Harris, Marvin. *Good to Eat: Riddles of Food and Culture*. New York: Simon and Schuster, 1985.

Health Canada. *State of Knowledge Report on Environmental Contaminants and Human Health in the Great Lakes Basin*. Ottawa: Minister of Public Works and Government Services, 1997.

Hearst, David, and Paul Brown. "Soviet Whaling Lies Revealed." *The Guardian*, 12 February 1994, 1.

Heck, C. "Collective Arrangements for Managing Ocean Fisheries." *International Organization* 29, no. 3 (1975).

Henke, Janice, and Stephen S. Boynton. "Legal Issues Regarding the International Whaling Commission (IWC) Under the International Convention for the Regulation of Whaling (ICRW)." 1995.

Henkin, Louis. *How Nations Behave: Law and Foreign Policy*. New York: Praeger, 1968.

Herscovici, Alan. *Second Nature: The Animal-Rights Controversy*. Montreal: CBC Enterprises, 1985.

Hirschman, Albert O. *Exit, Voice, and Loyalty*. Cambridge: Harvard University Press, 1972.

Hodge, Ian. *Environmental Economics: Individual Incentives and Public Choices*. New York: St. Martin's Press, 1995.

Hoel, Alf Hakon. *The International Whaling Commission, 1972–1984: New Members, New Concerns*. Lysaker, Norway: Fridtjof Nansen Institute, 1985.

Holt, Sidney. "Whale Mining, Whale Saving." *Marine Policy* 3 (1985): 192–214.

Homer-Dixon, Thomas F. "On the Threshold: Environmental Changes as Causes of Acute Conflict." *International Security* 16, no. 2 (fall 1991): 76–116.

———. "Environmental Scarcity and Global Security." *Foreign Policy Association Headline Series* 300 (fall 1993).

Homer-Dixon, Thomas, and Jessica Blitt, eds. *Ecoviolence: Links among Environment, Population, and Security*. Lanham, Md.: Rowman and Littlefield, 1998.

Homer-Dixon, Thomas F., Jeffrey H. Boutwell, and George W. Rathjens. "Environmental Change and Violent Conflict." *Scientific American*, February 1993, 38–47.

Hopkins, Raymond F., and Donald J. Puchala. *Global Food Interdependence: Challenge to American Foreign Policy*. New York: Columbia University Press, 1980.

Hoyt, Erich. *The Worldwide Value and Extent of Whale Watching, 1995*. Bath, England: Whale and Dolphin Conservation Society, 1995.

Hugill, Peter J. "Structural Changes in the Core Regions of the World Economy, 1830–1945." *Journal of Historical Geography* 14, no. 12 (1988): 111–27.

HWNNews. "CITES: Continues to Defer to IWC Decisions." www.highnorth.no, June 1997.

HWNNews. "Majority Vote for Downlisting Minke Whales." www.highnorth.no, June 1997.

Inglehart, Ronald. *Modernization and Postmodernization: Cultural, Economic, and Political Change in Forty-three Societies.* Princeton, N.J.: Princeton University Press, 1997.

Intergovernmental Panel on Climate Change. *Second Assessment Synthesis of Scientific-Technical Information Relevant to Interpreting Article 2 of the UN Framework Convention on Climate Change."* Geneva: World Meteorological Organisation, 1995.

International Convention for the Regulation of Whaling (ICRW). 62, Stat. 1716, 161 U.N.T.S. 74.

International Convention for the Regulation of Whaling (ICRW). Tables 1–3, Schedule. Cambridge, England: IWC, 1995.

International Fund for Animal Welfare (IFAW). "Report of the Workshop on the Socioeconomic Aspects of Whale Watching." Kaikoura, New Zealand: IFAW, 1998.

International Harpoon 5 (1995).

International Network for Whaling Research (INWR). *INWR Digest* 12, no. 1 (1997).

International Whaling Commission (IWC). *Proceedings of the 45th Meeting of the IWC.* Cambridge, England: IWC, 1993.

———. "Resignation of the Chairman of the Scientific Committee," Circular Communication to Commissioners, Contracting Governments, and Members of the Scientific Committee, 1 June 1993. Cambridge, England: IWC, 1993.

———. Intersessional Meeting of the Working Group on a Sanctuary in the Southern Ocean. *Proceedings of the 46th Meeting of the IWC* (IWC/46/19). Cambridge, England: IWC, 1994.

———. Joint Opening Statement by the Observers for ICC And IWGIA. *Proceeding of the 46th Meeting of the IWC.* Cambridge, England: IWC, 1994.

———. Resolution on the Unreliability of Past Whaling Data. *Proceedings of the 46th Meeting of the IWC.* Cambridge, England: IWC, 1994.

———. Chairman's Report. *Proceedings of the 47th Meeting of the IWC.* Cambridge, England: IWC, 1995.

———. Final Press Release. *Proceedings of the 48th Meeting of the IWC.* Cambridge, England: IWC, 1996.

———. *Provisional Estimate of Financial Contributions 1997/98.* Cambridge, England: IWC, 1997.

IUCN (World Conservation Union). *Report of the Fourth Global Biodiversity Forum,*

31 August–1 September, 1996, Montreal, Canada. Gland, Switzerland: IUCN and World Resources Institute, 1997.

IUCN/UNEP/WWF. *Caring for the Earth: A Strategy for Sustainable Living.* Gland, Switzerland: IUCN, 1991.

IUCN (World Conservation Union), World Commission on Protected Areas, and World Wide Fund for Nature. *Principles and Guidelines on Indigenous and Traditional Peoples and Protected Areas.* Gland, Switzerland: IUCN, 1996.

Ivarsson, Johann Vioas. *Science, Sanctions, and Cetaceans: Iceland and the Whaling Issue.* Reykjavík: Center for International Studies, University of Iceland, 1994.

Iwasaki-Goodman, M. "Social and Cultural Change in Ayukawa-hama (Ayukawa Shore Community)." Ph.D. diss., University of Alberta-Edmonton, 1994.

Iwasaki-Goodman, M., and Milton M. R. Freeman. *Social and Cultural Significance of Whaling in Contemporary Japan: A Case Study of Small-Type Coastal Whaling* (IWC/46/SEST3). Cambridge, England: IWC, 1993.

Japan. *Aboriginal Subsistence Whaling with Special Reference to the Alaska and Greenland Fisheries.* Cambridge, England: IWC: 1982.

———. *Report to the Working Group on Socio-Economic Implications of a Zero Catch Limit* (IWC/41/21). Cambridge, England: IWC, 1989.

———. *Socio-Economic Impact Countermeasures in the Four Japanese STCW Communities* (TC/42/SEST2). Cambridge, England: IWC, 1990.

———. *Age Difference in Food Preference with Regard to Whale Meat: Report on a Questionnaire in Oshika Township* (TC/43/SEST4). Cambridge, England: IWC, 1991.

———. *The Cultural Significance of Everyday Food Use* (TC/43/SEST1). Cambridge, England: IWC, 1991.

———. *A Critical Evaluation of the Relationship between Cash Economies and Subsistence Activities* (IWC/44/SEST5). Cambridge, England: IWC, 1992.

———. *The Importance of Everyday Food Use* (IWC/44/SEST4). Cambridge, England: IWC, 1992.

———. *Action Plan for Japanese Community-Based Whaling (CBW)* (IWC/45/SEST3). Cambridge, England: IWC, 1993.

———. *Action Plan for Japanese Community-Based Whaling (CBW): Distribution and Consumption of Whale Products* (IWC/46/31 Rev. 2). Cambridge, England: IWC, 1994.

———. *"Commercial" vs. "Subsistence," "Aboriginal" vs. "Nonaboriginal," and the Concept of Sustainable Development in the Context of Japanese Coastal Fisheries Management* (IWC/46/SEST1). Cambridge, England: IWC, 1994.

———. *Action Plan for Japanese Community-Based Whaling (CBW)* (IWC/47/SEST1). Cambridge, England: IWC, 1995.

———. *Action Plan for Japanese Community-Based Whaling (CBW): (Revised)* (IWC/47/46). Cambridge, England: IWC, 1995.

Jarvenpa, Robert, and Hetty-Jo Brumbach. "Occupational Status, Ethnicity, and Ecology: Metis Cree Adaptations in a Canadian Trading Frontier." *Human Ecology* 13, no. 3 (1985): 309–29.

Jasanoff, Sheila. "Science and Norms in Global Environmental Regimes." In *Earthly Goods: Environmental Change and Social Justice,* ed. Fen Osler Hampson and Judith Reppy. Ithaca: Cornell University Press, 1996.

Jasper, James M., and Dorothy Nelkin. *The Animal Rights Crusade: The Growth of a Moral Protest.* New York: Free Press, 1992.

Jensen, J., K. Adare, and R. Shearer, eds. *Canadian Arctic Contaminants Assessment Report.* Ottawa: Department of Indian Affairs and Northern Development, 1997.

Jones, Olive R. "Commercial Foods, 1740–1820." *Historical Archaeology* 27, no. 2 (1993): 25–41.

Kalland, Arne. *The Spread of Whaling Culture in Japan* (TC/41/STW3). Cambridge, England: IWC, 1989.

———. "Management by Totemization: Whale Symbolism and the Anti-Whaling Campaign." *Arctic* 46, no. 2 (1993): 124–33.

———. "Some Reflections after the Sendai 'Workshop.'" *Isana* 16 (1997): 11–15.

Kalland, Arne, and B. Moeran. *Japanese Whaling: End of an Era?* London: Curzon Press, 1992.

Kapel, Finn O., and Robert Petersen. "Subsistence Hunting-The Greenland Case." In *Aboriginal/Subsistence Whaling (with Special Reference to the Alaska and Greenland Fisheries),* ed. G. P. Donovan. Cambridge, England: IWC, 1982.

Kaufman, Wallace. *No Turning Back.* New York: Basic Books, 1994.

Kaul, Inge, Isabelle Grunberg, and Marc A. Stern, eds. *Global Public Goods: International Cooperation in the Twenty-first Century.* New York: Oxford University Press, 1999.

Kawakami, Rin'itsu. "Save Whaling." *Isana* 11 (1994).

Kempton, Willett, James S. Boster, and Jennifer A. Hartley. *Environmental Values in American Culture.* Cambridge: MIT Press, 1997.

Keohane, Robert O. *After Hegemony: Cooperation and Discord in the World Political Economy.* Princeton, N.J.: Princeton University Press, 1984.

Kimball, Lee. "Treaty Implementation: Scientific and Technical Advice Enters a New Stage." *Studies in Transnational Legal Policy* 28 (1996).

Kindleberger, Charles P. *The World in Depression, 1929–1939.* Berkeley and Los Angeles: University of California Press, 1973.

King, Lauriston R., and Kimberly McGar Stephens. "Politics and the Animal Rights

Bibliography

Movement in the United States." *Annual Meeting of the Southern Political Science Association,* 1991.

Kline, Gary. "Food as a Human Right." *Journal of Third World Studies* 10 (1993): 92–107.

Klinowska, Margaret. "How Brainy Are Cetaceans?" *Oceanus* 32, no. 1 (spring 1989): 19–20.

Knauss, John A. "The International Whaling Commission: Its Past and Possible Future." *Ocean Development and International Law* 28, no. 1 (1997): 79–99.

Kneese, Allen V. *Economics and the Environment.* Middlesex, England: Penguin, 1977.

Kowalewski, David. "Transnational Corporations and the Third World's Right to Eat: The Caribbean." *Human Rights Quarterly* 3, no. 4 (1981): 45–64.

Krasner, Stephen D. "Global Communication and National Power: Life on the Pareto Frontier." *World Politics* 43, no. 3 (1991): 336–66.

Krech, Shepard, III. "The Influence of Disease and the Fur Trade on Arctic Drainage Lowlands Dene, 1800–1850." *Journal of Anthropological Research* 39, no. 2 (1983): 123–46.

Lamothe, Rene M. J. *It Was Only a Treaty: A Historical View of Treaty 11 According to the Dene of the Mackenzie Valley.* Ottawa: Royal Commission on Aboriginal Peoples, 1993.

Langdon, Steve J. "Alaska Native Subsistence: Current Regulatory Regimes and Issues." *Alaska Native Review Commission Reports.* Anchorage: Alaska Native Review Commission, 1984.

Leonard, William R., Anne Keenleyside, and Evgueni Ivakine. "Recent Fertility and Mortality Trends among Aboriginal and Non-Aboriginal Populations of Central Siberia." *Human Biology* 69, no. 3 (1997): 403–17.

Levy, Marc A. "European Acid Rain: The Power of 'Tote-Board' Diplomacy." In *Institutions for the Earth: Sources of Effective International Environmental Protection,* ed. Peter M. Haas, Robert O. Keohane, and Marc A. Levy. Cambridge: MIT Press, 1994.

Levy, Marc A., et al. *The Study of International Regimes,* IIASA, Vienna, 1994.

Litfin, Karen T. *Ozone Discourses: Science and Politics in Global Environmental Cooperation.* New York: Columbia University Press, 1994.

Litfin, Karen T., ed. *The Greening of Sovereignty in World Politics.* Cambridge: MIT Press, 1998.

Littell, R. *Endangered and Other Protected Species: Federal Laws and Regulations.* Washington, D.C.: Bureau of National Affairs, 1992.

Los Angeles Times, 19 May 1997, A12.

Los Angeles Times, 22 October 1997, A12.

Lubchenco, Jane. "Entering the Century of the Environment: A New Social Contract for Science." *Science* 279 (1998): 491–97.

Ludwig, Donald, Ray Hilborn, and Carol Walters. "Uncertainty, Resource Exploitation, and Conservation: Lessons from History." *Science* 260 (2 April 1993): 17.

Lyke, M. L. "After Whale Hunt, Makah Celebrate New Sense of Unity; Annual Festival Marked by Fewer Protests." *Seattle Post-Intelligencer,* 30 August 1999.

Lynge, Finn. *Arctic Wars, Animal Rights, Endangered Peoples.* Trans. Marianne Stenbaek. Hanover, N.H.: University Press of New England, 1992.

Macnow, Alan. "A Whaling Moratorium Opposed by I.W.C.'s Own Scientists." *New York Times,* 29 September 1984, 22.

Makombe, K. ed. *Sharing the Land, Wildlife, People, and Development in Africa.* IUCN/ROSA Environmental Issues Series, no. 1. Harare, Zimbabwe: IUCN, 1994.

Malnes, R. "'Leader' and 'Entrepreneur' in International Negotiations: A Conceptual Analysis." *European Journal of International Relations* 1, no. 1 (1995): 87–112.

Manderson, Lenore, and Haruko Akatsu. "Whale Meat in the Diet of Ayukawa Villagers." *Ecology of Food and Nutrition* 30 (1993): 207–20.

Manderson, Lenore, and Helen Hardacre. "Small-Type Coastal Whaling in Ayukawa." Draft Report of Research: December 1988–January 1989 (IWC/41/SE3). Cambridge, England: IWC, 1989.

Marchione, Thomas J. "The Right to Food in the Post-Cold War Era." *Food Policy* 21, no. 1 (1996): 83–102.

Marquardt, Ole, and Richard A. Caulfield. "Development of West Greenlandic Markets for Country Foods since the Eighteenth Century." *Arctic* 49, no. 2 (1996): 107–19.

Martin, Gene S., Jr., and James W. Brennan. "Enforcing the International Convention for the Regulation of Whaling: The Pelly and Packwood-Magnuson Amendments." *Denver Journal of International Law and Policy* 17, no. 2 (1989): 300.

Mason, Jim. "Brave New Farm?" In *In Defence of Animals,* ed. Peter Singer, 89–107. Oxford: Basil Blackwell, 1985.

Matthews, Jessica Tuchman. "Redefining Security." *Foreign Affairs* 68, no. 2 (spring 1989): 162–77.

McDorman, Ted L. "Iceland, Whaling and the U.S. Pelly Amendment: The International Trade Law Context." *Nordic Journal of International Law* 66 (1997): 453–74.

———. "Canada and Whaling: An Analysis of Article 65 of the Law of the Sea Convention." *Ocean Development and International Law* 29 (1998): 179–94.

McDougal, Myres Smith, Harold D. Lasswell, and James C. Miller. *The Interpretation of Agreements and World Public Order: Principles of Content and Procedure.* New Haven, Conn.: Yale University Press, 1967.

McGoodwin, James R. *Culturally-Based Conflicts in the Use of Living Resources and Suggestions for Resolving or Mitigating Such Conflicts* (KC/FI/95/TECH/9). International Conference of Sustainable Contribution of Fisheries to Food Security, Kyoto, 4–9 December 1995.

McGrath, Ann. *"Born in the Cattle": Aborigines in Cattle Country.* London: Allen and Unwin, 1987.

McHugh, J. L. "The Role and History of the Whaling Commission." In *The Whale Problem: A Status Report,* ed. William E. Schevill. Cambridge: Harvard University Press, 1974.

McKeown, C. Timothy. "The Human Context of Scientific Expectations." *Futures* 22, no. 1 (1990): 46–56.

McLaughlin, Richard J. "UNCLOS and the Demise of the United States' Use of Trade Sanctions to Protect Dolphins, Sea Turtles, Whales, and Other International Marine Living Resources." *Ecology Law Quarterly* 21, no. 1 (1994): 1–78.

———. "U.S. Accession of the United Nations Convention on the Law of the Sea and the Loss of Unilateral Trade Sanctions to Protect Marine Living Resources." In *Ocean Yearbook 11,* ed. Elizabeth Mann Borgese, Norton Ginsburg, and Joseph R. Morgan. Chicago: University of Chicago Press, 1994.

McNeely, Jeffrey A. "The Great Reshuffling: How Alien Species Help Feed the Global Economy." In *Proceedings: Norway/UN Conference on Alien Species; The Trondheim Conference on Biodiversity, 1–5 July 1996,* ed. Odd Terje Sandlund, Peter Johan Schei, and Aslaug Viken. Trondheim: Norwegian Institute for Nature Research, 1996.

Merchant, Carolyn, ed. *Ecology.* Atlantic Highlands, N.J.: Humanities Press, 1994.

Messer, Ellen. "Anthropological Perspectives on Diet." *Annual Review of Anthropology* 13 (1984): 205–49.

Miles, Edward L. *Legal Aspects of International Whaling Commission Activities,* Tokyo: Institute of Cetacean Research, 1991.

Miles, Edward L., Arild Underdahl, Steinar Andresen, Jon Birger Skjaerseth, Jørgen Wettstad, Elaine Carlin. *Unlocking Effectiveness: Soft Institutions and Hard Environmental Problems.* Cambridge: MIT Press, forthcoming.

Mintz, Sidney W. "Zur Beziehung zwischen Ernahrung und Macht." *Jahrbuch fur Wirtschaftsgeschicte* 1 (1994): 61–72.

Mitchell, B. R. *Abstract of British Historical Statistics.* Cambridge: Cambridge University Press, 1962.

Mitchell, Ronald. "Forms of Discourse, Norms of Sovereignty: Interests, Science, and Morality in the Regulation of Whaling." In *The Greening of Sovereignty in World Politics,* ed. Karen T. Litfin, 141–71. Cambridge: MIT Press, 1998.

Bibliography

Mizroch, Sally A. "The Development of Balaenopterid Whaling in the Antarctic." *Cetus* 5, no. 2 (1984).

Moeran, Brian, T. Akimichi, Richard Caulfield, N. Doubleday, Milton Freeman, Arne Kalland, G. Palsson, H. Stefansson, and J. Takahashi. *Similarities and Diversity in Coastal Whaling Operations: A Comparison of Small-Scale Whaling in Greenland, Iceland, Japan, and Norway* (IWC/44/SEST6). Cambridge, England: IWC, 1992.

Murphy, Neil J., Cynthia D. Schraer, Maureen C. Thiele, Edward J. Boyko, Lisa R. Bulkow, Barbara J. Doty, and Anne P. Lanier. "Dietary Changes and Obesity Associated with Glucose Intolerance in Alaska Natives." *Journal of the American Dietary Association* 6 (1995): 672–76.

Naess, Arne. "The Shallow and the Deep, Long-Range Ecology Movement: A Summary." *Inquiry* 16 (1973): 95–100.

Nagasaki, Fukuzo. "On the Whaling Controversy." In *Whaling and Japan's Whale Research*, 5–20. Tokyo: Institute of Cetacean Research, 1993.

National Research Council. *Proceedings of the Conference on Common Property Resource Management.* Washington, D.C.: National Academy Press, 1986.

Nelson, Michael. "Social-Class Trends in British Diet, 1860–1980." In *Food, Diet, and Economic Change Past and Present,* ed. Catherine Geissler and Derek J. Oddy. Leicester, England: Leicester University Press, 1993.

"New Poll on Public Attitudes to Whaling." *INWR Digest (International Network for Whaling Research)* 14 (December 1997): 3.

Niimi, Reiko. "The Problem of Food Security." In *Japan's Economic Security,* ed. Nobutoshi Akao. New York: St. Martin's, 1983.

North Atlantic Marine Mammal Commission (NAMMCO). *Annual Report 1996.* Tromso, Norway: NAMMCO, 1997.

Norway. "Statement of the Norwegian Commissioner on Draft Resolution on Northeast Atlantic Minke Whales." *Proceedings of the 48th Meeting of the IWC.* Cambridge, England: IWC, 1996.

"NZ Bid to Win Rights for Apes Fails in Parliament." *Nature* 399 (1999).

Odell, John. *U.S. International Monetary Policy: Markets, Power, and Ideas as Sources of Change.* Princeton, N.J.: Princeton University Press, 1982.

———. *Policy Beliefs and International Economic Negotiation.* American Political Science Association Convention, 1995.

Ohnuki-Tierney, Emiko. *Rice as Self: Japanese Identities through Time.* Princeton, N.J.: Princeton University Press, 1993.

Olson, Mancur. *The Logic of Collective Action: Public Goods and the Theory of Groups.* Cambridge: Harvard University Press, 1965.

Bibliography

Ostrom, Elinor. *Governing the Commons: The Evolution of Institutions for Collective Action.* Cambridge: Cambridge University Press, 1990.

Parry, Jonathan, and Maurice Bloch, eds. *Money and the Morality of Exchange.* Cambridge: Cambridge University Press, 1989.

Pearce, David W., and R. Kerry Turner. *Economics of Natural Resources and the Environment.* Baltimore: Johns Hopkins University Press, 1990.

Pelto, Gretel H., and Pertti J. Pelto. "Diet and Delocalization: Dietary Changes since 1750." *Journal of Interdisciplinary History* 14, no. 2 (1983): 507–28.

Peluso, Nancy. "Coercing Conservation: The Politics of State Resource Control." In *The State and Social Power in Global Environmental Politics,* ed. Ronnie D. Lipschutz, and Ken Conca. New York: Columbia University Press, 1993.

Petersen, Robert. *Traditional and Present Distribution Channels in Subsistence Hunting in Greenland* (TC/41/22). Cambridge, England: IWC, 1989.

Peterson, M. J. "Whalers, Cetologists, Environmentalists, and the International Management of Whaling." *International Organization* 46, no. 1 (winter 1992): 149–87.

Peterson, Nicholas, and Toshio Matsuyama. *Cash, Commoditisation, and Changing Foragers.* Senri Ethnological Studies, no. 30 (Osaka: National Museum of Ethnology, 1991), 1–16.

Pimental, D., and M. Pimental. *Food, Energy, and Society.* New York: Wiley, 1979.

Pollack, Andrew. "Commission to Save Whales Endangered, Too." *New York Times,* 18 May 1993, C4.

Pontecorvo, Giulio, ed. *The New Order of the Oceans.* New York: Columbia University Press, 1986.

Povinelli, Daniel J. "What Chimpanzees (Might) Know about the Mind." In *Chimpanzee Cultures,* ed. Richard W. Wrangham, W. C. McGrew, Frans B. M. de Waal, and Paul G. Heltne. Cambridge: Harvard University Press, 1994.

Princen, Thomas, and Matthias Finger. *Environmental NGOs in World Politics.* London: Routledge, 1994.

Prosterman, Roy L., Tim Hanstad, and Li Ping. "Can China Feed Itself?" *Scientific American,* November 1996, 90–97.

Putnam, Robert D. "Diplomacy and Domestic Politics: The Logic of Two-Level Games." *International Organization* 42 (summer 1988): 427–60.

Rawls, John. *A Theory of Justice.* Cambridge: Harvard University Press, Belknap Press, 1971.

Reagan, Ronald. "Message to Congress: Whaling Activities of the U.S.S.R." *Public Papers of the President of the United States.* Washington, D.C.: GPO, 1985.

———. "Message from the President of the United States: Whaling Activities of

Norway." *Public Papers of the President of the United States.* Washington, D.C.: GPO, 1986.

Redclift, Michael, and David Goodman. *Refashioning Nature: Food, Ecology, and Culture.* New York: Routledge, 1991.

Redford, Kent H., and John G. Robinson. "The Game of Choice: Patterns of Indian and Colonist Hunting in the Neotropics." *American Anthropologist* 89, no. 4 (1987): 650–67.

Reichel-Dolmantoff, Gerardo. "Cosmology as Ecological Analysis: A View from the Rain Forest." *Man* 11, no. 3 (1976): 307–18.

Reischauer, Edwin O., and Albert M. Craig. *Japan: Tradition and Transformation.* Boston: Houghton Mifflin, 1978.

Renker, Ann. *Whale Hunting and the Makah Tribe: A Needs Statement* (IWC/49/AS5). Cambridge, England: IWC, 1997.

Rifkin, Jeremy. *Beyond Beef: The Rise and Fall of the Cattle Culture.* New York: Dutton, 1992.

Ritvo, Harriet. "Toward a More Peaceable Kingdom." *Technology Review* 95, no. 3 (1992): 56–61.

Robertson, Claire C. "Black, White, and Red All Over: Beans, Women, and Agricultural Imperialism in Twentieth-Century Kenya." *Agricultural History* 71, no. 3 (1997): 259–99.

Rosenne, Shabtai. *Developments in the Law of Treaties, 1945–1986.* Cambridge: Cambridge University Press, 1989.

Rossum, J. "The Negotiations over Reductions of Quotas in Antarctica: The International Whaling Commission, 1960–65." Master's thesis, University of Oslo, 1985.

Rothenberg, David. "Have a Friend for Lunch: Norwegian Radical Ecology versus Tradition." In *Ecological Resistance Movements: The Global Emergence of Radical and Popular Environmentalism,* ed. Bron Raymond Taylor. Albany: State University of New York Press, 1995.

Rousham, E. K., and M. Gracey. "Persistent Growth Faltering among Aboriginal Infants and Young Children in Northwest Australia: A Retrospective Study from 1969–1993." *Acta Paediatrica* 86, no. 1 (1997): 46–50.

Royal Commission on Seals and the Sealing Industry. *Seals and Sealing in Canada: Report of the Royal Commission.* Ottawa: Minister of Supply and Services, 1986.

Ruggie, John Gerard. "Territoriality and Beyond: Problematizing Modernity in International Relations." *International Organization* 47, no. 1 (winter 1993): 139–74.

Sabine, George H. *A History of Political Theory.* New York: Holt, 1937.

Sahlins, Marshall D. *Stone Age Economics.* New York: Aldine, 1972.

Said, Edward W. *Orientalism.* New York: Pantheon, 1978.

Sands, Phillipe. *Principles of International Environmental Law.* Manchester, England: Manchester University Press, 1995.

Sands, Phillipe, ed. *Greening International Law.* New York: New Press, 1994.

Sarewitz, Daniel. "Social Change and Science Policy." *Issues in Science and Technology* 13, no. 4 (1997): 31.

Scarff, James E. "The International Management of Whales, Dolphins, and Porpoises: An Interdisciplinary Assessment." *Ecology Law Quarterly* 6, no. 2–3 (1977): 323–638.

Scheiber, Harry N. "Historical Memory, Cultural Claims, and Environmental Ethics in the Jurisprudence of Whaling Regulation." *Ocean and Coastal Management* 38, no. 1 (1998): 5–40.

Scheiber, Harry N., and Chris Carr. "The Limited Entry Concept and the Pre-History of the ITQ Movement in Fisheries Management." In *Social Implications of Quota Systems in Fisheries,* ed. Gisli Palsson, and Gudrun Petursdottir. Copenhagen: Nordic Council of Ministers, 1997.

Schelling, Thomas C. *The Strategy of Conflict.* New York: Oxford University Press, 1963.

Schevill, William E., ed. *The Whale Problem: A Status Report.* Cambridge: Harvard University Press, 1974.

Schleifer, Harriet. "Images of Death and Life: Food Animal Production and the Vegetarian Option." In *In Defence of Animals,* ed. Peter Singer. Oxford: Basil Blackwell, 1985.

Schweder, Tore. *Intransigence, Incompetence, or Political Expediency? Dutch Scientists in the International Whaling Commission in the 1950s: Injection of Uncertainty,* SC/44/O 13. Cambridge, England: IWC, 1992.

Sen, Amartya. "Population: Delusion and Reality." *New York Review of Books* (22 September 1994): 62.

Sethi, S. Prakash, and James E. Post. "Public Consequences of Private Action: The Marketing of Infant Formula in Less Developed Countries." *California Management Review* 21, no. 4 (1979): 35–48.

Shammas, Carole. "The Eighteenth-Century English Diet and Economic Change." *Explorations in Economic History* 21, no..3 (1984): 254–69.

Shaw, Gareth. "Changes in Consumer Demand and Food Supply in Nineteenth-Century British Cities." *Journal of Historical Geography* 11, no. 3 (1985): 280–96.

Sheridan, Richard B. *Sugar and Slavery: An Economic History of the British West Indies, 1623–1775.* Barbados: Caribbean Universities Press, 1974.

———. "The Crisis of Slave Subsistence in the British West Indies during and after the American Revolution." *William and Mary Quarterly* 33, no. 4 (1976): 615–41.

Shima, Kazuo. *Letter to Ray Gambell*, Cambridge, England: IWC, 1994.

Sinclair, Ian. *The Vienna Convention on the Law of Treaties.* Manchester, England: Manchester University Press, 1984.

Singer, Peter. *Animal Liberation: A New Ethics for Our Treatment of Animals.* New York: Avon, 1975.

———. "The Singer Solution to World Poverty." *New York Times Magazine,* 5 September 1999.

Sklair, Leslie. *Sociology of the Global System.* Baltimore: Johns Hopkins University Press, 1991.

Small, George L. *The Blue Whale.* New York: Columbia University Press, 1971.

Soden, Dennis L., and Brent S. Steel, eds. *Handbook of Global Environmental Policy and Administration.* New York: Marcel Dekker, 1999.

Sokal, Alan, and Jean Bricmont. *Fashionable Nonsense: Postmodern Intellectuals' Abuse of Science.* New York: Picador, 1998.

Solway, Jacqueline S. "Foragers, Genuine or Spurious? Situating the Kalahari San in History." *Current Anthropology* 31, no. 2 (1990): 109–22.

Sorel, Georges. *Reflections on Violence.* New York: Collier, 1950.

Soule, Michael E., and Gary Lease, eds. *Reinventing Nature? Responses to Postmodern Deconstruction.* Washington, D.C.: Island Press, 1995.

Spector, Michael. "The Dangerous Philosopher." *New Yorker* (1999): 46–55.

Spencer, Leslie, Jan Bollwerk, and Richard C. Morais. "The Not So Peaceful World of Greenpeace." *Forbes,* 11 November 1991, 174.

Spencer, Paul D. "Optimal Harvesting of Fish Populations with Nonlinear Rates of Predation and Autocorrelated Environmental Variability." *Canadian Journal of Fisheries and Aquatic Sciences* 54, no. 1 (1997): 59–74.

Spencer, Valeria Neale. "Domestic Enforcement of International Law: The International Convention for the Regulation of Whaling." *Colorado Journal of International Environmental Law and Policy* 2: 109–27.

Stevenson, Marc G., Andrew Madsen, and Elaine Maloney. *The Anthropology of Community-Based Whaling in Greenland: A Collection of Papers Submitted to the International Whaling Commission,* Occasional Paper 42. Edmonton, Alberta: Canadian Circumpolar Institute, 1997.

Stewart-Smith, Jo. *In the Shadow of Fujisan: Japan and Its Wildlife.* Harmondsworth, England: Viking/Penguin, 1987.

Stiles, D. "The Hunter-Gatherer "Revisionist" Debate." *Anthropology Today* 8 (1992): 13–17.

Stoett, Peter J. "The International Whaling Commission: From Traditional Concern to an Expanded Agenda." *Environmental Politics* 14, no. 1 (1995): 130–35.

———. *The International Politics of Whaling.* Vancouver: University of British Columbia Press, 1997.

Stokke, Olaf Schram. "Transnational Fishing: Japan's Changing Strategy." *Marine Policy* (July 1991): 231–43.

———. "The Effectiveness of CCAMLR." In *Governing the Antarctic: The Effectiveness and Legitimacy of the Antarctic Treaty System,* ed. Olaf Schram Stokke, and Divor Vidas. Cambridge: Cambridge University Press, 1996.

Stokke, Olaf Schram, and Divor Vidas, eds. *Governing the Antarctic, the Effectiveness and Legitimacy of the Antarctic Treaty System.* Cambridge: Cambridge University Press, 1996.

Stone, Christopher D. *The Gnat Is Older Than Man: Global Environment and Human Agenda.* Princeton, N.J.: Princeton University Press, 1993.

———. "Deciphering "Sustainable Development"" *Chicago-Kent Law Review* 69 (1994): 201.

———. "Legal and Moral Issues in the Taking of Minke Whales." In *Report: International Legal Workshop, Sixth Annual Whaling Symposium,* ed. Robert L. Friedheim. Tokyo: Institute of Cetacean Research, 1996.

———. "Too Many Fishing Boats, Too Few Fish: Can Trade Laws Trim Subsidies and Restore the Balance of Global Fisheries?" *Ecology Law Quarterly* 24, no. 3 (1997): 505–44.

———. "Can the Oceans Be Harbored? A Four-Step Plan for the Twenty-first Century." Paper prepared for a conference, "Toward the International Protection of the Oceans: From Rules to Compliance," Lisbon, 17–19 September 1998.

Sullivan, Robert. "The Face of Eco-Terrorism." *New York Times Magazine,* 20 December 1998, 46ff.

Susskind, Lawrence E. *Environmental Diplomacy: Negotiating More Effective Global Agreements.* New York: Oxford University Press, 1994.

Susskind, Lawrence, Paul Levy, and Jennifer Thomas-Larmer. *Negotiating Environmental Agreements.* Washington, D.C.: Island Press, 1999.

Susskind, Lawrence E., and William Moomaw, eds. *New Directions in International Environmental Negotiation.* Cambridge, England: Program on Negotiation, 1999.

Suttles, Wayne P. *Coast Salish Essays.* Seattle and Vancouver: University of Washington Press and Talonbooks, 1987.

Takahashi, T., A. Kalland, B. Moeran, and T. C. Bestor. *Japanese Whaling Culture: Continuities and Diversities* (IWC/41/STW2). Cambridge, England: 1989.

Taylor, Bron Raymond, ed. *Ecological Resistance Movements: The Global Emergence of Radical and Popular Environmentalism.* Albany: State University of New York Press, 1995.

Thies, Carsten, and Teja Tscharntke. "Landscape Structure and Biological Control in Agroecosystems." *Science* 285 (1999): 893–95.

Thomas, Brindley. "Feeding England during the Industrial Revolution: A View from the Celtic Fringe." *Agricultural History* 56, no. 1 (1982): 328–42.

"Threat of Global Warming Is for 'Real' Clinton Declares." *Los Angeles Times,* 25 July 1997, A21.

Tønnessen, Johan Nicolay, and Arne Odd Johnsen. *The History of Modern Whaling.* Berkeley and Los Angeles: University of California Press, 1982.

Turner, B. L., II, and Karl W. Butzer. "The Columbian Encounter and Land-Use Change." *Environment* 34, no. 8 (October 1992): 16ff.

Underdal, Arild. *The Politics of International Fisheries Management: The Case of the Northeast Atlantic.* Oslo: Universitetsforlaget, 1980.

———. "Solving Collective Problems: Notes on Three Models of Leadership." In *Challenges of a Changing World.* Lysaker, Norway: Fridtjhof Nansen Institute, 1991.

———. "The Concept of Regime Effectiveness." *Cooperation and Conflict* 27, no. 3 (1992): 227–40.

Underdal, Arild, et al. *Science and the Environment, Integrity, and Involvement.* Manchester, England: Manchester University Press, 1998.

United Nations. *Convention on the Law of the Sea.* New York: UN, 1983.

———. Agreement for the Implementation of the Provisions of the United Nations Convention on the Law of the Sea of 10 December 1982 Relating to the Conservation and Management of Straddling Fish Stocks and Highly Migratory Fish Stocks. 1995. UN Doc. A/CONF.164/37.

———. *Kyoto Protocol to the United Nations Framework Convention on Climate Change.* New York: UN, 1997.

United Nations Center on Transnational Corporations. *Transnational Corporations in World Development: Trends and Prospects.* 1988. UN Doc. ST/CTC/89.

United Nations Environment Programme (UNEP). *Convention on Biological Diversity: Text and Annexes.* Geneva: UN, 1994. UN Doc. UNEP/CDB/94/1.United Nations Food and Agricultural Organization (FAO). *Global Fishery Production in 1994.*

United Nations General Assembly. *Report of the United Nations Conference on Environment and Development; Vol. I, Resolutions Adopted by the Conference."* New York: UN, 1992. UN Doc. No. A/CONF.151/26 Rev.1. (Vol. 1).

United States. "United States Opening Statement." *Proceedings of the 46th Annual Meeting of the IWC.* Cambridge, England: IWC, 1994.

Bibliography

Vagts, Detlev F. "Taking Treaties Less Seriously." *American Journal of International Law* 92 (1998): 458–62.

Van den Brink, Rogier, Daniel W. Bromley, and Jean-Paul Chavas. "The Economics of Cain and Abel: Agro-Pastoral Property Rights in the Sahel." *Journal of Development Studies* 31, no. 3 (1995): 373–99.

Van Dyke, Jon, Durwood Zaelke, and Grant Hewison, eds. *Freedom for the Seas in the Twenty-first Century.* Washington, D.C.: Island Press, 1993.

Verity, C. William. Letter to President Ronald Reagan, 14 April 1988.

Victor, David, Kal Raustiala, and Eugene B. Skolnikoff, eds. *"The Implementation and Effectiveness of International Environmental Commitments: Theory and Practice,* Cambridge: MIT Press, 1998.

Victor, David, and Julian Salt. "Keeping the Climate Treaty Relevant: An Elaboration." *International Institute for Applied Systems Analysis* (April 1995): 22–23.

Vietmeyer, Noel D. "Lesser-Known Plants of Potential Use in Agriculture and Forestry." *Science* 232 (13 June 1986): 1379–84.

Walsh, Virginia. "Eliminating Driftnets from the North Pacific Ocean: U.S.-Japanese Cooperation in the International North Pacific Fisheries Commission, 1953–1993." *Ocean Development and International Law* 29, no. 4 (1998): 295–322.

Warnock, John W. *The Politics of Hunger: The Global Food System.* London: Methuen, 1987.

Weber, P. "Abandoned Seas: Reversing the Decline of the Oceans." *World Watch Paper* 116 (November 1993).

Webster, David J. "The Political Economy of Food Production and Nutrition in Southern Africa in Historical Perspective." *Journal of Modern African Studies* 24, no. 3 (1986): 447–63.

Weiss, Edith Brown. *In Fairness to Future Generations: International Law, Common Patrimony, and Intergenerational Equity.* Dobbs Ferry, N.Y.: Transnational, 1989.

———. "Our Rights and Obligations to Future Generations for the Environment." *American Journal of International Law* 84, no. 1 (1990): 198–206.

———. *Environmental Change and International Law.* Tokyo: United Nations University Press, 1992.

Wenz, Peter S. *Environmental Justice.* Albany: State University of New York Press, 1988.

Wenzel, George. *Animal Rights, Human Rights: Ecology, Economy, and Ideology in the Canadian Arctic.* Toronto: University of Toronto Press, 1991.

Wettestad, Jørgen. *Designing Effective International Environmental Institutions: The Conditional Key.* Cheltenham: Edward Elgar, 1998.

Wettestad, Jørgen, and Steinar Andresen. "The Effectiveness of International Resource Cooperation: Some Preliminary Findings." *International Challenges* 11, no. 3 (1991): 55.

Whiten, A., et al. "Cultures in Chimpanzees." *Nature* 399 (1999): 682–85.

Williams, Michael. *Americans and Their Forests: A Historical Geography.* Cambridge: Cambridge University Press, 1989.

Wilson, C. Anne. *Food and Drink in Britain from the Stone Age to the Nineteenth Century.* London: Constable, 1973.

Wilson, Michael Clayton, and Ineke J. Dijks. "Land of No Quarter: The Palliser Triangle as an Environmental-Cultural Pump." In *The Palliser Triangle: A Region in Space and Time,* ed. R. W. Barendregt, M. C. Wilson, and F. J. Jankunis, 37–61. Lethbridge, Canada: Lethbridge University Press, 1993.

Woods, Ngaire. "Economic Ideas and International Relations: Beyond Rational Neglect." *International Studies Quarterly* 39 (1995): 162.

World Commission on Environment and Development. *Our Common Future.* New York: Oxford University Press, 1987.

World Wide Fund for Nature (w w f). *Integrating Economic Development with Conservation.* Gland, Switzerland: w w f, 1993.

Yablokov, Alexy V. "On the Soviet Whaling Falsification, 1947–1972." *Whales Alive* 6, no. 4 (1997).

Yost, James A., and Patricia M. Kelly. "Shotguns, Blowguns, and Spears: The Analysis of Technological Efficiency." In *Adaptive Responses of Native Amazonians,* ed. Raymond B. Hames, and William T. Vickers. New York: Academic Press, 1983.

Young, H. Peyton. " Dividing the Indivisible." *American Behavioral Scientist* 38, no. 6 (1995): 904–20.

Young, Nina M., ed. *Examining the Components of a Revised Management Scheme.* Washington, D.C.: Center for Marine Conservation, 1993.

Young, Oran R. *Resource Regimes: Natural Resources and Social Institutions.* Berkeley and Los Angeles: University of California Press, 1982.

———. *International Cooperation: Building Regimes for Natural Resources and the Environment.* Ithaca, N.Y.: Cornell University Press, 1989.

———. "Political Leadership and Regime Formation: On the Development of Institutions in International Society." *International Organization* 45, no. 3 (summer 1991): 281–308.

———. *International Governance: Protecting the Environment in a Stateless Society.* Ithaca: Cornell University Press, 1994.

Young, Oran R., Milton M. R. Freeman, Gail Osherenko, Raoul R. Andersen, Richard A. Caulfield, Robert L. Friedheim, Steve J. Langdon, Mats Ris, and Peter J. Usher.

Bibliography

"Subsistence, Sustainability, and Sea Mammals: Reconstructing the International Whaling Regime." *Ocean and Coastal Management* 23 (1994): 117–27.

Young, Oran R., and Gail Osherenko, eds. *Polar Politics: Creating International Environmental Regimes.* Ithaca: Cornell University Press, 1993.

Zartman, I. William. *The 50% Solution.* Garden City, N.Y.: Doubleday, Anchor, 1976.

Zemsky, V. A., A. A. Bergin, Y. A. Mikhaliev, and D. D. Tormsov. *Report of the Subcommittee on Southern Hemisphere Baleen Whales, Appendix 3: Soviet Antarctic Pelagic Whaling after WWII: Review of Actual Catch Data.* Report of the International Whaling Commission. Cambridge, England: I W C, 1994.

CONTRIBUTORS

STEINAR ANDRESEN is a political scientist from the University of Oslo, Norway. His specialty is international relations, more specifically international environmental and resources regimes. Andresen has been affiliated with the Fridtjof Nansen Institute since 1979, as a senior research fellow since 1987. He was research director from 1992 though 1997. Andresen was a visiting research fellow at the School of Marine Affairs, University of Washington, 1987–1988. From 1994 through 1996 he was associated on a part-time basis with the International Institute for Applied Systems Analysis (IIASA), Austria. Recently (1997–1998) Andresen was a visiting research fellow at the Center of International Studies, Woodrow Wilson School, Princeton University, where he worked on a book project on the International Whaling Commission. Andresen has published widely on this issue as well as on other international regimes.

WILLIAM ARON received his Ph.D. from the University of Washington in 1960. He was director of the Office of Limnology and Oceanography at the Smithsonian Institution, 1967–1971. He also served as director of the Office of Ecology and Environmental Conservation, National Oceanic and Atmospheric Administration (NOAA), 1971–1978; director of the Office of Marine Mammals and Endangered Species, NOAA-NMFS, 1978–1980; director of the Alaska Fisheries Science Center, NOAA-NMFS, 1980–1996. He was a member of the Scientific Committee of the International Whaling Commission, 1972–1977, was U.S. commissioner to the IWC in 1977, and served as an advisor to the U.S. commissioner, 1972–1980. As director of the Alaska Fisheries Science Center, he provided a supervisory overview of the National Marine Mammal Laboratory, which is part of the center. Currently, Aron is an affiliate professor in the College of Ocean and Fisheries Sciences at the University of Washington.

RUSSEL LAWRENCE BARSH has worked with indigenous communities throughout North America as a lawyer, ecologist, and development planner and has been a consultant for the International Labour Organization, United Nations Development Programme, and Canada's Royal Commission on Indigenous Peoples. He was a member of Canada's Royal Commission on Seals and the Sealing Industry and assessed Native American customary harvesting for the U.S. Department of the Interior's Outer Continental Shelf program. He is currently teaching law at New York University and directing an NYU-based research program on indigenous peoples and the biotechnology industry.

WILLIAM T. BURKE is a professor of law emeritus and professor of marine affairs emeritus at the University of Washington, where he taught from 1968 to 1999. His teaching and research interests center upon the international law of the sea. Since 1980 his writings have dealt mainly with living marine resources in international law. For information about affiliations and selected publications from 1982–1996, see the Web page *http://www2.law.washington.edu/burke*.

ELIZABETH R. DESOMBRE is an assistant professor of environmental studies and government at Colby College. She has published research on fishery and ocean issues, the Montreal Protocol Multilateral Fund, the use of sanctions and aid in international environmental politics, and the environmental impact of the international trading system.

MILTON M. R. FREEMAN is Senior Research Scholar, Canadian Circumpolar Institute, University of Alberta. An ecologist (Ph.D., McGill University, 1965) with research interests in maritime societies and resource use, Freeman has served as advisor to aboriginal groups, conservation organizations, and governments, and he has served as a senior science advisor to the Department of Indian Affairs and Northern Development in Ottawa. Freeman is a fellow of the American Anthropological Association, the Society for Applied Anthropology, and the Arctic Institute of North America.

ROBERT L. FRIEDHEIM is a professor of international relations at the University of Southern California (USC). He also taught at Purdue University and served as a staff member of the Center for Naval Analyses. At USC, Friedheim has served as the associate director of the Institute for Marine and Coastal Studies, director of the USC Sea Grant Institutional Program, and director of the School of International Relations. He is the author, or author and editor, of eight books and over fifty articles and studies, mostly dealing with ocean policy.

JON L. JACOBSON is Bernard B. Kliks Professor of Law Emeritus and codirector emeritus of the Ocean and Coastal Law Center at the University of Oregon School of Law, where he taught contracts, international law, and law of the sea. He has twice been a Fulbright scholar at the University of Oslo's Scandinavian Institute of Maritime Law and served as Stockton Chair of International Law at the U.S. Naval War College during 1982–1983. He is coauthor of the textbook *Coastal and Ocean Law* (3d ed., 1998) and has published several articles and book chapters on ocean law and policy. Until his retirement, Jacobson served as editor-in-chief of *Ocean Development and International Law,* a leading journal on marine affairs.

CHRISTOPHER D. STONE is the J. Thomas McCarthy Trustee Professor of Law at the University of Southern California. He received his undergraduate degree in philosophy from Harvard, a law degree from Yale, and was a fellow in law and economics at the University of Chicago. At the University of Southern California he teaches international environmental law, as well as courses in moral philosophy, property, and corporations. He has served as rapporteur for the American Bar Association's Special Committee on International Law and the Environment in preparation for the 1992 UNCED meeting in Rio. Stone is a trustee of the Center for International Environmental Law (ICEL/USA) and an advisor to the Foundation for International Environmental Law and Development (FIELD/London). Among his books are the environmental classic *Should Trees Have Standing? Toward Legal Rights for Natural Objects.* Stone's current interests include examining strategies for underwriting the defense and repair of the global environment in the context of North-South tensions.

DAVID G. VICTOR is the Robert W. Johnson Jr. Fellow for Science and Technology at the Council on Foreign Relations in New York City. At the council his activities include leading a research project on ways to limit emissions of greenhouse gases through policies that spur the development and deployment of new energy technologies. Victor previously directed a three-year multinational research project on implementation of international environmental treaties at the International Institute for Applied Systems Analysis (IIASA) in Laxenburg, Austria. His research focused on how the international system monitors, verifies, and enforces compliance with environmental treaties. He has a Ph.D. in political science (international relations) from the Massachusetts Institute of Technology and a B.A. in history and science from Harvard University. His publications include an edited book of case studies on the implementation of international environmental agreements (MIT Press, 1998) and articles in various journals, including *Climatic Change, International Journal of Hydrogen Energy, Nature,* and *Scientific American.*

INDEX

Aboriginal whaling, 120

Advocates, animal rights, 12, 13, 306

Agenda 21, 58, 136, 169

provisions on whaling, 57

Alaska Eskimo Whaling Commission, 172

Alaskan Inuit, 167

Allen, K. Radway, 105, 108, 110

Amazonia, 155, 157, 160

Andresen, Steinar, 10, 29, 278, 279

Animal rights, 14, 16, 19, 37, 269, 276, 296, 299, 307, 313. *See also* Advocates, animal rights

Antarctic Treaty System, 241

Antigua and Barbuda, 15, 187, 188

Anti-whaling coalition and forces, 7, 8, 10, 22, 34, 36, 98, 99, 215, 245–56 passim, 284, 296, 305, 308, 312, 320, 327, 330

breakup fostered, 219

creativity of, 244

focus on monetized exchange, 134

impact on indigenous people, 219

need to negotiate, 226

stability of, 223

Anti-whaling states, 17, 36, 51, 58, 83–86, 97, 98, 183, 192–93, 200, 216, 222–24, 244, 248, 251–54, 306, 317, 328

majority in IWC, 298

Arctic, 6, 158, 163–65

Argentina, 156

Aron, William, 4, 29, 278

Artisanal whalers, 6, 7

Australia, 12, 59, 105, 115, 117, 137, 139, 151, 156–59, 204, 221, 251

desire to end whaling, 215

Austria, 251

Bailey, Ronald, 140

Baleen whales, 4, 31, 33, 169

Ban on bowhead whaling, 167

Barber, Benjamin R., 19

Barsh, Russel Lawrence, 29, 283, 284, 285, 286

BATNA. *See* Best Alternative to a Negotiated Agreement

Beluga whales, 11

Best Alternative to a Negotiated Agreement, 226, 245

Biodiversity, 24, 81, 155, 250, 277
 and conservation efforts, 161
 convention, 170
 and European agrosystems, 158
 maintenance of, 160

Blue whales, 6, 107–10, 243, 296

Blue Whale Unit, 4, 109, 110, 116, 139, 295
 as currency, 296
 dropped after 1972, 118
 elimination of, 33, 107
 value of, 31

Bowhead whales, 36, 120, 126
 aboriginal quotas, 302
 reproductive continuity of, 148

Boycotts, 8, 115, 305, 322

Brazil, 9, 154, 155, 188

Britain, 150, 151, 152, 154

Brown, Lester, 24

Brundtland, Gro Harlem, 305

Brundtland Commission Report, 13, 20, 319

Burke, William, 4, 7, 29, 83, 94, 125, 270, 273–75, 302, 316

BWU. *See* Blue Whale Unit

Canada, 11, 14, 15, 139, 151, 155, 160, 171

Canny, Michael, 36, 37, 200, 216, 316, 327

CCAMLR. *See* Convention on the Conservation of Antarctic Marine Living Resources

CCLMR. *See* Convention for Conservation of the Living Marine Resources

CCSBT. *See* Commission for the Conservation of Southern Bluefin Tuna

Center for Marine Conservation, 12

Chapman, Douglas, 108, 110, 111

Charismatic megafauna, 5, 14, 25, 83, 313

Chayes, Abram and Antonia, 195

Chile, 183, 188, 190

China, 15

Chukchi people, 280, 281

CITES. *See* Convention on International Trade in Endangered Species

Clark, Colin, 129, 131

Clinton, William Jefferson, 22

Cold War, 18, 71, 92, 149, 217, 225, 296

Colombia, 154

Commercial whaling, 32, 36, 59, 81–85, 97, 98, 127, 129, 168, 183, 188, 195, 200, 211, 244, 245, 246, 250, 252–56, 292, 293, 297, 302, 305, 306, 312, 313, 317, 326–28, 330
 compared to aboriginal whaling, 120
 effect of IWC on, 237
 in Japan, 5, 132
 moratorium on, 5, 34, 65, 87, 94, 111, 186–87, 191–92, 201, 203, 301
 outlawed in U.S., 189
 permitted by Irish Proposal, 248
 resumed by Iceland, 307
 resumed by Norway, 299
 and Scientific Committee, 116–17, 118
 and stock recovery, 185

Commission for the Conservation of Southern Bluefin Tuna, 275

Common pool resources, 18, 320
 management of, 30
 sustainable use, 295

Compliance mechanisms, 28

Consensus, formal, 22

Consent, 26

Conservation, 5, 13, 14, 32, 37, 52–56, 59, 72, 108, 149, 160

cooperation, 57
obligations, 53
and quotas, 161
Convention for Conservation of the
Living Marine Resources, 220
Convention for the Regulation of
Whaling (1931), 282
Convention on Biological Diversity, 136,
169, 170, 171
Convention on the Conservation and
Management of Pollock Resources
in the Central Bering Sea (1994). *See*
North Pacific Pollock Agreement
Convention on International Trade in
Endangered Species (CITES), 19, 20,
37, 171, 184, 195
Appendix I, 313
Appendix II, 313
downlisting whales, 313
Iceland joins, 313
lessons, 316
Convention on International Trade in
Species of Wild Fauna and Flora
(1973), 184
Convention on the Conservation of
Antarctic Marine Living Resources,
239
Convention on the High Seas (1958), 85,
86, 88
Costa Rica, 188
Cote d'Ivoire, 154
Cree, 160
Cronon, William, 25
CRW. *See* Convention for the Regula-
tion of Whaling
Cuba, 149
Culture, 29, 96, 154, 211
conflict, 123

dependence on wildlife, 23
diversity, 23, 24, 162
food and, 162, 167
homogenization, 18
rights of, 16, 19, 148
sensitivity to, 139
traditional, 7, 326

Dene people, 160
Denmark, 15, 204, 221, 252
DeSombre, Elizabeth, 8, 10, 29, 213, 235,
277, 278
Diamond, Jared, 285
Dominica, 15, 188, 204
Douglas, Mary, 133

EEZ. *See* Exclusive Economic Zone
Egypt, 187
EIA. *See* Environmental Investigation
Agency
Elephants, 5, 13, 25
endangered status of, 37
Endangered Species Act, 315
Environment, governance of, 20, 184, 196
Environmental Investigation Agency, 12,
280
Epistemic community, 21, 205
Ethics, 12, 14
European Bureau for Conservation and
Development, 15
European Community, 158, 170, 172
Exclusive Economic Zone, 11, 35, 36, 54,
62, 70, 84–86, 208, 324–30 passim
whaling within, 53, 193

Faeroe Islands, 23, 281
FAO. *See* Food and Agriculture
Organization

FCCC. *See* Framework Convention on Climate Control

Fin whales, 6, 109–10
 BWU equivalence, 107
 quota, 109, 190

Fisher, Roger, 214, 221, 226

Fisherman's Protective Act, 189

Fishing, freedom of, 53, 62, 63, 84–85

Flipper, 5, 25

Food
 hegemony, 152, 153, 156, 158, 159, 285
 security, 24, 29, 147–51, 161, 166, 271, 328

Food and Agriculture Organization, 220, 242

Framework Convention on Climate Change, 22, 270, 277

France, 12, 139, 204, 215, 251

Freeman, Milton M. R., 4, 29, 257, 277, 279–82, 303

Free Willy, 25, 95

Friedheim, Robert L., 10, 29, 245, 247, 252–53, 257, 269, 277–78, 294, 300

Friends of the Earth, 250

Gambell, Ray, 33, 121

GATT, 225

Germany, 4, 12, 139, 204, 208, 251

Governance
 environmental, 20, 196
 without government, 184

Gray whales, 6, 36, 120, 166, 168

Greenland, 11, 128, 252, 281

Greenpeace, 5, 12, 36, 156, 187, 208, 213, 249, 277, 296
 influence on whaling issue, 250

Grenada, 15, 188, 248

Gulland, John, 109, 110

Hague Declaration, 10, 238

Hammond, Philip, 22, 35, 117

Harpoon, 30

Harvard Negotiating Project, 221

High North Alliance, 15

Hoel, Alf Hakon, 257

Holt, Sidney, 106, 108, 110, 250

Humpback whales, 6, 31, 109–10
 BWU equivalence, 107
 songs, 95

Iceland, 6, 9, 11, 15, 23, 35, 36, 191, 221, 245, 247, 253–54, 278, 292, 297–300, 303
 benefits from whaling, 215, 298
 and blue whale prohibition, 108
 boycott of fish from, 208
 Cold War bargaining chip, 225
 disputes with Norway, 246
 joins CITES, 313
 poor conditions under IWC, 293
 resumption of whaling, 307, 313
 role in establishing NAMMCO, 246
 and smallholder whaling, 281

ICES. *See* International Committee for the Exploration of the Seas

ICESCR. *See* International Covenant on Economic, Social and Cultural Rights

ICRW. *See* International Convention on the Regulation of Whaling

Ikle, Fred Charles, 201

India, 151, 154, 168

Indian Ocean, 70

Indigenous people, 6, 10, 23, 24, 35, 138, 147, 154, 160, 162, 169, 170–71, 219, 280–81
 and food systems, 152, 158
 disintegration of, 159

and European agrosystems, 158
physical and cultural sustainability,
 159
protecting way of life, 148
rights of, 172
and whaling industry, 241
Indonesia, 11, 154
Inglehart, Ronald, 24
Initial Management Stock, 6, 21
Inter-generational responsibility, 13
Intergovernmental Panel on Climate
 Change, 277
International Commission for Northwest Atlantic Fisheries, 59
International Committee for the Exploration of the Seas, 249
International Convention for the Regulation of Whaling, 4, 9, 11, 20, 23, 24,
 31, 32, 51–52, 57–59, 62–70, 72, 82–88
 passim, 91, 93, 97, 98, 125, 134, 186,
 190, 201, 240–41, 244, 270, 272, 292,
 295, 303, 308, 318, 330,
 adoption of, 81
 alleged violations of, 61
 Article V, 36, 56, 63
 authority over parties, 274
 and charter, 56
 flexibly interpreted provisions, 94
 function in international law, 304
 IWC exceeding its powers under, 274
 lack of dispute settlement provisions, 95
 purposes, 63, 171
 revision of, 54
 right to leave, 275
 similarities to UN Charter, 91
 ultra vires moratorium, 275
 whaling states leaving, 273

International Court of Justice, 69, 89,
 93, 94, 95, 274
International Covenant on Economic,
 Social and Cultural Rights, 149, 168
International Dolphin Protection
 Program, 70
International Dolphin Watch, 12
International Fund for Animal Welfare,
 12
International law, 7, 8, 16, 26, 51, 52, 54,
 65, 72, 80, 86, 87, 95–98, 140, 149, 171,
 190, 194, 247, 273, 276, 300, 301–3,
 307–8
 customary, 84
 environmental protection under, 97
 function of ICRW, 304
 IWC resolutions and, 206
 support for indigenous rights, 147
 treaty termination doctrine, 84
 whaling outlawed under, 84
International organization, 56
International regimes, 141, 244, 248, 277
 defined, 27
 effect on behavior, 238
 increased participation, 241
 legal inconsistencies in, 171
 veto provisions of, 239
International Union for the Conservation of Nature, 136, 137
International Whaling Commission,
 325, 327. *See also* Scientific
 Committee
 and aboriginal peoples, 302, 325
 adopting RMS, 327
 annual budget, 243
 annual meetings, 36, 37, 109, 111, 187,
  204, 248
 Articles V and VIII, 85, 207

International Whaling Commission
(*cont.*)
  bandwagoning, 209
  and bribery, 8, 186, 187, 189, 194, 223,
    241, 277, 314
  and bullying, 8, 185–87, 194, 223,
    241, 277, 293, 302, 314
  committees, 10, 108–10, 115, 117,
    119
  decision-making, 4, 29, 186, 239
  logrolling, 209
  membership, 34, 67, 115, 243
  and NGOs, 213
  objections, 9, 195
  open design, 241
  purpose, 80, 83, 140, 271, 275
  and quotas, 330
  regional alternatives, 55
  RMP accepted, 116, 118, 297
  Schedule, 6, 9, 21, 33, 36, 51, 64, 85,
    206–7, 219, 225, 238–39, 244, 272,
    296, 317, 318, 321–23, 329
  Secretariat, 242, 243
  Technical Committee, 32, 33, 321–22
  voting procedures, 32
  Working Groups, 10
International Wildlife Coalition, 12
International Wildlife Management
  Consortium, 15
Inuit, 23, 30, 36, 136, 159, 166, 280
  and dietary changes, 158, 165
Inuit Circumpolar Conference, 15
IPCC. *See* Intergovernmental Panel on
  Climate Change
Iraq, 149
Ireland, 151, 251
Irish proposal, 72
Italy, 12, 251

IUCN. *See* International Union for the
  Conservation of Nature
Ivory, 20
IWC. *See* International Whaling
  Commission

Jacobson, Jon, 7, 80, 270, 273, 276, 316
Jamaica, 188
Japan, 4, 6, 139, 149, 190, 275
  and commerciality of whaling, 134
  food security, 166, 271
  history of whaling in, 30, 31, 132
  and ICRW, 9, 195, 301
  financial influence over IWC states,
    187–88
  and IWC, 15, 191, 204, 210, 243,
    245–47, 300
  minke whaling, 132–33, 297
  and RMS, 317
  salience of whaling, 215, 277
  scientific whaling, 11, 36, 193, 207,
    256, 292, 307, 312
  smallholder whaling, 23–24, 281
  and Southern Ocean sanctuary, 5,
    35, 66, 271, 327–30
  subsistence whaling, 136, 303
  and whale stocks, 16, 110
Japan Small-Type Whaling Association,
  15
Japan Whaling Association, 15

Kalland, Arne, 279
Keiko, 95
Kellogg, Remington, 106
Kenya, 187
Knauss, John, 3, 327
Koh, Tommy, 217

Index

Korea, 15
Kyoto Protocol, 22

Magnuson Fishery Conservation and
    Management Act, 189, 208
Makah tribe, 36, 147, 168–69, 225, 276,
    279, 280, 282, 283
    aboriginal subsistence patterns, 166
    anti-whaling campaign, 167, 169
    appeal of whaling, 166
    intermediate case between Japan
        and Inuit, 166
    IWC permits whaling, 166
    offered bribe not to whale, 40
    significance of whaling, 167
Marine Mammal Commission, 189
Marine mammal meat, 158, 164
Marine Mammal Protection Act, 128,
    189, 315
Marx, Karl, 130, 281
Maximum Sustainable Yield, 21, 34, 111,
    115, 295, 296, 319
Minke whales, 6, 11, 36, 83, 94, 98, 111,
    132–35, 141, 293, 297, 298, 312–13, 327
    quota objected, 190
Mitchell, Ronald, 205
Monaco, 36, 204
MSY. See Maximum Sustainable Yield

Namibia, 15
NAMMCO. See North Atlantic Marine
    Mammal Commission
Narwhal, 11
National Oceanic and Atmospheric Ad-
    ministration, 3, 315
Netherlands, 12, 107, 139, 251
New Management Procedure, 4, 5, 6,
    116, 118

application problems, 296
effective date, 34
flawed procedure in, 115
improvements, 296
New Zealand, 12, 59, 137, 168, 204, 251
desire to end whaling, 215
NGOs. See Non-governmental
    organizations
Nicaragua, 149
NMP. See New Management Procedure
Non-governmental organizations, 8, 10,
    13–15, 17, 34, 97, 147, 171, 200, 203–5,
    210, 215, 218, 241, 242, 251, 278, 327
    as international actors, 156
    and IWC, 185, 213, 247
North Atlantic Marine Mammal Com-
    mission, 11, 58, 223, 226, 246, 292
    as alternative to IWC, 245, 302, 312,
        313
    develops expertise, 312
    Russia an observer, 252
North Atlantic Salmon Conservation
    Organization, 59
North Pacific Pollock Agreement, 318
Norway, 6, 9, 11, 14–7, 23, 35–7, 191, 193,
    195, 204, 243, 245, 253–254, 278, 297,
    300, 312, 317
    and whaling, 215, 255–56, 292,
        298–99, 305
    disputes with Iceland, 246
    formal compliance with ICRW, 301
    lack of support for NAMMCO, 246
    smallholder whaling, 281
Norwegian Whalers Union, 15, 248

Ohsumi, 111
Oman, 37, 204, 242, 248
    joins IWC, 187

Packwood-Magnuson Amendment, 189, 191
Pacta sunt servanda, 26, 95
Panama, 191
Pelly Amendment, 37, 189, 190, 191, 208, 225
Peru, 9, 188, 190
Philippines, 11, 188
Pilot whales, 11
Precautionary principle, 21, 52, 69, 272, 318, 320, 329
    defined, 319
    requirements of, 319
    whaling, 98
Preservation regime, 3, 4, 7, 13, 14, 17, 28, 34, 82
    as new norm, 10
Protection Stocks, 6, 21

Quotas, 4, 6, 8, 33, 36, 110, 121, 160

Regional Fisheries Management Orga-
    nizations, 273–74
Remote regions, 23, 25, 30, 321, 325
    whaling necessary for habitation, 162
Revised Management Procedure, 5, 11, 21, 22, 35, 36, 69, 116–19, 219–20, 240, 245, 249, 297, 316–19, 321
    false data, 312
Revised Management Scheme, 5, 6, 14, 35, 36, 117, 200, 219–20, 245, 297, 317–19, 327–29
    development of, 316
    issue of contention, 244
R F M Os. *See* Regional Fisheries Manage-
    ment Organizations
Right whales, 6, 295, 296
Ripeness, 201

R M P. *See* Revised Management
    Procedure
R M S. *See* Revised Management Scheme
Rule of law, 27, 95, 96, 98, 99, 301, 308
    and the I W C, 304
Russia, 9, 15, 31, 32, 97, 110, 139, 170, 190–91, 204, 243
    passive role in I W C, 252

St. Kitts and Nevis, 15, 188
St. Lucia, 15
St. Vincent and the Grenadines, 15, 183, 188, 204, 252
Salt, Julian E., 214
Sanctions, binding legal, 10
S B T. *See* Southern Bluefin Tuna
Schweder, Tore, 106, 107
Scotland, 151
Sei whales, 6, 31, 110
    B W U equivalence, 107
Seychelles, 187
Shamu, 5
Siberia, 159
Singer, Peter, 161
Single negotiating text, 322
Skjaerseth, Jon B, 257
Smallholders, 23, 138
S N T. *See* Single negotiating text
Soft law, 7, 26, 206
Solomon Islands, 15
South Africa, 139, 152
South Korea, 190
Southern Ocean Sanctuary, 5, 10, 31, 35, 65–67, 81–84, 87, 94, 111, 188, 201, 204–5, 209, 247, 251, 270–71, 273, 275, 292, 298, 300, 311, 316, 327, 329–30
    creation of, 36
    scientific grounds for, 12

Sovereignty, 9, 184, 207
  aboriginal, 280
  violation of, 27
Soviet Union. *See* Russia
Sperm whales, 6, 87, 95, 192
Stockholm Conference. *See* United
  Nations Conference on the
  Environment
Stone, Christopher, 8, 24, 29
Straddling Stocks Agreement. *See*
  United Nations Agreement for
  the Implementation of the
  Provisions of 1982 UNCLOS
  Relating to the Conservation and
  Management of Straddling Fish
  Stocks and Highly Migratory Fish
  Stocks
Strong, Maurice, 217
Subsistence, 30, 126
  defined, 127
  integrated with cash economy, 128
  peoples' means of, 149
  similarity between Japanese and
    Inuit, 136
Supreme Commander for the Allied
  Powers, 31
Sustainability, 4, 16, 20, 23, 115, 156, 237
Sustainable development, 82, 169, 294,
  305
Sustainable Use Program of the World
  Conservation Union, 137
Sustained Management Stocks, 6, 21
Sweden, 251
Switzerland, 251
Symbolic politics, 17

Taiwan, 190
Thomas, Erica Keen, 257

Tillman, Michael, 257
Toothed whales, 4
Treaties, self-enforcing, 26, 32
Treaty of Westphalia, 207

UN. *See* United Nations
Underdal, Arild, 257
UNEP. *See* United Nations Environ-
  mental Program
UNHCR. *See* United Nations High
  Commissioner for Refugees
UNICEF, 148, 149
United Kingdom, 12, 59, 68, 139, 193,
  204, 208, 251
  Charter, 89, 91, 93
  desire to end whaling, 215
  Security Council, 91, 92, 93
  United Nations, 4, 52, 53, 111, 148–50,
    168, 172, 223
United Nations Agreement for the Im-
  plementation of the Provisions of
  1982 UNCLOS Relating to the Con-
  servation and Management of
  Straddling Fish Stocks and Highly
  Migratory Fish Stocks (1995), 21, 56,
  274, 318, 319
United Nations Commission on Sustainable Development, 170
United Nations Conference on Environment and Development, 57, 136,
  169
United Nations Conference on the Environment (1972), 4, 10, 34, 111
United Nations Convention on Biological Diversity, 165
United Nations Covenant on Economic, Social and Cultural Rights,
  140

United Nations Convention on the Law
of the Sea, 52–61 passim, 67, 85, 165,
171, 217, 273–75
Annex VI, 64, 65, 66, 69
Article 116, 53, 62, 63, 68
Article 117, 53
Article 118, 53, 54
Article 119, 53, 72
Article 120, 11, 53, 54, 55
Article 287, 62, 64
Article 56, 53, 324
Article 61, 53
Article 64, 54, 55, 58
Article 65, 11, 37, 53, 54, 55, 56, 57, 324
Article 87, 53, 62, 63
dispute settlement, 62, 68
Part V, 53
Part XV, 61, 62, 64, 67, 68
regional organizations consistent
with, 273
Sections 2 and 3, 62
Tribunal, 64, 65, 66, 67, 68, 274, 275
United Nations Environmental Pro-
gram, 136, 242
United Nations High Commissioner for
Refugees, 148, 149
United Nations Law of the Sea. *See*
United Nations Convention on the
Law of the Sea
United Nations World Commission on
Environment and Development, 136
United States, 9, 151, 154, 221, 241, 255,
270, 273, 316
aboriginal/indigenous whaling, 7,
36, 128–29, 137, 147, 166–68, 210,
222, 252, 276, 280, 282–83
and BWU quota, 108
Constitution of, 89–90, 275–76

cultural dominance, 19
dolphin protection, 70
economic threats by, 186, 189–92,
208
and EEZ, 323
High Seas Convention, 85–86
and ICESCR, 149
and ICRW, 32, 56
and IWC, 12, 37, 94, 97, 189, 204, 213
and Japan, 11, 15, 195, 245, 328, 330
Kyoto protocol, 22
ocean management, 315
opposition to commercial whaling,
83, 193, 253
public opinion on whaling, 140–41
and Scientific Committee, 121, 240
whaling industry, 139, 215
whaling policies, 224–25, 253–54,
256, 277
and WTO, 170–71
Universal Declaration of Human Rights
(1948), 149
Ury, William, 226
USSR. *See* Russia

Victor, David, 29, 214, 256, 257, 275, 276,
278, 311, 312
Vienna Convention on the Law of
Treaties (1969), 60, 84, 85, 94
emphasis on language, 87
types of treaties, 88

Washington Conference, 241, 270
Wettestad, Jorgen, 257
Whale and Dolphin Conservation Soci-
ety, 12
Whales, 5, 7, 13, 25, 124, 148
Antarctic, 109, 243

as food, 31, 147

conservation of, 4, 33, 57, 81, 82, 85, 87, 117, 123, 185, 187, 192, 237, 239, 241

endangered status of, 37, 161

and extinction, 5, 16, 18, 87, 295, 296, 299, 302

harvesting, 54, 57

humane killing, 204

management before IWC, 106

management of, 25, 28–30, 34, 105, 120, 226, 237, 243, 298, 307, 311, 315

overharvesting of, 30, 31, 108, 118, 186, 192

preservation of, 7, 82, 84, 85, 87, 88, 97, 298

quotas, 118, 120

remote sensing, 71

stock survival, 6, 16, 18, 30, 123, 131

substitutes for, 14

symbolic meaning of, 16

totemization of, 279

Whale sanctuary. *See* Southern Ocean Sanctuary

Whales and science, 8, 11, 32, 33, 36, 55, 58, 64, 65, 72, 85, 108, 125, 191, 193, 203, 207, 245, 249, 250, 272, 293, 312, 316, 322, 329, 330

Whale stocks, 16

assessment of, 34, 116, 292

Whale watching, 7, 135, 204, 222, 256, 271

economic advantages of, 320

Whalers, 7, 295

aboriginal, 60, 126, 127, 129, 130, 137

Alaskan, 128

artisanal, 15, 23, 30, 147, 312, 326, 327

Bequian, 167

Chukchi, 167

community-based, 137, 138

Russian, 31

subsistence, 126, 128–30, 137, 139, 326

Yankee, 32

Yupiit, 167

Whaling, 128, 280, 299

aboriginal, 23, 36, 69, 120, 129, 137, 249–52, 293, 302–3, 306, 326

aboriginal subsistence, 126, 127, 128, 138, 204, 294, 317

artisanal, 16, 131, 326

community-based, 132, 134, 136, 138–39

costs, 320

cultural arguments, 30, 127, 167

dominant moral arguments, 148

economic interest, 60

enforcement system, 70, 71

governance of, 53

industrial, 30, 138, 139

international management, 123

Inuit, 167

pelagic, 16, 30, 32, 139, 191

regional management, 274

Whaling, commercial. *See* Commercial whaling

Whaling Committee of the International Council for the Exploration of the Sea, 31

Whaling industry, 5, 8, 32, 64, 80, 82, 85, 86, 94, 107, 115, 139, 183, 193, 205, 215, 237, 244, 271, 292

development of, 4

Whaling olympics, 33, 220, 240, 296, 316, 320, 327

WHO. *See* World Health Organization

World Conservation Strategy, 136

World Council of Whalers, 15, 172, 313
World Food Program, 149
World Health Assembly, 89
World Health Organization, 89, 93, 95
World Trade Organization, 170, 171,
   208
World War II, 15, 31, 148, 328

World Wildlife Fund, 12, 136, 156, 250
w t o. *See* World Trade Organization

Young, Oran, 106, 116, 257, 258

Zartman, I. William, 38, 203, 222
Zimbabwe, 15, 20